A Guide to
Genetic Counseling

A Guide to Genetic Counseling

Edited by

Diane L. Baker
University of Michigan
Department of Human Genetics
Ann Arbor, Michigan

Jane L. Schuette
University of Michigan
Division of Pediatric Genetics
Ann Arbor, Michigan

Wendy R. Uhlmann
University of Michigan
Division of Molecular Medicine and Genetics
Ann Arbor, Michigan

A John Wiley & Sons, Inc., Publication
New York · Chichester · Weinheim · Brisbane · Singapore · Toronto

Copyright © 1998 by Wiley-Liss, Inc. All rights reserved.

Published simultaneously in Canada.

Library of Congress Cataloging-in-Publication Data:

A guide to genetic counseling / edited by Diane L. Baker, Jane L.
Schuette, Wendy R. Uhlmann.
    p. cm.
  "A Wiley-Liss publication."
   ISBN 0-471-18541-8 (cloth : alk. paper). — ISBN 0-471-18867-0
(pbk. : alk. paper)
   1. Genetic counseling. I. Baker, Diane L. (Diane Lynn), 1951– .
II. Schuette, Jane L., 1956– . III. Uhlmann, Wendy R., 1961– .
   [DNLM: 1. Genetic counseling. QZ 50 G946 1998]
RB155.7.G85 1998
616'.042—dc21
DNLM/DLC
for Library of Congress   98-5906
                                   CIP

Printed in the United States of America.

10 9 8 7 6 5 4

*To our patients and their families, from whom we receive the
inspiration to pursue our work
and
to our students, who provide the
opportunity for us to look at our work in new ways.*

∞

*Colleague, friend, and contributing author Beth Fine Kaplan
died before this book went to press. She gave
richly and vigorously to the field of genetic counseling.*

Diane L. Baker, Jane L. Schuette, and Wendy R. Uhlmann

Contents

Foreword   xv

Preface   xix

Contributors   xxi

1 The Practice of Genetic Counseling   1
   Ann Platt Walker

   Historical Overview / 1

   Models of Genetic Counseling / 2

   Definition and Goals of Genetic Counseling / 5

   Components of the Genetic Counseling Interaction / 9

   Counseling Contexts and Situations / 11

   Providers of Genetic Counseling / 12

   Professional and Educational Landmarks / 15

   Directions for Professional Growth and Skill Acquisition / 18

   References / 18

   Appendix 1.1: Practice-Based Competencies / 21

2 **Lessons in History: Obtaining the Family History and Constructing a Pedigree**   **27**

Jane L. Schuette and Robin L. Bennett

The Evolution of the Pedigree / 28

Family History Basics / 29

Gathering the Information and Constructing a Pedigree / 32

Interpreting the Family History and Pedigree Analysis / 43

Psychosocial Aspects of Obtaining a Family History / 47

Summary / 50

References / 50

Appendix 2.1: Example History and Pedigree / 52

3 **Interviewing Techniques**   **55**

Diane L. Baker

Goals of Interviewing / 56

Getting Started / 57

Interviewing Tools / 59

Listening Skills / 65

Interviewing or Counseling? / 68

Counselor–Client Dynamics / 68

Some Common Pitfalls of the New Interviewer / 71

Conclusions / 72

References / 73

4 **The Medical Genetics Evaluation**   **75**

Elizabeth Petty

A Historical Perspective / 75

The Making of a Clinical Medical Geneticist / 76

Purposes of the Medical Genetics Examination / 77

The Medical Genetics Examination: Basic Components of a Medical History / 79

The Physical Examination / 85

Documentation of the Medical Genetics Evaluation / 91

Diagnostic Studies / 93

Tools and Resources of the Clinical Geneticist / 95

Clinical Case Conferences and Outside Expert Consultants / 96

Summary / 97

References / 97

5 Patient Education **99**
Ann C. M. Smith

Adult Learners / 100

Tenets of Health Education and Promotion: Application to
Genetic Counseling / 102

The Genetic Counseling Session: A Vehicle for Patient Education / 105

Additional Aspects of Genetic Counseling / 111

Barriers to Client Understanding / 113

Application of Instructional Aids for Patient Education / 115

Designing Informational Materials / 117

Future Directions / 118

Summary / 119

References / 119

Appendix 5.1: Design and Development Model: A Case Example / 122

6 Psychosocial Counseling **127**
Luba Djurdjinovic

The Genetic Counselor as Psychotherapist / 128

Psychosocial Assessment and the Structure of a Session / 129

The Patient's Story / 130

The Working Relationship / 131

Disruptions in the Working Relationship / 136

Discussing Difficult Issues and Giving Bad News / 139

Counselees' Coping Styles / 145

Counselor's Self-Awareness / 146

Theories That Surround Our Work / 146

Psychosocial Case Discussion / 155

Stages of Ongoing Genetic Counseling Practice / 156

The Unfolding of a Genetic Counseling Case / 157

Conclusions / 162

References / 163

Appendix 6.1: Major Models of Family Therapy / 167

7 Multiculturalism and the Practice of Genetic Counseling   **171**
Anne Greb

Case Example / 172

What Are Culture and Ethnicity? / 174

What Is a Minority Group? / 175

Developing Multiculturalism / 177

Beginning to Understand What Happened in the Case Example / 177

Considering the Perspective of the Family in the Case Example / 192

Developing Skills / 196

References / 196

8 A Guide to Case Management   **199**
Wendy R. Uhlmann

The Initial Intake / 200

Obtaining Family History Information / 202

Obtaining Medical Records / 202

Preparing a Case / 205

Seeking Information on Genetic Conditions / 207

Performing a Risk Assessment / 208

Coordinating Genetic Testing / 211

Finding Support Groups and Patient Resources / 219

Formulating a Plan for the Clinic Visit / 221

Managing a Session / 222

Communicating Results / 225

Communicating with Other Specialists and Researchers / 226

Case Documentation / 227

Is a Genetics Case Ever Complete? / 228

Summary / 228

References / 228

9 Medical Documentation   **231**
Deborah Lochner Doyle

Types of Medical Documentation / 232

The Importance of Medical Documentation / 234

Recommendations for Medical Documentation / 235

Disclosure of Information Contained in Medical Records / 239

Documentation That Is Subject to External Review / 242

The Medical Records Audit / 243

Legal Issues / 244

Retention of Medical Records / 245

Summary / 246

References / 247

10 Ethical and Legal Issues   **249**
Susan Schmerler

Ethical Issues / 249

Theoretical Influences on Genetic Counseling / 252

Codes of Ethics: The NSGC Code / 257

Case Analysis / 257

Legal Issues / 260

Areas of Practice Raising Ethical and Legal Questions / 267

Conclusions / 273

References / 273

11 Relationally Based Professional Conduct for Genetic Counselors   **277**
Deborah Lee Eunpu

Development of the NSGC Code of Ethics / 277

Relationally Based Concepts of Professional Conduct / 278

Developing a Personal Professional Conduct Code / 285

The Genetic Counselor's Conduct in Relationships with Colleagues and Society / 288

Conclusions / 289

References / 289

Appendix 11.1: The Code of Ethics of the National Society of Genetic Counselors / 291

12 Student Supervision   **295**

Patricia McCarthy and Bonnie S. LeRoy

Definitions of Supervision / 296

Genetic Counselors as Clinical Supervisors: Responsibilities / 297

The Goals of Supervision / 298

Supervision Discussion Topics / 300

Supervisor and Student Roles / 301

Student Responsibilities / 304

Methods in Supervision / 306

Student Assessment / 309

Ethical and Legal Issues / 313

Other Issues in Supervision / 315

Considerations for Supervising Novice and Advanced Students / 316

Conclusions / 318

References / 318

Appendix 12.1: Example Consent Form for Clinical Supervision in the Genetic Counseling Setting / 320

Appendix 12.2: Supervisor Evaluation of Student Skills and Performance / 323

Appendix 12.3: Student Evaluation of the Clinical and Supervisory Experience / 328

13 Professional Development   **331**

Beth A. Fine and Karen Greendale

The Genetic Counseling Profession and Role Expansion / 332

Lifelong Learning Practices / 334

Opportunities in One's Own Institution / 338

Participation in Professional Organizations / 338

References / 339

Appendix 13.1: Clinical Genetics–Related Professional Societies / 341

14 New and Evolving Technologies: Implementation Considerations for Genetic Counselors   **347**
Patricia A. Ward

The Technology Transfer Process: An Overview / 348

Laboratory Issues: Certification and Quality / 352

Practical Issues: Establishing Policy and Practice Guidelines / 354

Complexities of New Technologies in the Genetic Counseling Session: Illustrative Cases / 358

Conclusions / 365

References / 365

Appendix 14.1: Sample DNA Laboratory Report / 368

15 Computer-Based Resources for Clinical Genetics   **371**
Debra L. Collins

Clinical Diagnostic Resources / 372

Private Databases / 376

Cytogenetic Databases / 376

Genetic Testing Laboratories / 377

Teratogen Databases / 377

Pedigree-Drawing Programs / 379

Internet Resources / 380

Funding Agencies / 383

Summary / 384

References / 384

Appendix 15.1: Genetics Computer-Based Resources / 385

Appendix 15.2: Computer Terms / 389

16 Putting It All Together: Three Case Examples   **391**
Vivian J. Weinblatt, Robin L. Bennett, and Elsa Reich

References / 418

Index   **423**

Foreword

What's in a name? About twenty-five years ago I was foolish enough to get embroiled in a semantic argument about who should do genetic counseling and, by extension, to whom the term "genetic counselor" should apply. To be quite explicit, I thought that it should apply to me! Well, have no fear—the term truly belongs to the genetic counselors as we now know them, health professionals extensively trained in both genetics and counseling. There is no question that they are the ones who are doing most, if not all of the genetic counseling in the United States and perhaps elsewhere as well. Where would those of us engaged in clinical genetics be without them?

The success of the profession of genetic counseling is certainly a tribute to the vision of the early pioneers who were bold enough to initiate genetic counseling training programs, generally in colleges and universities removed from the usual sites of clinical genetics activity in medical schools, and to recognize the importance of the *counseling* aspects of genetic counseling. Although the physicians and Ph.D.'s engaged in the genetic counseling of those days had a reasonably good grasp of the genetics, which was much simpler then than it is now (because we knew so much less), we were, sad to say, not as well versed in counseling as we thought we were. It really took the new breed of genetic counselors to show us what should and could be done. As a result, medical genetics became one of the first specialties in medicine to be able to put into action the proposition that patients and their families, if well enough informed and guided, are truly capable of making very good decisions on their own.

Things were not always easy between genetic counselors and medical geneticists, whose fates have been inextricably intertwined. However, genetic counselors are now members of a recognized and well-organized profession. They have developed their own professional organizations and are engaged in the certification of genetic counselors and the accreditation of the genetic counseling training programs. With the latter, genetic counselors have truly taken responsibility not only for their own destinies but also for the future of their profession.

By seeking to improve the education of the genetic counselors of the future, as well as to expand the horizons of those who have already completed their training programs, the writing and publication of this book by Baker, Schuette, and Uhlmann represents another example of the assumption of responsibility for counselors by counselors. It fills a great void in the present literature on genetic counseling. It is a text written by active and engaged genetic counselors to systematize what they have learned by virtue of their own training, research, and experience about the essential components of genetic counseling as it is actually practiced. It is not a theoretical treatise about counseling or counseling techniques or an exposition of the principles of human genetics. Rather, it treats in depth all the practical aspects of genetic counseling, from the first contact of a patient or family with medical genetics and geneticists to the ultimate disposition of the case. It provides a valuable guide to how genetic counselors function within the medical environment in which their activities are embedded, right down to what kinds of paperwork need to be done.

As the mapping, cloning, and sequencing of the human genome proceed at an accelerated pace and as new genes responsible for disease or susceptibility to disease are being identified almost daily, the application of human genetics to all aspects of health and disease is rapidly increasing. Genetic counselors are being called on to play a critical role in making the benefits of this research available to those who are in need of them. However, to do so they will have to be able to adapt their approaches to the ever changing requirements imposed by the particular conditions with which they are concerned. This book provides a strong foundation on which to build.

CHARLES J. EPSTEIN, M.D.

University of California
Department of Pediatrics
San Francisco

As director since the 1970s of the first graduate program in genetic counseling at Sarah Lawrence College, I faced the challenge of demonstrating that this new field of nonphysician genetic counselors could improve the quality of care provided to individuals and families affected by, or at risk for, a genetic condition. The concept of a master's-level genetic counselor was initially greeted with doubt. There was a vigorous and challenging response from the medical community, questioning whether the training was comprehensive enough to provide the necessary medical and scientific expertise.

Today, over 25 years later, these are no longer issues of concern. Many hurdles were surmounted in establishing this new field, and we owe much to the committed, creative, and pioneering efforts of the early graduates of the 1970s. The impressive evolution in the acceptance and utilization of nonphysician genetic counselors has been followed by increased opportunities for genetic counselors to join medical genetics programs at academic centers throughout North America.

One of the many challenges now facing the field of genetic counseling is the need to train an adequate number of qualified genetic counselors. Genetic counseling is a science-based profession whose best practice is definitely an art. Scientific competence, knowledge, and strength of character are the hallmarks of professionals in genetic counseling.

The publication of *A Guide to Genetic Counseling* is an important step in the growth of the profession. The contributors, all experienced genetic counselors, describe in depth important areas of counseling practice that can enrich the understanding of beginning and experienced counselors alike. The collection of chapters illustrates the breadth of the field and its diverse components. The counseling dimension of caring for patients facing a genetic crisis is addressed through discussion of the emotional conflicts inherent in carrying and transmitting a genetic disease. A clear message is given that exclusively information-based counseling will be a disservice to the client.

To prepare counselors capable of working with patients on an emotional as well as an informational level, training programs need to motivate students to examine their own personal issues with regard to helping others. This timely guide serves these needs well.

JOAN H. MARKS

Sarah Lawrence College

Preface

It all began with a breakfast meeting and a letter asking if we would be interested in writing a textbook on genetic counseling. Initially, we were hesitant; a book seemed like a daunting task. However, as we considered the request, our enthusiasm grew. Many textbooks address medical genetics and have chapters on genetic counseling, but until now there has not been an entire book devoted to describing in depth the components of genetic counseling and how it is provided. As genetic counselors actively involved in the teaching and supervising of genetic counseling students at the University of Michigan, we recognized that the primary method of teaching in this field has been through sharing our clinical experiences, creating guidelines, and assigning articles from the primary literature. A textbook would codify this knowledge and facilitate academic instruction.

This book provides a comprehensive overview of genetic counseling, focusing on the components, theoretical framework, goals, and unique approach to patient care that are the basis of this profession. Our emphasis is on describing the fundamental principles and practices that are broadly applicable to all areas of genetic counseling, rather than focusing on specialty areas. Throughout the book, cases are used to illustrate concepts and applications. Major topics provide the reader with a strong foundation of what genetic counseling encompasses, and include obtaining a family history, conducting case preparation, developing approaches for patient education and psychosocial exploration, and identifying the components of the medical genetics evaluation and case management. Important ethical, legal, and cultural considerations in providing genetic counseling are also addressed. A chapter on student supervision serves both as a resource for clinical supervisors and as a set of tips to help students make the most of their training. Information on computer-based resources in clinical genetics and a listing of clinical genetics professional societies are also included.

As more gene mutations are identified, the availability of genetic testing and the need for genetic counseling services will rapidly increase. This book will provide both students and health professionals with a comprehensive foundation for learning about the exciting field of genetic counseling. We acknowledge the valuable contributions of our authors for putting into words their rich and varied experiences. We are grateful to John Wiley & Sons, Inc., for providing us with this opportunity and to Ann Boyle, Ph.D., for her confidence and enthusiasm for this project, from initiation to completion.

DIANE L. BAKER, M.S., C.G.C.
JANE L. SCHUETTE, M.S., C.G.C.
WENDY R. UHLMANN, M.S., C.G.C.

Ann Arbor, Michigan
June 1998

Contributors

DIANE L. BAKER, M.S., University of Michigan, Department of Human Genetics, Ann Arbor, Michigan

ROBIN L. BENNETT, M.S., University of Washington Medical Center, Medical Genetics, Seattle, Washington

DEBRA L. COLLINS, M.S., University of Kansas Medical Center, Department of Endocrinology and Genetics, Kansas City, Kansas

LUBA DJURDJINOVIC, M.S., Genetic Counseling Program, Binghamton, New York

DEBORAH LEE EUNPU, M.S., Beaver College, Genetic Counseling Program, Glenside, Pennsylvania

BETH A. FINE, M.S., Northwestern University Medical School, Graduate Program in Genetic Counseling, Chicago, Illinois

ANNE GREB, M.S., Wayne State University, Center for Molecular Medicine and Genetics, Detroit, Michigan

KAREN GREENDALE, M.A., New York State Department of Health, Albany, New York

BONNIE S. LEROY, M.S., University of Minnesota Hospitals, Institute of Human Genetics, Minneapolis, Minnesota

DEBORAH LOCHNER DOYLE, M.S., Washington State Department of Health, Genetic Services Section, Seattle, Washington

PATRICIA MCCARTHY, Ph.D., University of Minnesota, Department of Educational Psychology, Minneapolis, Minnesota

ELIZABETH PETTY, M.D., University of Michigan, Division of Molecular Medicine and Genetics, Ann Arbor, Michigan

ELSA REICH, M.S., New York University School of Medicine, Division of Genetics, New York, New York

SUSAN SCHMERLER, M.S., J.D., University of Medicine and Dentistry of New Jersey, New Jersey Medical School, Paterson, New Jersey

JANE L. SCHUETTE, M.S., University of Michigan, Division of Pediatric Genetics, Ann Arbor, Michigan

ANN C. M. SMITH, M.A., Translational Research and Laboratory Support Unit, National Institutes of Health/National Human Genome Research Institute, Medical Genetics Branch, Bethesda, Maryland

WENDY R. UHLMANN, M.S., University of Michigan, Division of Molecular Medicine and Genetics, Ann Arbor, Michigan

ANN PLATT WALKER, M.A., University of California, Irvine Medical Center, Division of Human Genetics and Birth Defects, Orange, California

PATRICIA A. WARD, M.S., M.D. Anderson Cancer Center, Department of Behavioral Sciences, Houston, Texas

VIVIAN J. WEINBLATT, M.S., Thomas Jefferson University Hospital, Division of Genetics, Philadelphia, Pennsylvania

1

The Practice of Genetic Counseling

Ann Platt Walker

HISTORICAL OVERVIEW

Until the beginning of the twentieth century, there existed little scientifically based information for people concerned about the chances of an apparently familial disorder or birth defect occurring (or recurring) in themselves or their offspring. Observations of such conditions had sometimes led to correct interpretations of the pattern of inheritance, as in the understanding of hemophilia evidenced by the Talmudic proscription against circumcising brothers of bleeders, in Broca's report of a seemingly dominant breast cancer predisposition in five generations of his wife's family (Broca, 1866), and in societal taboos against marriages between close relatives. Often, however, birth defects and familial disorders were attributed to exogenous causes—punishment (or perhaps, favor) by a deity, a misdeed on the part of a parent (usually the mother), a fright, a curse, or some natural phenomenon such as an eclipse. Indeed, similar beliefs are still widespread in many cultures and may even figure subliminally in irrational fears of people who are otherwise quite scientifically and medically informed.

Throughout the late 1700s and the 1800s, investigators attempted to deduce mechanisms for the transmission of traits. Lamarck's theories regarding the inheritance of *acquired* characteristics persisted into this century. Darwin recognized that traits that became advantageous in particular circumstances might increase the likelihood of survival and reproduction—leading to the eventual development of a population sufficiently different from its ancestors to constitute a new species. Darwin's cousin, Galton,

A Guide to Genetic Counseling, Edited by Diane L. Baker, Jane L. Schuette, and Wendy R. Uhlmann.
ISBN 0-471-18541-8 (cloth), 0-471-18867-0 (paper).   Copyright © 1998, Wiley-Liss, Inc.

by studying families and twin pairs, attempted to develop mathematical models to tease out the relative contributions of environment and heredity. At the turn of this century, Bateson and Garrod each recognized that the familial occurrence of alcaptonuria (described by Garrod in 1899) and other recessive "inborn errors of metabolism" could be explained by the neglected and recently rediscovered laws of Mendel (Garrod, 1902). These observations ushered in a new era in which the pattern of inheritance of certain genetic conditions—hence their risks of recurrence—could be deduced.

Over this century, understanding of genetic disorders, variability, mechanisms, and contributions to common diseases has grown exponentially. Less dramatic but equally important advances have occurred in the study of human behavior, public health policy, ethics, counseling theory, and in empowering people to assume greater responsibility for their health care decisions. The *activity* of genetic counseling has developed and changed accordingly over this time. However it is only in the last three decades that a *profession* specifically devoted to genetic counseling has arisen. The education and practice of a genetic counselor encompass all the elements above, allowing such professionals, as members of a genetics health care team, to bridge such diverse disciplines as research scientist, clinical geneticist, social worker, and hospital administrator. More importantly, today's genetic counselor provides a service that is unique—distinct from the contributions of these other individuals—for patients and families who seek to understand and to cope with the genetic and psychosocial issues of the disorders they confront.

In the fewer than 30 years since the first master's degrees were awarded in genetic counseling, these new professionals have achieved a prominent place in genetic health care delivery, education, and public policy development; formed professional organizations in the United States and Canada; helped start training programs in five countries; become board-certified as distinct subspecialists; and initiated accreditation for genetic counseling graduate programs. This chapter gives an overview of how these developments have occurred—and perhaps a glimpse of the challenges and excitement to come.

MODELS OF GENETIC COUNSELING

Eugenic Model

Sheldon Reed is credited with introducing the term "genetic counseling" in 1947 (Reed, 1955). However, the practice of "advising" people about inherited traits had actually begun about 1906, shortly after Bateson suggested that this new medical and biological study of heredity be called "genetics." By then the public (and many geneticists) had become intrigued by the thought that this new science might identify hereditary factors contributing not only to medical conditions, including mental retardation, but to social and behavioral conditions such as poverty, crime, and mental illness. Galton himself had suggested in 1885 that "eugenics" (a word he coined from the Greek ευγενης, meaning "well-born") become the study of "agencies under social control that may improve or impair racial qualities of future generations, either physically or mentally" (Carr-Saunders, 1929).

Enthusiasm over the possibility that genetics might be used to improve the human condition gave rise, for example, to the Eugenics Records Office at Cold Spring Harbor (a section of the Carnegie Institution of Washington's Department of Genetics), and to the establishment of a chair of eugenics (by a bequest of Galton, himself) at University College, London. Not only did scientists in these institutions collect data on human traits, but they also sometimes provided information to affected families— usually with the aim of persuading them not to reproduce. Unfortunately, at least at the Eugenics Records Office, the collection of data was often scientifically unsound, or was biased and tainted by social or political agendas.

The eugenics movement, of course, ultimately led to horrendous excesses. By 1926, 23 of the 48 United States had laws mandating sterilization of the "mentally defective," and over 6000 people had been sterilized (presumably involuntarily) (Carr-Saunders, 1929). Astoundingly, it has recently come to light that this practice persisted up until the 1960s and 1970s in some Scandinavian countries (Wooldridge, 1997). Laws also were passed at the federal level in the United States instituting quotas that limited immigration by various "inferior" ethnic groups. In Germany, euthanasia for the "genetically defective" was legalized in 1939—leading to the deaths of over 70,000 people with hereditary disorders in addition to Jews and others killed in the Holocaust (Neel, 1994). The specter of these past abuses in the name of mandatory eugenics is at the heart of the "non directive" approach to genetic counseling that prevails today.*

Medical/Preventive Model

Revulsion at the outcomes of what had started out as legitimate scientific inquiry caused most geneticists to retreat from "advising" families about potentially hereditary conditions for at least a decade. However, by the mid-1940s, heredity clinics had begun functioning at the Universities of Michigan and Minnesota and at the Hospital for Sick Children in London (Harper, 1988). A decade later, as medicine began to focus on prevention, several additional genetics clinics had been started. Information about risks was offered—based almost entirely on empirical observations—so that families could avoid recurrences of disorders that had already occurred. But in 1956, few diagnostic tests were available. Knowledge of the physical structure of DNA was only three years old, and there was no way to prospectively identify unaffected *carriers* of genetic conditions. Given that it was still thought that there were 48 chromosomes in the human genome and that the mechanism of sex determination was the same in humans and *Drosophila* (Therman and Susman, 1993), the basis for chromosomal syndromes was completely unknown. Even with the goal of preventing genetic disorders, "counseling" was limited to offering families information, sympathy, and the option to avoid childbearing. For most geneticists, the presumption was that "rational" families would want to prevent recurrences.

Decision-Making Model

The capabilities of genetics changed dramatically over the next 10 years as the correct human diploid complement of 46 was reported by Tjio and Levan (1956) and

---

*Robert G. Resta has written an excellent essay reviewing the complex issues around eugenics and non directiveness for the *Journal of Genetic Counseling* (1997, 6:255–258).

the cytogenetics of Down (Lejeune et al., 1959), Klinefelter (Jacobs and Strong, 1959), and Turner (Ford et al., 1959) syndromes, and trisomies 13 (Patau et al., 1960) and 18 (Edwards et al., 1960; Patau et al., 1960; Smith et al., 1960) were elucidated. Over this decade it also became possible to identify heterozygotes for α- and β-thalassemia (Kunkel et al., 1957; Weatherall, 1963); a host of abnormal hemoglobins; galactosemia (Hsia, 1958); Tay–Sachs disease (Volk et al., 1964); and G6PD deficiency (Childs et al., 1958), among others. Amniocentesis was first utilized for prenatal diagnosis—initially for sex determination using Barr body analysis (Serr et al., 1955)—and then for karyotyping (Steele and Breg, 1966), with the first diagnosis of a fetal chromosome anomaly being reported by Jacobson and Barter (1967).

These advances in genetics meant that families had some new options for more specifically assessing their risks and possibly avoiding a genetic disorder. The choices were by no means straightforward. Tests were not always informative. Prenatal diagnosis was novel, and its potential pitfalls were incompletely understood. Explaining the technologies and the choices was time-consuming. However, the clinical genetics tenet of nondirective counseling was beginning to evolve toward emphasizing patient autonomy in decision making. The emphasis in genetic counseling shifted too . . . from providing information toward a more interactive process in which individuals were not only *educated* about risks, but helped with the complex tasks of exploring issues related to the disorder in question and making decisions about reproduction, testing, or management that were consonant with their own needs and values.

Psychotherapeutic Model

Although families often come to genetic counseling seeking information, they cannot effectively process or act on what they learn until they have dealt with the powerful reactions such information can evoke. For this reason, exploring with clients their experiences, emotional responses, goals, cultural and religious beliefs, financial and social resources, family and interpersonal dynamics, and coping styles has become an integral part of the genetic counseling process. Genetic disorders and birth defects often catch individuals completely off guard—raising anxiety about the unfamiliar, assaulting the self-image, provoking fears for one's own future and that of other family members, and generating guilt. Even when a client brings a lifetime of experience with a disorder, or has known about his or her own reproductive risk for some time, certain cognitive or emotional dimensions may need to be addressed before they can participate in genetic decision making. A skilled genetic counselor must be able to recognize and elicit these factors, identify normal and pathological responses, reassure clients (when appropriate) that their reactions are normal, prepare them for new issues and emotions that may emerge in the future, and help them marshal intrinsic and extrinsic resources to promote coping and adjustment. Some genetic counselors have chosen to develop these skills even further by obtaining additional training so that they are able to provide long-term therapy for individuals or families who seek assistance in coping with genetic conditions and/or genetic risk.

DEFINITION AND GOALS OF GENETIC COUNSELING

Genetic Counseling as Defined in 1975

In 1975 a committee of the American Society of Human Genetics (ASHG) proposed a definition of genetic counseling that was subsequently adopted by the society. No textbook of genetic counseling would be complete without this often-cited paragraph:

> Genetic counseling is a communication process which deals with the human problems associated with the occurrence or risk of occurrence of a genetic disorder in a family. This process involves an attempt by one or more appropriately trained persons to help the individual or family to: (1) comprehend the medical facts including the diagnosis, probable course of the disorder, and the available management, (2) appreciate the way heredity contributes to the disorder and the risk of recurrence in specified relatives, (3) understand the alternatives for dealing with the risk of recurrence, (4) choose a course of action which seems to them appropriate in view of their risk, their family goals, and their ethical and religious standards and act in accordance with that decision, and (5) to make the best possible adjustment to the disorder in an affected family member and/or to the risk of recurrence of that disorder. (American Society of Human Genetics, 1975)

This definition has held up quite well, articulating as it does several central features of genetic counseling. The first is the two-way nature of the interaction—quite different from the "advice giving" of the eugenic period or the primarily information-based counseling characteristic of the midpart of this century. The second is that genetic counseling is a *process,* ideally taking place over a period of time so that the client can gradually assimilate complex or potentially distressing information regarding diagnosis, prognosis, and risk, and be able to formulate decisions or strategies. The third is the emphasis on the client's autonomy in decision making related to reproduction, testing, or treatment, and the recognition that such decisions will *appropriately* be different depending on the personal, family, and cultural contexts in which they are made. The fourth acknowledges that because the occurrence or risk for occurrence of a genetic disorder can have a family-wide impact different from that of other kinds of disease, there should be a psychotherapeutic component of genetic counseling to help people cope with reproductive and other implications of a rare disorder. The phrase "appropriately trained persons" implies that because of these particular features, genetic counseling requires special knowledge and skills distinct from those needed in other medical and counseling interactions.

How Genetic Counseling Has Changed Since 1975

More Indications for Genetics Services and Counseling The ASHG definition focuses mostly on the reproductive implications of genetic counseling. In the two decades since this definition was proposed, however, genetic counseling's purview has expanded to include conditions that are not entirely, and sometimes not at all, genetic. Genetic counselors now provide information about potentially teratogenic or mutagenic exposures; about birth defects that may have little or no genetic basis; and about common diseases of adulthood that have heterogeneous causes. It is not inconceivable that in the future, patients will receive counseling about otherwise inconsequential genetic polymorphisms that may affect their response to therapeutic drugs or environmental pollutants, or even about the genetics of "normal" behavioral and physical traits.

Changes in Clients and Health Care Delivery   As individuals seeking genetic counseling have become more diverse and the technology ever more powerful and complex, new elements have gained prominence in the genetic counseling process. In 1975 one could not have predicted that access to various evaluations or interventions would frequently be limited by lack of insurance or by the constraints inherent in managed care, making advocating for funding a new part of the genetic counseling process. The counselor today needs to inform clients not only about the nature of the disorder, the risks, and the testing and reproductive options, but about ethical dilemmas that might arise as a result of testing, such as possible discrimination in employment or insurance. Additionally, the genetic counseling "process" sometimes has to be accomplished in a period as short as half an hour. Counseling with a recently arrived immigrant might be severely compromised by passage through two translations and perhaps by the client's unfamiliarity with rudimentary concepts of biology. While the basic tenets and goals may remain as they were in 1975, the face of genetic counseling will continue to change.

Philosophy and Ethos of Genetics Services and Counseling

Voluntary Utilization of Services   Genetic counseling operates on a number of assumptions or principles. Among these is the belief that the decision to utilize genetics services should be entirely voluntary. Society at large and other entities, such as insurance companies, clearly have economic and eugenic interests in promoting the prevention of genetic disease. However, in North America and western Europe at least, the prevailing philosophy is that information should be made available and tests offered when appropriate, but that patients and families should have the right to make decisions—particularly about genetic testing and reproduction—unencumbered by pressure or any intimation that a particular course of action is fiscally or socially irresponsible.

In reality, of course, some patients are referred for genetics services not at their own request, but by virtue of a care providers' fear of litigation, or because they have been identified through a screening program about which they were not adequately educated. Furthermore, decisions about testing or reproduction must often be influenced by financial considerations. Genetic disorders are usually associated with additional health care costs, which may or may not be covered by health insurance or public assistance. In some cases, third-party payers consider the newer genetic tests "experimental" or see genetic counseling as unnecessary outside the context of pregnancy. To assume that families can make voluntary decisions about utilizing genetic services or about reproduction based solely on their preferences, personal goals, and moral views is unrealistic.

Equal Access   Ideally, genetic services, including counseling, diagnosis, and treatment, should be equally and readily available to all who need and choose to use them. Perhaps even more than with other medical specialties, however, genetics care is likely to be accessed by those who live in heavily populated areas, have some sort of health coverage and enough education or medical sophistication to know that the services exist, and are able to act as their own advocates in the health care system. Even for patients who are lucky enough to have these attributes, certain newer genetic technologies (e.g., preimplantation diagnosis) may be out of reach just because of their costs, novelty, or limited availability. As capabilities continue to expand, it will be ever more important to assess expensive genetic services, both in terms of how likely they are to be available to *all* those who might benefit and in terms of their costs relative to other public health care needs.

Client Education   A central feature of genetic counseling is a firm belief in the importance of patient education. Expanding on the ASHG definition, typical patient education regarding a particular disorder includes information about (1) the features, natural history, and range of variability of the condition in question, (2) its genetic (or nongenetic) basis, (3) how it can be diagnosed and managed, (4) the chances that it will occur or recur in various family members, (5) the economic, social, and psychological impacts—positive as well as negative—it may have, (6) resources that are available to help families deal with the challenges the disorder presents, and (7) strategies for amelioration or prevention the family may wish to consider.

Complete Disclosure of Information   In providing education about diagnosis and related issues, most geneticists and genetic counselors subscribe to the belief that all relevant information should be disclosed. Being selective in what one tells a patient is viewed as paternalistic and disrespectful of an individual's autonomy and competence. However, there is philosophical and practical disagreement on what geneticists or counselors view as "relevant." Most would probably concur that competent patients should be given the facts about their own diagnoses—even in a challenging situation such as informing a phenotypic female with androgen insensitivity syndrome about her XY karyotype. But there is likely to be less consensus on such issues as disclosing nonpaternity revealed through DNA testing when no risk is entailed. As genetic testing has become widely available for a host of carrier states, there has also been considerable debate about whether all possible tests (e.g., screening for cystic fibrosis) need to be offered and discussed. And it is not clear whether a counselor is obligated to address issues of potential genetic significance that are not related to the reason for referral.

As our testing capabilities and understanding of genetic mechanisms have become more complex, clients have become increasingly diverse—both culturally and educationally. Often, the time available for counseling them has also decreased. Two decades ago, when a typical session might last an hour and a half and the majority of clients were college-educated, middle class, and fluent in English, a "genetics lesson" was a prominent feature of genetic counseling. We thought that to make informed decisions, clients needed a basic understanding of genes, chromosomes, and how testing was done. Now, however, with the incredible complexity of genetic knowledge and technology, the pressure to see clients more rapidly, and the growing need to work through interpreters, achieving this level of client education is often impractical. Moreover, full disclosure of all "relevant information" could overwhelm even the most sophisticated patient. Despite these pressures, however, it will always be critical for the counselor to disclose any information *relevant to decision making* in ways that the client can interpret and act on.

Nondirective Counseling Although the counselor can use clinical judgment in choosing what information is most likely to be important and helpful in a client's adjustment to a diagnosis or decision making, it should be presented fairly and even-handedly—not with the purpose of encouraging a particular course of action. Adherence to a nonprescriptive (often referred to as "nondirective") approach is perhaps the most defining feature of genetic counseling. The philosophy stems from a firm belief that genetic counseling should—insofar as is possible—be devoid of any eugenic motivation. While this is a time-honored tradition, for the counselor to avoid expressing *any* opinions in all instances could be at times counterproductive. A client should expect a genetics professional to be able to provide guidance in some cases—particularly in the presence of complex genetic and medical issues, conflicting data, or choices that raise problematic moral issues. Being entirely nondirective under such circumstances leaves clients to flounder.

Attention to Psychosocial and Affective Dimensions in Counseling Just presenting information does not necessarily promote client autonomy. To succeed in empowering individuals to cope with a genetic condition or risk, or to make difficult decisions with which they can live, the counselor needs to encourage clients to see themselves as competent and to help them anticipate how various events or courses of action could affect them and their families. This cannot be done without knowing something of their social, cultural, educational, economic, emotional, and experiential circumstances. A client's ability to hear, understand, interpret, and utilize information will be influenced by all these factors. An effective counselor will be attuned to affective responses and able to explore not only clients' understanding of information, but what it means to them, and what impact they feel it will have in their respective social and psychological frameworks.

Confidentiality and Protection of Privacy Genetic counseling raises special issues with regard to confidentiality and privacy protection. Information about an individual's family history, carrier, status, diagnosis, or risk of genetic disease in self or offspring is potentially stigmatizing and can lead to discrimination in employment or inability to obtain insurance. For these reasons it is especially critical that such information be kept confidential. On the other hand, knowing a person's genotype can sometimes provide important information not about the individual's own risk, but that of family members. When this risk is substantial, or serious, and when options are available to prevent harm, the client—and in some cases, the counselor—may have an ethical duty to warn relatives. In only a few other situations in medicine (a serious infectious disease or a threat to another's safety disclosed in the course of psychotherapy) may a breach of patient confidentiality be warranted.

Recently, with the advent of computerized databanks and storage of samples containing DNA, concerns have been raised about the *privacy* of genetic information. Genetic material obtained for one purpose (e.g., genetic linkage studies, newborn screening, military identification) can also reveal information that may be both unwanted and damaging (e.g., risk for late-onset disease, nonpaternity). The privacy of genetic information increasingly will become a cause for both litigation and legislation.

COMPONENTS OF THE GENETIC COUNSELING INTERACTION

Information Gathering

An integral part of a genetic evaluation is, of course, the family history. This is usually recorded in the form of a pedigree to clarify relationships and phenotypic features that may be relevant to the diagnosis. Additional family history of potential genetic significance (ethnicity, consanguinity, infertility, birth defects, late-onset diseases, mental disability) should also be obtained as a matter of course. Adherence to conventional symbols notating gender, biologic relationships, pregnancy outcomes, and genotypic information, when known (Bennett et al., 1995), will assure that a pedigree can be readily and accurately interpreted.

Medical history is routinely obtained, as is information about previous and current pregnancies, including complications and possible teratogenic exposures that might have a bearing on the outcome. Often, clinical features or history potentially relevant to a diagnosis must be confirmed—even before the visit—by obtaining medical records not only on the proband, but on any family members previously evaluated or treated for possibly related symptoms.

Of equal importance to the success of counseling is ascertaining the client's or family's understanding about the reason for referral and expectations about what will be gained through the genetics consultation. Assessment of the family's beliefs about causation, and of emotional, experiential, social, educational, and cultural issues that may affect their perception of information given in the course of their evaluation, must be an ongoing process throughout counseling.

Establishing or Verifying Diagnosis

Although a genetic diagnosis can sometimes be established or ruled out solely by reviewing medical records, evaluation usually involves at least one clinic visit. This might be for a diagnostic procedure—as in the case of prenatal testing—or for a physical examination by a clinical geneticist or another specialist experienced with the condition. Confirmation of a clinically suspected diagnosis often requires additional evaluations, such as imaging studies, assessments by other specialists, or examinations of particular family members. Increasingly, cytogenetic or molecular genetic testing alone may be sufficient, not only to diagnose an affected individual or carrier, but to provide important clues to prognosis or severity (if the phenotypic correlates of specific mutations are known). The commercial availability of many of these tests means that certain genetic diagnoses can now be confirmed by a primary care physician or genetic counselor without involving a clinical geneticist. This capability challenges the traditional team approach to genetic evaluation and counseling.

Risk Assessment

In many cases, the client's concerns center on assessing future reproductive or personal health risks. The counselor can sometimes make such an assessment by analyzing the pedigree—taking into account the pattern of inheritance and the client's relationship to individuals with the condition. Mathematical calculations can sometimes be used to modify a risk by incorporating additional information (e.g., carrier frequencies, numbers of affected and unaffected individuals, client's age). Questions about carrier status may be resolved with appropriate laboratory tests. However, when a condition is multifactorial or genetically heterogeneous, risks used in counseling may have to be derived by means of data from observations of other families with affected individuals. Answering concerns about potentially mutagenic or teratogenic exposures usually requires having empirical data on the agent, and assessing timing and dose of the exposure.

Information Giving

Once a diagnosis or risk has been determined, the family needs to understand how the result was arrived at and what the implications are for the affected person and other family members. This includes describing the condition, its variability, and its natural history—making sure that the family's prior view of the disorder (if any) is still appropriate in light of current understanding of the genetics and treatment. Medical, surgical, social, and educational interventions that can correct, prevent, or alleviate symptoms should be described, and there should be thorough discussions of financial and social resources available for support and of reproductive options that would reduce risk or provide information during pregnancy.

Psychological Counseling

Being given a diagnosis or learning about a personal or reproductive risk is likely to generate powerful emotional responses that must be acknowledged and dealt with if the information is to be assimilated. Part of counseling is preparing clients for these responses and helping them cope, often over a period of months or years. In some cases, as in a fetal or neonatal diagnosis, critical decisions must be made rapidly on the basis of new and distressing data. In other situations, carrier or presymptomatic testing may reveal that a person is *not* at increased risk for developing a disease or having affected children. If this new knowledge overturns long-held beliefs, it can be quite disorienting. Clients often need help in trying possible scenarios "on for size" to help them imagine how various courses of action—even the decision just to undergo diagnosis—may affect them and their family. In addition to knowing about re-sources that can help families adjust to the reality of a condition or risk, the counselor must be alert to pathological reactions that are beyond her skills to treat and be able to make an effective referral when necessary.

COUNSELING CONTEXTS AND SITUATIONS

Genetic Counseling for Reproductive Issues

As genetics has become increasingly relevant to all areas of medicine, the contexts in which genetic counseling occurs also have expanded. Genetic counselors once worked mostly in pediatrics, prenatal diagnosis, and a few specialty clinics. Today, however, clients may seek counseling *before* they conceive, because of concerns about the reproductive implications of their own or their family's history. Similarly, many seek counseling as a result of having been screened for carrier status for con-ditions that occur more frequently in individuals of their ethnic background. Others are referred to genetic counseling in the course of an evaluation for infertility or fetal loss, or for donor screening if they are considering using assisted reproductive techniques. With the advent of preimplantation genetic diagnosis, "prenatal coun-seling" may actually occur before conception. Genetic counseling regarding prena-tal diagnosis of birth defects and genetic disorders not only has become more complicated as techniques have improved, but also has moved from large university genetics units into private obstetricians' offices, commercial laboratories, and pri-vate hospitals.

Genetic Counseling in Pediatrics

Most genetic conditions and birth defects appear without warning. Genetic coun-selors have an important role to play after the birth or stillbirth of an abnormal baby or upon the death of an infant with a genetic condition. The counselor not only can help the family understand the cause of the problem (if it is known) but also can help them grieve for the baby's death or the "loss" of the normal child they had antici-pated. At a time when families can feel abandoned by trusted professionals or friends who are made uncomfortable by a baby's death or a birth defect, the counselor can provide not only information, but ongoing emotional support that can continue through subsequent (and usually successful) pregnancies.

Many genetic conditions are not suspected until later in childhood, or even in ado-lescence or adulthood. In some situations, such as delayed physical or cognitive de-velopment, problems may have become evident over time. In others, a newly recognized health problem or feature of a disorder may prompt concerns about a par-ticular diagnosis. Genetic counseling in these circumstances includes gathering in-formation relevant to establishing the diagnosis, anticipating its impact on the patient or family, addressing the clients' fears or distress, educating them about the condition and its implications, and ensuring that they access necessary medical and social services. Because genetic counselors understand the unique medical, genetic, and psychosocial issues that surround many chronic conditions, they are often part of the team of professionals called on to provide ongoing management for diseases such as cystic fibrosis, neuromuscular disorders, and hemoglobinopathies.

Genetic Counseling for Adult-Onset Diseases

A more recent arena for genetic counseling is genetic testing for conditions that develop later in life. As molecular tests have become available for disorders such as Huntington disease, familial amyotrophic lateral sclerosis, and certain hereditary cancer predispositions, healthy individuals who are at risk may consider learning their genotype—to diminish anxiety, to remove uncertainty, or to be able to make informed personal or medical decisions. However, a host of complex genetic and psychosocial issues are inherent in helping families consider testing or interpret and cope with the results. Many physicians who traditionally have cared for these patients feel ill-equipped to provide the education and counseling that should surround testing. Consequently, genetic counselors increasingly are finding themselves in settings such as cancer centers, dialysis units, and adult neurology clinics that historically may not have had close relationships with genetics.

These different work settings present a challenge in that genetic counselors work apart from the traditional "genetics team" and sometimes are looked to for diagnostic expertise that formerly was provided by clinical geneticists. Up until recently, diagnosing most genetic conditions has required the assistance of a clinical geneticist skilled in physical diagnosis and able to synthesize complex historical and laboratory information. For many genetic conditions, this undoubtedly will always be the case. Genetic counselors are trained to understand genetic test results, but their training does not typically include the development of clinical diagnostic skills. However, our increasing ability to diagnose certain disease predispositions and adult-onset genetic disorders solely through molecular testing raises the possibility that in appropriate clinical situations, future genetic counselors may function even more autonomously.

PROVIDERS OF GENETIC COUNSELING

Geneticists

Elements of genetic counseling—risk assessment, education about genetic disorders and reproductive options, and treatment of psychological distress related to these issues—are provided in diverse settings by a variety of health professionals. However, since genetics programs are capable of providing these services in a comprehensive setting and are now recognized as comprising a distinct medical specialty, those with a genetic condition or birth defect ideally should be seen at some point by individuals with special training in genetic diagnosis and counseling. In many centers, the services are provided by a team composed of geneticists who have distinct disciplinary backgrounds, roles, and areas of expertise. The subsections that follow discuss these team members.

being sensitive to inherent ethical dilemmas, and (5) knowing when it is necessary to refer to a geneticist.

PROFESSIONAL AND EDUCATIONAL LANDMARKS

Development of Training Programs and Curricula for Genetic Counseling

In 1971, the year in which the first 10 "genetic associates" graduated from Sarah Lawrence College, a report of the National Institute of General Medical Services predicted that by 1988, 68% more geneticists would be needed to provide appropriate services. By 1973, four additional genetic counseling programs had been started. The next year various faculty, students, and graduates from four of the five programs met at the California state conference grounds in Asilomar to discuss training goals and expectations for this new profession. A second Asilomar meeting, sponsored by the March of Dimes in 1976, was attended by representatives from state and federal health care agencies, genetics centers, volunteer health organizations, and legislators, in addition to those from counseling programs. The importance of overall planning for genetic counseling training and for evaluating programs and graduates was emphasized by the participants.

A much more comprehensive meeting, involving about 50 participants, was sponsored by the Office of Maternal and Child Health (MCH) in Williamsburg, Virginia, in the spring of 1979. The constituencies represented were similar to those from 1976, but participants also included planners from health maintenance organizations and the health insurance industry, nurses and social workers who provided genetics services, and representatives of the NSGC, which had been formed just a year before. Four panels were assigned different tasks: evaluating the curricula of existing programs in light of graduates' experience, exploring how genetic counselors' services could be reimbursed, making recommendations for assuring continuing education opportunities for genetic counselors, and drafting suggestions for a means of evaluating the quality of programs and the competence of their graduates (Dumars et al., 1979). Of all the recommendations to come from the Williamsburg meeting, the guidelines that were established for the curricular content and structure of genetic counseling training had the most lasting impact.

The Williamsburg guidelines influenced planning for the many new training programs that were started over the next decade. Then, in 1989, a third conference was held in Asilomar—this time under sponsorship of the NSGC as well as MCH. The purpose of the 1989 meeting was to reevaluate the Williamsburg recommendations for the programs' curriculum and clinical training and to explore innovative ways for addressing newer features of genetics practice, such as cross-cultural counseling and molecular genetics. Additional issues that were discussed included the pros and cons of instituting a doctoral degree in genetic counseling and possible solutions for shortfalls in genetics "person-power"—including the possibility of more limited training for "genetics assistants" or "aides" who would assume routine tasks and perhaps bring more diverse cultural perspectives to counseling (Walker et al., 1990).

The National Society of Genetic Counselors

A milestone in the evolution of any profession is the formation of its own society. Genetic counselors mark 1979 as the year when the National Society of Genetic Counselors was incorporated. The goals of the new society were "to further the professional interest of genetic counselors, to promote a network of communication within the genetic counseling profession and to deal with issues related to human genetics (Heimler, 1979). Over the years, the society has, in fact, achieved significant progress toward all these goals. In 1980 the newly formed NSGC—then numbering only about 200 members—lobbied successfully for genetic counselors to be included among the subspecialties that would be certified by the ABMG. The NSGC has helped achieve representation by genetic counselors on the boards of directors of the ABMG and the American Society of Human Genetics and on numerous committees of the ASHG, the American College of Medical Genetics (ACMG), and various government advisory boards. The NSGC sponsors an annual meeting to provide continuing education for its members and a forum for discussing research and clinical issues of interest. Since 1991, it has published a journal that is now indexed in multiple databases. By 1991, the NSGC had developed and adopted a code of ethics for the profession. Most importantly, the society is now recognized as the voice of the profession and a resource for information about genetic counseling issues by the media, the public, and other health and genetics professionals.

Accreditation of Genetic Counseling Training Programs

In 1992 the American Board of Genetic Counseling was incorporated to take over the role of certifying genetic counselors that had been the province of the ABMG since it began in 1980. This change had been necessary to make the ABMG eligible for recognition by the American Board of Medical Specialties, which does not allow member boards to certify non-doctoral-level professionals. Formation of the ABGC followed a majority vote by ABMG diplomates (including counselors) that occurred following a year of tense debate. The ABGC's establishment, however, was an important landmark in the evolution of the genetic counseling profession, inasmuch as it opened the way not only for more autonomous decision making regarding the certification of genetic counselors but also for developing a mechanism to accredit their training programs. The ABMG has always accredited fellowship programs for Ph.D. and M.D. geneticists, but these are not degree-granting programs. The ABMG had chosen to limit its oversight of *counseling programs* to approving the sites in which genetic counseling students acquired clinical experiences and logbook cases for board eligibility, not to reviewing academic programs.

The newly formed accreditation committee of the ABGC explored ways to approach accreditation with consultants and with boards that accredited other allied health programs. From these investigations, it became clear that flexibility, variety, and innovation in training were more likely to occur if the accreditors evaluated a program's ability to develop various *professional competencies* in its graduates, rather than simply determined how well the program adhered to prescribed guidelines for curriculum and clinical experience. However, expectations for "entry-level" competencies in genetic counseling had never been clearly defined. To this end, the ABGC sponsored a 1994 meeting that included the directors of all existing genetic counseling programs, the ABGC directors, and consultants with expertise in clinical supervision and accreditation. The goal was to develop consensus about what new graduates should be able to do. By analyzing the counselor's role in various clinical scenarios, participants identified the areas of required knowledge and skills (Fiddler et al., 1996). From these analyses, 27 "competencies" were described. Competencies were further refined by the ABGC (Fine et al., 1996) to form the basis of a document that would be used to guide nascent programs and those seeking accreditation (American Board of Genetic Counseling, 1996). Since helping to develop these competencies is what this book is about, the ABGC description of them is appended to this chapter.

The ABGC knew that it would take time to develop a mechanism for accreditation and to review all the genetic counseling programs. Moreover, since there had been a proliferation of new and proposed programs, the board felt that there should be interim standards to guide new programs and to help potential students evaluate a program's ability to provide appropriate didactic and clinical training experiences. So in 1995, existing programs were invited to apply to the ABGC for a circumscribed review to obtain "interim accreditation," which would remain in effect (provided a program remained in compliance) until the time of eligibility for full accreditation. Beginning in 1996, genetic counseling programs could apply for full accreditation review in order of seniority.

Requirements for Certification in Genetic Counseling

When the ABMG first offered certification in 1981, participating genetic counselors came from a variety of training backgrounds. Many had graduated from genetic counseling programs, but others were nurses, social workers, or simply holders of a degree in genetics. The ABMG established that eligibility for certification in genetic counseling required at least a master's degree in a relevant discipline, provision of genetic counseling in 50 diverse cases described in a logbook submitted with the application, and letters of references from three other geneticists.

Since the early 1980s, requirements for certification have become more stringent, and the ABGC now requires applicants to have graduated from an accredited genetic counseling training program. Logbook cases need to be acquired in clinical sites that have been approved either via accreditation of the program(s) with which they are affiliated or by board review of an application submitted for a site that is utilized by one particular trainee over a specified time period. The nature of the candidate's involvement in each logbook case needs to be clearly documented, and cases must demonstrate a variety of counseling roles and clinical situations. Clinical supervision of all cases must have been provided by one or more ABGC- or ABMG-certified individuals. The ABGC has certified genetic counselors since 1993 and publishes a bulletin describing requirements and providing application forms and instructions for each examination cycle. Examinations are given jointly by the ABGC and ABMG every three years. Achieving certification requires that a candidate's application be approved by the ABGC Credentials Committee (credentialing) to establish "active candidate status." All candidates for certification take a common "general examination" and then a second examination in the appropriate subspecialty. To be certified, an applicant must pass both the general and subspecialty examinations.

DIRECTIONS FOR PROFESSIONAL GROWTH AND SKILL ACQUISITION

While genetic counselors enter the field with an impressive armamentarium of skills and knowledge, there is no way that two years of training can prepare them for all counseling situations or for burgeoning new developments in genetics. Ongoing self-education is critical, and to provide quality service, counselors must stay abreast of the literature, routinely attend professional meetings, and communicate with genetics colleagues. Some counselors elect to obtain additional formal training to enhance their ability to do psychotherapy, research, or administration, or perhaps to enable them to function in a new domain, such as cancer genetics. Maintaining membership in professional societies and being active on their committees afford opportunities to work with colleagues from around the country and to develop leadership skills. Involvement in education and advocacy, and political activism, can also bring rewards and lead to recognition in the community and beyond.

Being a member of a relatively small profession that deals with issues at the cutting edge of science, medicine, and ethics requires a commitment to continued growth and to the assumption of responsibility for helping other health professionals, policy makers, and clients understand genetics and its implications. The challenges are many, but the personal and professional rewards are enormous.

REFERENCES

American Board of Genetic Counseling (1996) "Requirements for Graduate Programs in Genetic Counseling Seeking Accreditation by the American Board of Genetic Counseling. Bethesda, MD: American Board of Genetic Counseling.

American Society of Human Genetics Ad Hoc Committee on Genetic Counseling (1975) Genetic counseling. *Am J Hum Genet* 27:240–242.

Bennett RL, Steinhaus KL, Uhrich SB, O'Sullivan CK, Resta RG, Lochner-Doyle D, Markel DS, Vincent V, Hamanishi J (1995) Recommendations for standardized human pedigree nomenclature. *Am J Hum Genet* 56:745–752.

Broca PP (1866) *Traité des Tumeurs,* Vol. 1, p. 80. Paris: P. Asselin.

Carr-Saunders AM (1929) Eugenics. In: *The Encyclopedia Britannica,* 14th ed., Vol. 8, p. 806. London: Encyclopedia Britannica Company.

Childs B, Zinkham W, Browne EA, Kimbro EL, Torbert JV (1958) A genetic study of a defect in glutathione metabolism of the erythrocyte. *Bull Johns Hopkins Hosp* 102:21–37.

Dumars KW, Burns J, Kessler S, Marks J, Walker AP (1979) Genetics associates: Their training, role and function. A conference report. Washington, DC: U.S. Department of Health, Education and Welfare.

Edwards JH, Harnden DF, Cameron AH, Crosse VM, Wolff OH (1960) A new trisomic syndrome. *Lancet* I:787–790.

Fiddler MB, Fine BA, Baker DL, and ABGC Consensus Development Consortium (1996) A case-based approach to the development of practice-based competencies for accreditation of and training in graduate programs in genetic counseling. *J Genet Counsel* 5:105–112.

Fine BA, Baker DL, Fiddler MB, and ABGC Consensus Development Consortium (1996) Practice-based competencies for accreditation of and training in graduate programs in genetic counseling. *J Genet Counsel* 5:113–121.

Ford CE, Jones KW, Polani PE, Almeida JC de, Briggs JH (1959) A sex-chromosome anomaly in a case of gonadal dysgenesis (Turner's syndrome). *Lancet* I:711–713.

Garrod AE (1899) A contribution to the study of alcaptonuria. *Proc R Med Chir Soc,* n.s., 2:130.

Garrod AE (1902) The incidence of alkaptonuria: A study in chemical individuality. *Lancet* 2:1616.

Harper PS (1988) *Practical Genetic Counselling,* 3rd ed. London: Wright, Butterworth and Co., p. 4.

Heimler A (1979) From whence we've come: A message from the president. *Perspect Genet Counsel* 1:2.

Hsia DY, Huang I, Driscoll SG (1958) The heterozygous carrier in galactosemia. *Nature* 182:1389–1390.

Jacobs PA, Strong JA (1959) A case of human intersexuality having a possible XXY sex-determining mechanism. *Nature* 183:302–303.

Jacobson CB, Barter RH (1967) Intrauterine diagnosis and management of genetic defects. *Am J Obstet Gynecol* 99:795–805.

Kunkel HG, Cappellini R, Müller-Eberhard U, Wolf J (1957) Observations on the minor basic hemoglobin component in the blood of normal individuals and patients with thalassemia. *J Clin Invest* 35:1615.

Lejeune J, Gautier M, Turpin R (1959) Étude des chromosomes somatiques de neuf enfants mongoliens. *Compt Rend* 24:1721–1722.

Lochner-Doyle D (1996) The 1996 professional status survey. *Perspect Genet Counsel* suppl. 18:1–8.

Marks JH, Richter ML (1976) The genetic associate: A new health professional. *Am J Public Health* 66:388–390.

Neel JV (1994) *Physician to the Gene Pool. Genetic Lessons and Other Stories.* New York: Wiley.

Patau K, Smith DW, Therman E, Inhorn SL, Wagner HP (1960) Multiple congenital anomaly caused by an extra autosome. *Lancet* I:790–793.

Reed S (1955) *Counseling in Medical Genetics.* Philadelphia: Saunders.

Serr DM, Sachs L, Danon M (1955) Diagnosis of sex before birth using cells from the amniotic fluid. *Bull Res Counc Israel* 5B:137.

Smith DW, Patau K, Therman E, Inhorn SL (1960) A new autosomal trisomy syndrome: Multiple congenital anomalies caused by an extra chromosome. *J Pediatr* 57:338–345.

Steele MW, Breg WR Jr (1966) Chromosome analysis of human amniotic fluid cells. *Lancet* I:383–385.

Therman E, Susman M (1993) In: *Human Chromosomes. Structure, Behavior, and Effects,* 3rd ed., p. 1. New York: Springer-Verlag.

Tjio JH, Levan A (1956) The chromosome number in man. *Hereditas* 42:1–6.

Volk BW, Aronson SM, Saifer SM (1964) Fructose-1-phosphate aldolase deficiency in Tay–Sachs disease. *Am J Med* 36:481.

Walker AP, Scott JA, Biesecker BB, Conover B, Blake W, Djurdjinovic L (1990) Report of the 1989 Asilomar meeting on education in genetic counseling. *Am J Hum Genet* 46:1223–1230.

Wooldridge A (1997) "Eugenics: The secret lurking in many nations' past," *Los Angeles Times,* Sept. 7.

Weatherall DJ (1963) Abnormal haemoglobins in the neonatal period and their relationship to thalassemia. *Br J Haematol* 9:625–677.

*Practice-Based Competencies**

An entry-level genetic counselor must demonstrate the practice-based competencies listed below to manage a genetic counseling case before, during, and after the clinic visit or session. Therefore, the didactic and clinical training components of a curriculum must support the development of competencies that are categorized into the following domains: Communication Skills; Critical-Thinking Skills; Interpersonal, Counseling, and Psychosocial Assessment Skills; and Professional Ethics and Values. Some competencies may pertain to more than one domain. These domains represent practice areas that define activities of a genetic counselor. The italicized facet below each competency elaborates on skills necessary for achievement of each competency. These elaborations should assist program faculty in curriculum planning, development, and program and student evaluation.

DOMAIN I: COMMUNICATION SKILLS

(a) **Can establish a mutually agreed upon genetic counseling agenda with the client.**

The student is able to contract with a client or family throughout the relationship; explain the genetic counseling process; elicit expectations, perceptions and knowledge; and establish rapport through verbal and nonverbal interaction.

(b) **Can elicit an appropriate and inclusive family history.**

The student is able to construct a complete pedigree; demonstrate proficiency in the use of pedigree symbols, standard notation, and nomenclature; structure questioning

---

*Reproduced from American Board of Genetic Counseling (1996).

for the individual case and probable diagnosis; use interviewing skills; facilitate recall for symptoms and pertinent history by pursuing a relevant path of inquiry; and in the course of this interaction, identify family dynamics, emotional responses, and other relevant information.

(c) Can elicit pertinent medical information including pregnancy, developmental, and medical histories.

The student is able to apply knowledge of the inheritance patterns, etiology, clinical features, and natural history of a variety of genetic disorders, birth defects, and other conditions; obtain appropriate medical histories; identify essential medical records and secure releases of medical information.

(d) Can elicit a social and psychosocial history.

The student is able to conduct a client or family interview that demonstrates an appreciation of family systems theory and dynamics. The student is able to listen effectively, identify potential strengths and weaknesses, and assess individual and family support systems and coping mechanisms.

(e) Can convey genetic, medical, and technical information including, but not limited to, diagnosis, etiology, natural history, prognosis, and treatment/management of genetic conditions and/or birth defects to clients with a variety of educational, socioeconomic, and ethnocultural backgrounds.

The student is able to demonstrate knowledge of clinical genetics and relevant medical topics by effectively communicating this information in a given session.

(f) Can explain the technical and medical aspects of diagnostic and screening methods and reproductive options including associated risks, benefits, and limitations.

The student is able to demonstrate knowledge of diagnostic and screening procedures and clearly communicate relevant information to clients. The student is able to facilitate the informed-consent process. The student is able to determine client comprehension and adjust counseling accordingly.

(g) Can understand, listen, communicate, and manage a genetic counseling case in a culturally responsive manner.

The student can care for clients using cultural self-awareness and familiarity with a variety of ethnocultural issues, traditions, health beliefs, attitudes, lifestyles, and values.

(h)   Can document and present case information clearly and concisely, both orally and in writing, as appropriate to the audience.

The student can present succinct and precise case-summary information to colleagues and other professionals. The student can write at an appropriate level for clients and professionals and produce written documentation within a reasonable time frame. The student can demonstrate respect for privacy and confidentiality of medical information.

(i)   Can plan, organize, and conduct public and professional education programs on human genetics, patient care, and genetic counseling issues.

The student is able to identify educational needs and design programs for specific audiences, demonstrate public speaking skills, use visual aids, and identify and access supplemental educational materials.

DOMAIN II: CRITICAL-THINKING SKILLS

(a)   Can assess and calculate genetic and teratogenic risks.

The student is able to calculate risks based on pedigree analysis and knowledge of inheritance patterns, genetic epidemiologic data, and quantitative genetics principles.

(b)   Can evaluate a social and psychosocial history.

The student demonstrates understanding of family and interpersonal dynamics and can recognize the impact of emotions on cognition and retention, as well as the need for intervention and referral.

(c)   Can identify, synthesize, organize and summarize pertinent medical and genetic information for use in genetic counseling.

The student is able to use a variety of sources of information including client/family member(s), laboratory results, medical records, medical and genetic literature and computerized databases. The student is able to analyze and interpret information that provides the basis for differential diagnosis, risk assessment and genetic testing. The student is able to apply knowledge of the natural history and characteristics/symptoms of common genetic conditions.

(d)   Can demonstrate successful case management skills.

The student is able to analyze and interpret medical, genetic, and family data; to design, conduct, and periodically assess the case management plan; arrange for test-

ing; and follow up with the client, laboratory, and other professionals. The student should demonstrate understanding of legal and ethical issues related to privacy and confidentiality in communications about clients.

(e) Can assess client understanding and response to information and its implications to modify a counseling session as needed.

The student is able to respond to verbal and nonverbal cues and to structure and modify information presented to maximize comprehension by clients.

(f) Can identify and access local, regional, and national resources and services.

The student is familiar with local, regional, and national support groups and other resources, and can access and make referrals to other professionals and agencies.

(g) Can identify and access information resources pertinent to clinical genetics and counseling.

The student is able to demonstrate familiarity with the genetic, medical, and social-science literature, and on-line databases. The student is able to review the literature and synthesize the information for a case in a critical and meaningful way.

DOMAIN III: INTERPERSONAL, COUNSELING, AND PSYCHOSOCIAL ASSESSMENT SKILLS

(a) Can establish rapport, identify major concerns, and respond to emerging issues of a client or family.

The student is able to display empathic listening and interviewing skills, and address clients' concerns.

(b) Can elicit and interpret individual and family experiences, behaviors, emotions, perceptions, and attitudes that clarify beliefs and values.

The student is able to assess and interpret verbal and nonverbal cues and use this information in the genetic counseling session. The student is able to engage clients in an exploration of their responses to risks and options.

(c) Can use a range of interviewing techniques.

The student is able to identify and select from a variety of communication approaches throughout a counseling session.

(d) Can provide short-term, client-centered counseling and psychological support.

The student is able to assess clients' psychosocial needs and recognize psychopathology. The student can demonstrate knowledge of psychological defenses, family dynamics, family theory, crisis-intervention techniques, coping models, the grief process, and reactions to illness. The student can use open-ended questions; listen empathically; employ crisis-intervention skills; and provide anticipatory guidance.

(e) Can promote client decision-making in an unbiased, non-coercive manner.

The student understands the philosophy of non-directiveness and is able to recognize his or her values and biases as they relate to genetic counseling issues. The student is able to recognize and respond to dynamics, such as countertransference, that may affect the counseling interaction.

(f) Can establish and maintain inter- and intradisciplinary professional relationships to function as part of a health-care delivery team.

The student behaves professionally and understands the roles of other professionals with whom he or she interacts.

DOMAIN IV: PROFESSIONAL ETHICS AND VALUES

(a) Can act in accordance with the ethical, legal, and philosophical principles and values of the profession.

The student is able to recognize and respond to ethical and moral dilemmas arising in practice and to seek assistance from experts in these areas. The student is able to identify factors that promote or hinder client autonomy. The student demonstrates an appreciation of the issues surrounding privacy, informed consent, confidentiality, real or potential discrimination, and other ethical/legal matters related to the exchange of genetic information.

(b) Can serve as an advocate for clients.

The student can understand clients' needs and perceptions and represent their interests in accessing services and responses from the medical and social service systems.

(c) Can introduce research options and issues to clients and families.

The student is able to critique and evaluate the risks, benefits, and limitations of client participation in research; access information on new research studies; present this information clearly and completely to clients; and promote an informed-consent process.

(d) Can recognize his or her own limitations in knowledge and capabilities regarding medical, psychosocial, and ethnocultural issues and seek consultation or refer clients when needed.

The student demonstrates the ability to self-assess and to be self-critical. The student demonstrates the ability to respond to performance critique and integrates supervision feedback into his or her subsequent performance. The student is able to identify and obtain appropriate consultative assistance for self and clients.

(e) Can demonstrate initiative for continued professional growth.

The student displays a knowledge of current standards of practice and shows independent knowledge-seeking behavior and lifelong learning.

2

Lessons in History: Obtaining the Family History and Constructing a Pedigree

Jane L. Schuette and Robin L. Bennett

Genetic counseling is very much dependent on the gathering of accurate, detailed, and relevant information. The family history, which is essentially a compilation of information about the physical and mental health of an individual's family, is *the fundamental component* of this process. Obtaining a family history provides a basis for making a diagnosis, determining risk, and assessing the needs for patient education and psychosocial support. It is therefore of utmost importance that genetic counselors possess the skills necessary to gather an accurate and relevant family history.

The pedigree is the diagram that records the family history information, the tool for converting information provided by the client and/or obtained from the medical record into a standardized format. It demonstrates the biological relationships of the client to his or her family members through the use of symbols, vertical and horizontal lines, and abbreviations. When complete, the pedigree stands as a quick and accurate visual record that assists in providing genetic counseling. An analysis of the pedigree may reveal the pattern of inheritance of a disorder within a family. It may provide information that aids in making a diagnosis, as well as information about the

A Guide to Genetic Counseling, Edited by Diane L. Baker, Jane L. Schuette, and Wendy R. Uhlmann.
ISBN 0-471-18541-8 (cloth), 0-471-18867-0 (paper).   Copyright © 1998, Wiley-Liss, Inc.

natural history of a disorder and its variable expression among family members. The pedigree offers a means of identifying family members at risk for being affected with a disorder as well as estimating risks for recurrence in future offspring. The pedigree may also indicate a history of other conditions for which an evaluation and/or counseling are recommended.

The process of obtaining a family history and constructing a pedigree may reveal the social relationships of the client to his or her family members. Information about adoption, divorce, separation, and "blended families" may be obtained. Pregnancy loss, infertility, or death of family members is also recorded. The family history may provide information that suggests the extent of the medical, emotional, and social impact of a disorder for a family. Myths developed by family members, explaining who in the family is at risk and why, may be revealed. And finally, because the family history is often obtained at the beginning of the genetic counseling process, it is a critical mechanism for establishing a productive relationship with the patient and family.

This chapter reviews the components of gathering a family history and constructing a pedigree. In addition, we explore opportunities for psychosocial assessment and patient education that present themselves during the process of taking a history.

THE EVOLUTION OF THE PEDIGREE

> *A complete pedigree is often a work of great labour, and in its finished form is frequently a work of art.*
> *—Pearson, 1912*

Interest in family origins has existed for thousands of years as evidenced in many historical texts, the Bible being a prime example. The pedigree, as a diagram using lines to connect an individual to his or her offspring, was developed in the fifteenth century as one of several techniques of illustrating ancestry (Resta, 1993). However, the use of the pedigree to demonstrate inheritance of traits is a more recent convention, dating back to the mid-nineteenth century, when the inheritance of colorblindness was documented in a publication by Pliny Earle and the "inheritance" of genius and artistic ability was demonstrated by Francis Galton. Earle's pedigree utilized anonymous symbols (circles and squares) to represent the members of a family, while Galton's included the names of family members (Resta, 1993).

Throughout the history of the pedigree, a variety in styles and symbols has been commonplace, reflecting differences in individual preferences, professional training, and national styles. The use of squares and circles for males and females, respectively, by American geneticists and the astronomical symbols for Mars and Venus by English geneticists was one major difference in symbols evident in the early twentieth century (Resta, 1993). More recent surveys of pedigrees recorded in clinical practice and in professional publications have continued to demonstrate a wide variation in the use of pedigree symbols and nomenclature. A survey of genetic counselors published in 1993 showed discrepancies even in common symbols such as those used to indicate pregnancy and miscarriage (Bennett et al., 1993). A review of medical genetic textbooks and human genetics journals also identified inconsistencies in the use of symbols (Steinhaus et al., 1995). If pedigree symbols and abbreviations cannot be interpreted, the value of the family history in establishing an accurate diagnosis and risk assessment is diminished. To address inconsistencies in pedigree symbols and nomenclature, the Pedigree Standardization Task Force (PSTF) was formed in 1991 through the National Society of Genetic Counselors in conjunction with the Pacific Northwest Regional Genetics Group (PacNoRGG) and the Washington State Department of Health. Recommendations for standardized human pedigree nomenclature were developed and subjected to peer review. These recommendations, which have helped to set a more universal standard, were published in the *American Journal of Human Genetics* (Bennett et al., 1995a) and the *Journal of Genetic Counseling* (Bennett et al., 1995b), and have since been reproduced internationally.

The pedigree has also been an invaluable research tool, historically providing evidence for the establishment of the inheritance pattern of particular disorders, and more recently, as an essential component of linkage analysis. For studying complex disorders long considered to be polygenic or multifactorial, such as depression and epilepsy, extended pedigrees and the pooling of many pedigrees can be used in the testing of hypotheses of single-gene versus multifactorial transmission, and environmental contributions (Spence and Hodge, 1996).

The value of family history information to an individual's health care has also been reported in the popular press. Numerous articles have been published in magazines (Cowley, 1993; Adato, 1995) and newspapers (Weiss, 1994), and entire books are devoted to the subject (Krause, 1995). Publications in professional journals from many diverse areas of health and medicine are further examples of the ever increasing awareness of the importance of family history among health professionals (Gordon, 1972; Juberg, 1972; Gelehrter, 1983; Green, 1983; Stang, 1985; Gotto and Wittels, 1987; Papazian, 1994; Wolpert, 1994).

FAMILY HISTORY BASICS

A family history should be obtained from all clients seeking genetic evaluation and/or counseling. This includes the construction of a standard three-generation pedigree containing information on the client, the client's first-degree relatives (children, siblings, and parents), second-degree relatives (half-siblings, aunts, uncles, nieces, nephews, grandparents, and grandchildren), and third-degree relatives (first cousins). Genetics professionals differ, however, as to whether a complete or abbreviated pedigree should be drawn for every patient. This often depends primarily on the nature of the visit. For example, some genetics professionals do not draw complete pedigrees for patients referred for advanced maternal age but, rather, ask a series of general questions designed to elicit any family history of mental retardation, birth defects, or known inherited disorders. A complete pedigree is then obtained for those who indicate such a history (Schuette and Spergel, 1989).

The first considerations when gathering family history information are Who, What, Where, When, and How.

Who?

In general the pedigree begins with the individual for whom an evaluation is being performed or for whom genetic counseling is being provided. The consultand (or client) is the individual(s) seeking genetic evaluation, counseling, or testing, and may or may not be affected. "Proband" is the term that designates the affected family member who brings the family to medical attention (Bennett et al., 1995a, 1995c; Marazita, 1995). Consultand and proband may be the same person, and there may be more than one consultand seeking genetic services.

What?

The nature of the referral or reason for the visit should be clarified. For example, is this a referral for genetic counseling because of a family history of a particular disorder? Is this a diagnostic evaluation, and if so, for what reason? Or, is this a reproductive genetics consultation, and what is the indication?

The nature of the visit to the genetics clinic provides a focus for the process of obtaining a family history. During a consultation for advanced maternal age, for example, the counselor would obtain a standard three-generation pedigree and would use direct questioning to determine whether there are additional risks that need to be addressed or considered. Typical questions might include the following:

Are there any individuals in your family with mental retardation, birth defects, and/or inherited disorders?
Is there any history of stillbirth, multiple pregnancy loss, or infant death?

During a diagnostic evaluation triggered by findings suggesting the diagnosis of neurofibromatosis type I (NF 1), the counselor would obtain a standard three-generation pedigree that includes a series of focused questions about the presence of symptoms and signs of neurofibromatosis, especially in first-degree family members. This information is extremely important to the patient's evaluation because the presence of a first-degree affected family member is one of the criteria considered in establishing a diagnosis of NF 1 (Gutmann et al., 1997). This example of a targeted family history includes questions that focus on the gathering of information that weighs in favor of or against a particular diagnosis. It could be argued that all family histories are targeted, based on the indication for the genetics evaluation. The distinction probably relies more on the specificity of the indication for the evaluation and the degree of confidence that is given to the differential diagnosis provided. For example, a referral for a family history of mental retardation is fairly nonspecific, whereas a referral for a family history of tuberous sclerosis is very specific. Certain general questions should be asked during the course of obtaining any history, while more specific questions depend on the reason for genetic evaluation and counseling.

Inquiring about the presence of physical features *associated* with a particular diagnosis is also important when one is obtaining a family history; this may help to establish a diagnosis, and information about potentially affected family members may be revealed. In the case of NF 1, for example, asking about scoliosis and/or learning disabilities, problems associated with NF 1, may indicate a family member who may be affected. Or, if the indication for the visit is a history of fetal loss, the counselor would inquire specifically not only about fetal loss, but also about infant death, infertility, mental retardation, and birth defects in extended family members, seeking clues about the possibility of an inherited chromosome translocation, X-linked condition associated with male lethality, or other inherited conditions.

Where?

The family history should be obtained in an environment that is comfortable and free of distractions. It should also be obtained in a setting that preserves confidentiality.

When?

The pedigree is usually drawn in the presence of the client(s). Alternatives to the traditional face-to-face method include the family history questionnaire sent to patients in advance of their appointment and the telephone interview. Questionnaires and phone interviews offer the advantages of saving time and enabling advanced case preparation. However, the face-to-face interview is an opportunity for the counselor to make important observations, obtain psychosocial information from the client, and set the stage for a trusting relationship. Complex relationships, such as marriages between biological relatives (consanguinity) or a history of multiple partners, may more likely be revealed in person. The family history and pedigree may be more accurate when the client is present during the recording; in fact, the details of a family history obtained through a questionnaire or phone interview should be confirmed with the client in person.

In most instances, the family history is obtained during the information-gathering portion of the genetic counseling visit, before the examination, risk assessment, and/or counseling. In some instances, however, the family history may be deferred until the very end of the visit, or even until the time of a follow-up appointment. This may be the case when there is a newly established diagnosis such as trisomy 21 in an infant or in an ongoing pregnancy, and the family has very urgent questions and issues that are given priority.

How?

A pedigree is part of a client's medical record. Pedigrees should be drawn on official paper or specific forms designated by the institution for inclusion in the medical record. Preprinted forms that serve as a template from which to construct the pedigree can be useful because they are efficient. Space constraints in the recording of a large family history can be a disadvantage, however.

All medical documentation, including the pedigree, should be recorded in black pen. However, taking a pedigree in pencil can sometimes be useful, since it is not unusual for the client to remember additional information after the form has been com-

pleted. The penciled pedigree can be redrawn in ink later. Although redrawing a pedigree originally done in pencil or one that is simply messy improves clarity, there is the
potential for omitting important information in the transcription. The most efficient
method is to master the skill of drawing an accurate and legible pedigree in ink the
first time. A hint for fixing mistakes before a pedigree is placed in the medical record
is to have on hand a fine-point white correction fluid pen, available at most office
supply stores. Plastic drawing templates with varying sizes of squares, circles, trian-
gles, and arrows are available at art supply centers and university bookstores. Such
templates can help keep pedigree symbols neat and uniform, although some genetic
counselors find the use of templates to be awkward and prefer sketching the pedigree
symbols freehand. Generating pedigrees via a computer for publication, presentation,
or as part of the patient record is feasible through a number of pedigree programs (see
Chapter 15).

GATHERING THE INFORMATION AND CONSTRUCTING A PEDIGREE

Gathering the family history and constructing a pedigree is a process best conducted
in step-by-step fashion. Standardized symbols and nomenclature should be used.

Overview of Pedigree Construction and Symbols

Some of the commonly used pedigree symbols, definitions, and abbreviations are
summarized in Table 2.1. A male is designated by a square and, if possible, placed to
the left of the female partner; a female is designated by a circle.

The proband or consultand is identified with an arrow (Table 2.1); the proband is
distinguished from the consultand by the use of the letter P. It is extremely important
to identify the proband; otherwise, someone looking at a large pedigree may be unable to determine to whom the pedigree pertains.

There are four "line definitions" to orient generations within a pedigree (Table
2.2). A **relationship line** connects two partners (conventions for same-sex relationships are discussed shortly, in connection with assisted reproductive technology). A
break in the relationship line (a double slash) indicates separation or divorce. The **line
of descent** extends vertically (or sometimes diagonally if there are space constraints),
from the relationship line and connects to the horizontal **sibship line**. Each sibling is
attached to the sibship line by an **individual's line**. For pregnancies not carried to
term, the individual line is shortened. Twins share the same line of descent but have
different individual lines. If twins are known to be monozygotic, a horizontal line is
drawn above the symbols (not between the symbols, since it is a relationship line).

A number placed inside a symbol is an indication of how many males or females
are in a sibship. For example, a square with a 5 inside means five males. If a person
has had children with multiple partners, it is not always necessary to show each partner, especially if such information is not relevant to the family history. For example,
a line of descent can extend directly from a parent without including the partner
(Table 2.2).

TABLE 2.1   Common Pedigree Symbols, Definitions, and Abbreviations

	Male	Female	Sex Unknown
Individual	b. 1925	30 y	4 mo
Affected individual (define shading in key)			
Affected individual (more than one condition)			
Multiple individuals, number known	5	5	5
Multiple individuals, number unknown	n	n	n
Deceased individual	d. 35 y	d. 4 mo	
Stillbirth (SB)	SB 28 wk	SB 30 wk	SB 34 wk
Pregnancy (P) (light shading can be used for affected)	LMP: 7/1/94	P 20 wk	P
Spontaneous abortion (SAB) (ectopic = ECT)	male	female	ECT
Affected SAB	male	female	16 wk
Termination of Pregnancy (TOP)	male	female	
Affected TOP	male 16 wk	female	
Proband	P	P	p
Consultand			

Source: Reprinted, by permission, from Bennett et al., *Am J Hum Genet* 56:746. Copyright © 1995 by The University of Chicago Press.

A current pregnancy is symbolized by a square or a circle (if the fetal sex is known) or by a diamond, with the letter P inside. The "age" of the pregnancy is recorded by listing the first date of the last menstrual period (LMP), gestational age (e.g., 20 wk), or estimated date of confinement (EDC). Symbols for pregnancies that resulted in a miscarriage or elective termination are smaller (Table 2.1). Rarely is the gender known in a miscarriage, but if it is, "male" or "female" should be written below the symbol.

TABLE 2.2   Pedigree Line Definitions

Definitions	Comments
1. relationship line 2. line of descent 3. sibship line 4. individual's line	If possible, male partner should be to left of female partner on relationship line. Siblings should be listed from left to right in birth order (oldest to youngest). For pregnancies not carried to term (SABs and TOPs), the individual's line is shortened.

1. Relationship line (horizontal)

a. Relationships		A break in a relationship line indicates the relationship no longer exists. Multiple previous partners do not need to be shown if they do not affect genetic assessment.
b. Consanguinity		If degree of relationship not obvious from pedigree, it should be stated (e.g., third cousins) above relationship line.

2. Line of descent (vertical or diagonal)

a. Genetic				Biological parents shown.
- Twins	Monozygotic	Dizygotic	Unknown	A horizontal line between the symbols implies a relationship line.
- Family history not known/available for individual.	?	?		
- No children by choice or reason unknown		or vasectomy tubal		Indicate reason, if known.
- Infertility		or azoospermia endometriosis		Indicate reason, if known.
b. Adoption	in	out	by relative	Brackets are used for all adoptions. Social vs. biological parents denoted by dashed and solid lines of descent, respectively.

Source: Reprinted, by permission, from Bennett et al., *Am J Hum Genet* 56:748. Copyright © 1995 by The University of Chicago Press.

A diagonal line drawn through a symbol indicates that the person represented is deceased. This is a very visual way of recording who is alive or deceased on a pedigree, and some clients find it offensive to have a deceased relative "slashed out" when a genetic counselor is recording the pedigree in their presence. When obtaining a family history from a client who is adopted, it is essential to distinguish between the adoptive (nonbiological family) and the biological or birth family. In either situation, brackets are placed around the symbol for the adopted individual. If the nonbiological family is included, a dotted line of descent is used. Otherwise, a solid line of descent is used, just as for any other biological relationship (Table 2.2).

The need to represent pregnancies conceived through assisted reproductive technology (ART), such as artificial insemination by donor, is becoming more common. The conventions for symbolizing the biological and social relationships involved in ART within a pedigree are outlined in Table 2.3. Some general rules include placing a "D" inside the symbol for the egg or sperm donor. An "S" inside the female symbol represents a surrogate. If this female is both the ovum donor and a surrogate, she is referred to only as a donor (in the interest of genetic assessment). The relationship line is between the couple (same-sex or heterosexual), and the line of descent extends from the woman who is actually carrying the pregnancy. By using these rules, any method of ART can be clearly illustrated. Documenting elective sterilization (e.g., tubal ligation or vasectomy) is useful if reproductive counseling is considered and as a factor in assessing risks for recurrence.

A double horizontal relationship line is drawn to represent a consanguineous couple. If the degree of relationship is not obvious from the pedigree (e.g., third cousins), it should be indicated above the relationship line.

A shaded symbol indicates an individual affected with a condition that is known or suspected to be genetic. More than one genetic condition can be shown by partitioning the symbol into three or four sectors and filling in the sectors or using different patterns of fill. As long as the shading is defined in a key (see below), any pattern can be used.

Even when standardized pedigree symbols are used, it is essential to include a key (also called a legend). The key provides information that is vital to interpreting the pedigree. It is used to identify less commonly used symbols (e.g., adoption) or to identify shaded symbols, particularly when several disorders are represented by different symbols within a family.

The Standard Information Recorded on the Pedigree

The counselor should record ages or dates of birth of the client and family members (especially first-degree relatives) on the pedigree, below the symbol and to the right if necessary, regardless of whether these individuals are reported to be affected or unaffected with a disorder. The age reached by an apparently unaffected family member may have important implications not only for that particular family member but for the client as well. For individuals reported to be affected with a particular disorder, the age of onset or age at diagnosis should be obtained. This is particularly important for many adult-onset conditions. For example, a client whose maternal grandmother and maternal aunt developed breast cancer before menopause may be at greater risk for breast cancer than the general population. A client whose mother was reported to be unaffected at age 65, however, may have a risk assessment very similar to that of the general population.

The units used to measure ages should be included after each number using standard abbreviations [e.g., 35 y (years), 4 mo (months), 20 wk (weeks), 3 dy (days)]. It

TABLE 2.3   Assisted Reproductive Technologies (ART) Symbols and Definitions

Definitions:
- Egg or sperm donor (D)
- Surrogate (S)
- If the woman is both the ovum donor and a surrogate, in the interest of genetic assessment, she will only be referred to as a donor (e.g., 4 and 5).
- The pregnancy symbol and its line of descent are positioned below the woman who is carrying the pregnancy.
- Family history can be taken on individuals, including donors, where history is known.

Possible Reproductive Scenarios		
1. Sperm donor		Couple in which woman is carrying pregnancy using donor sperm. No relationship line is shown between the woman carrying the pregnancy and the sperm donor. For a lesbian relationship, the male partner can be substituted with a female partner.
2. Ovum donor		Couple in which woman is carrying pregnancy using donor egg(s) and partner's sperm.
3. Surrogate only		Couple whose gametes are used to impregnate another woman (surrogate) who carries the pregnancy.
4. Surrogate ovum donor		Couple in which male partner's sperm is used to inseminate a) an unrelated woman or b) a sister who is carrying the pregnancy for the couple.
5. Planned adoption		Couple contracts with a woman to carry a pregnancy using ovum of the woman carrying the pregnancy and donor sperm.

Source: Reprinted, by permission, from Bennett et al., *Am J Hum Genet* 56:749. Copyright © 1995 by The University of Chicago Press.

is important to note that the ability of clients to recall dates of birth or ages is extremely variable, especially with respect to extended family members. Unless precise information is required, the counselor may wish to encourage the client to provide close estimates by asking, for instance, whether the family member is in her 50s, 60s, or 70s.

The physical and mental health status of the client and each family member (or pregnancy) needs to be recorded succinctly. This includes information about the presence of birth defects, developmental delay and mental retardation, inherited disorders, and chronic illness. Specific and accurate information is best. For example, if a family member is reported to be affected with "muscular dystrophy," the type of muscular dystrophy should be specified. In most circumstances, if a family member at age 70 is reported to have had "heart problems requiring medication, multiple hospitalizations, and triple bypass surgery," it is adequate to record coronary artery disease or heart disease.

Although information recorded on a pedigree should not be too wordy, the use of multiple abbreviations can be confusing. For example "CP" may be an abbreviation for cleft plate or cerebral palsy, and "TS" is sometimes used for Tay–Sachs disease as well as tuberous sclerosis. If abbreviations are used, they should be defined in the key.

If a family member is deceased, cause and age at time of death are recorded on the pedigree. It is also important to inquire whether any family members, especially first-degree relatives, had pregnancies that resulted in miscarriages, stillbirths, or infant deaths. Often clients will omit information about unsuccessful pregnancies, as well as data on siblings, and other relatives who are not living or who died at or around the time of birth. It is also important to distinguish whether individuals who have no children have remained childless by choice or because of a known biological reason (infertility).

The Step-by-Step Process

The counselor should begin by providing a brief explanation to the client about the purpose of gathering the family history information. The counselor then asks sequential questions while, at the same time, drawing the pedigree. Usually the counselor begins with the client, drawing him or her on the pedigree, and obtaining the information described above (e.g., health status, age). If the client is an adult and has a partner, the partner may also be placed on the pedigree at this point, which helps with the positioning of the pedigree appropriately on the paper (usually in the center).

It simply makes sense to obtain family history information in chronological order. The counselor guides the client by proceeding through each first-degree family member, usually by first asking whether the client has had children and/or pregnancies. As a rule, it is easier to obtain information on the children before finding out about any siblings, simply because of the practical consideration that offspring are drawn on the pedigree on a line below the client. (If a child is the patient, it is usually easier to obtain information on his or her siblings before information is requested on the parents). Ages and information about the physical and mental health of each living child should be obtained. Cause of death and age at time of death should be obtained for those who are deceased.

It should be ascertained consistently whether all pregnancies were conceived with the same partner and whether all brothers and sisters within a sibship have the same parents. Otherwise, such information may not be revealed until the pedigree is complete. The counselor needs to indicate appropriately on the pedigree any pregnancies conceived with different partner(s) or any siblings who have a different parent (Table 2.2).

The counselor then asks whether the client has brothers or sisters and whether such siblings have had any pregnancies; if so, the outcomes must be elicited. The counselor should next inquire about the client's mother and father (whether living or deceased; their physical and mental health status), continuing through each family member until a three-generation pedigree has been constructed.

It is important to obtain both the maternal and paternal sides of a family history, even if the visit is for risk assessment and counseling regarding a history of a particular disorder on one side of the family. A full family history allows the counselor to determine whether there are additional factors that may impact on risk assessment, for psychosocial considerations (a focus on only one side of the family may potentially imply support of an individual's sense of responsibility or guilt), and for completeness. Not uncommonly, additional risks are identified during the course of obtaining a family history, necessitating discussion and/or evaluation (Schuette and Spergel, 1989; Holsinger et al., 1992). A question mark is placed above the line of descent in instances in which little is known about the family history to indicate that the appropriate inquiries were made.

A three-generation pedigree is usually adequate and is considered standard, unless the nature of the visit is such that information on extended family members may provide useful information. If a positive history of mental retardation, birth defects, or symptoms and signs of a genetic disorder in an extended family member is identified, documentation in the pedigree may be necessary. For example, a family history obtained for a patient in whom physical findings suggest the diagnosis of Marfan syndrome may very well include pertinent information beyond three generations. A positive history of sudden death in early adulthood in one or more distant family members may contribute toward the establishment of a diagnosis.

When the pedigree appears complete, many counselors ask a series of general questions about the presence of birth defects, mental retardation, and inherited disorders. More specific questions about the existence of associated anomalies or specific signs and symptoms may be asked when warranted, as in a targeted family history. This effort may seem redundant, but often clients recall additional important information after the pedigree has been completed, especially if the counselor makes a final extra attempt to elicit such recollections.

Additional Considerations

Several key points for gathering a family history and constructing an accurate pedigree bear highlighting. All have an impact on genetic risk assessment and counseling.

Ethnic Background   The counselor must inquire about the ethnic background of both the paternal and maternal sides of the family, including the family's country of origin and religion. Responses to these inquiries may yield information that is useful for diagnostic purposes as well as for identifying couples at increased risk for being carriers of certain autosomal recessive disorders.

The question about ethnic background should be posed in a manner that is clear to the client(s). The counselor may need to use alternative words or phrases such as "country of origin," "family's nationality," or "family's traditional religion," to ensure that clients thoroughly understand what information is sought.

Consanguinity The counselor should inquire whether key family members are related to one another. Determining whether consanguinity is present is important for the biological parents of the patient for whom an evaluation is being performed. Inquiring about consanguinity is also important for purposes of preconception counseling and when there are concerns about an ongoing pregnancy.

When inquiring about consanguinity the counselor may need to ask the question in a variety of ways (Are you and your partner blood relatives? Is there any chance that you and your partner are related to one another other than by marriage?). In some instances (e.g., when a rare autosomal recessive disorder is a diagnostic possibility), it may be necessary to obtain additional information related to the possibility of consanguinity. The counselor may wish to obtain maiden and surnames of the biological mother and father, and both sets of grandparents, as well as the town, city, or village

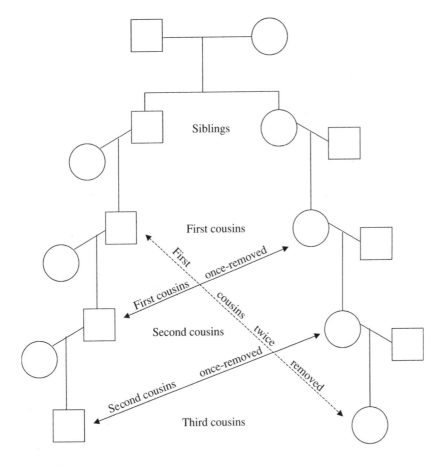

Figure 2.1 Relationships.

(in addition to country) of origin to search for possible distant consanguinity or to document a lack of any evidence. It is also important to note that clients sometimes describe biological relationships inaccurately, and therefore the counselor must carefully establish the exact nature of every relationship that is a potential source of ambiguity. For example, a client may report that she and her partner are second cousins when, in fact, they are first cousins once removed (Figure 2.1).

Verification of Pedigree Information and Documentation of Affected Status

Documenting which family members are known to be affected in a pedigree is essential. In addition, it is often important to *verify* family history information for purposes of ensuring an accurate diagnosis and providing counseling. Verification can be accomplished by obtaining the medical records of the proband or other affected family member(s), genetic tests or other laboratory results, pathology reports, autopsy reports, and in some instances, by performing examinations of key individuals. For example, when a couple is counseled regarding a family history of cystic fibrosis, it is important to obtain documentation of the diagnosis and DNA studies of the affected family member (if performed). The accuracy of risk assessment depends on confirmation of the diagnosis; the information regarding the sensitivity and specificity of CF mutation analysis depends on whether the affected family member's CF mutations are known. If the affected family member has had testing that revealed identifiable CF mutations, then negative carrier studies in the client would have greater predictive value. For a client who is concerned about a family history of cancer, pathology reports that document tumor types and medical records that confirm diagnoses are necessary components of an accurate risk assessment.

When the pedigree is being constructed, information about affected status that has not been documented (e.g., diagnoses reported by the client) should be distinguished from a diagnosis that was established by an examination, laboratory studies, and/or a review of medical records. A diagnosis documented by an evaluation is indicated by an E and an asterisk (*) (Table 2.4). If more than one evaluation has been done (e.g., MRI and DNA tests) subscripts are used (E_1, E_2) and defined in the key. If results of an evaluation are not known or unavailable, a "?" can be used.

As Table 2.4 shows, the situation of a client who has a high likelihood of developing a condition (e.g., because of a positive DNA test), but was asymptomatic at the time the family history was obtained, is represented by a vertical line down the center of the symbol. If the client develops symptoms, the symbol is shaded when the pedigree is updated. This designation is different from that of an individual who is known to carry a gene mutation by virtue of family history or through DNA testing but is not expected to develop symptoms. For example, a woman with normal intelligence whose child has a diagnosis of fragile X syndrome is represented in a pedigree by a circle with a dot inside.

Identifying Information

It is critical that the pedigree include the name and professional background (e.g., M.S., M.D.) of the person who recorded the pedigree. It is also important to identify the historian (i.e., the person providing the family history information). Information about an extended family provided by a foster parent may be more subject to questions of accuracy than information provided by a close relative. Recording the date the pedigree was obtained is also important, particularly if the pedigree gives ages of family members instead of years of birth.

TABLE 2.4   Pedigree Symbolization of Genetic Evaluation/Testing Information

Instructions:

-Evaluation (E) is used to represent clinical and/or test information on the pedigree.

  a. E is to be defined in key/legend.

  b. If more than one evaluation, use subscript (E_1, E_2, E_3) and define in key. May be written side by side or below each other depending on available space.

  c. Test results should be put in parentheses or defined in key/legend.

  d. If results of exam/family study/testing not documented or unavailable, may use a question mark (e.g., E?).

- Documented evaluation (.)

  a. Asterisk is placed next to lower right edge of symbol.

  b. Use *only* if examined/evaluated by *you* or *your* research/clinical team or if the outside evaluation has been personally reviewed and verified.

- A symbol is shaded only when an individual is clinically symptomatic.

- For linkage studies, haplotype information is written below the individual. The haplotype of interest should be on the left and appropriately highlighted.

- Repetitive sequences, trinucleotides and expansion numbers are written with affected allele first and placed in parentheses.

- If mutation known, identify and place in parentheses.

- Recommended order of information:

  1) age/date of birth or age at death

  2) evaluation information

  3) pedigree number (e.g., I-1, I-2, I-3)

Definition	Symbol	Scenario	Example
1. Documented evaluation (.)		Woman with normal physical exam and negative fragile X chromosome study (normal phenotype and negative test result).	E-
2. Obligate carrier (will **not** manifest disease).		Woman with normal physical exam and premutation for fragile X (normal phenotype and positive test result).	E+(100n/35n)
3. Asymptomatic/ presymptomatic carrier (clinically unaffected at this time, but could later exhibit symptoms)		Man age 25 with normal physical exam and positive DNA test for Huntington disease (symbol filled in if/when symptoms develop).	25 y E+(45n/18n)
4. Uninformative study (u)	Eu	Man age 25 with normal physical exam and uninformative DNA test for Huntington disease (E_1) and negative brain MRI study (E_2).	25 y E_1u(36n/18n) E_2-
5. Affected individual with positive evaluation (E+)	E+	Individual with cystic fibrosis and positive mutation study, although only one mutation has currently been identified.	E+(ΔF508/u)
		18 week male fetus with abnormalities on ultrasound and a trisomy 18 karyotype.	18 wk E+(tri 18)

Efficiency, Accuracy, and Time Constraints Obtaining a family history can be a lengthy undertaking especially in instances of large, extended families, multiple affected individuals, or numerous disorders or health problems. If the client is an overly enthusiastic participant with an affinity for details, the completion of this task could potentially occupy most of the time allotted for the clinic visit. How then does the counselor obtain a family history that is complete, accurate, and concise in a timely fashion?

1. Be prepared. This goes back to some of the first considerations of obtaining a family history: who and what. The counselor needs to be prepared for each case.

2. Prepare the client. Explain the purpose of obtaining a family history to the client and indicate clearly the kind of information needed. This will assist the client in reporting relevant data.

3. Control the process; keep the client focused. The counselor needs to provide guidance to the client if he or she loses sight of what is relevant. The counselor should ask direct and specific questions.

4. Be aware of time. It can be helpful to communicate time constraints to a client at the start of the visit.

5. Listen. Listening is a complex skill and it must be done efficiently in a multitasking environment: the counselor will have to listen attentively while simultaneously drawing an accurate pedigree, framing directed questions, sorting through family history data, and interpreting information. Listening carefully is important for accuracy.

6. Be aware of accuracy issues. The counselor needs to develop the ability to quickly assess what information is relevant, what information is suspect, and what information is likely to be inaccurate. The more distant a family member, the more likely the medical information provided about him or her is unreliable. For example, a report of a second cousin who is mentally retarded as a result of birth trauma should not be taken as a definitive diagnosis.

7. Practice! Many genetic counseling students acquire facility in obtaining family histories and constructing pedigrees by practicing on friends and fellow students.

8. Know when to use shortcuts. For large families that include little relevant history, use abbreviation symbols for drawing sibships and extended family members (Table 2.1).

Pedigree Updating   Updating the family history is an important component of a follow-up evaluation for clients with or without an established diagnosis. Additional information may be obtained that provides further clues about a possible diagnosis; or, in the instance of an established diagnosis, the births of additional family members may indicate other at-risk individuals for whom evaluations are indicated.

When the pedigree is updated, the date and recorder should be noted. Some genetic counselors record new information on a colored photocopy of the original pedigree to ensure easy recognition of the updated version.

Issues of Confidentiality   The recording of names on a pedigree makes it possible for the counselor to refer by name to a family member when asking questions. In addition, recording last names may help to uncover previously unknown instances of consanguinity. In some cases, genetic counselors see many family members from the same extended family over time, and having names enables the connection of members of the same family who otherwise would appear to be isolated family groups. Names of family members are necessary, as well, when medical records are requested for purposes of verifying diagnoses so that when records arrive they can be matched to the case to whom they pertain.

When names of persons not present are recorded, however, the resulting pedigree contains genetic information about some family members whose permission to record was not obtained. Using initials may mitigate such concerns, although family members commonly have the same initials.

Individuals in a large family can also be identified by numbers. Each generation can be recorded with a roman numeral to the far left of the pedigree. Each individual within a particular generation is then assigned an arabic number, from left to right, in ascending order (e.g., I-1, I-2). (When this identification method is used, the names of family members for whom medical records have been requested can be recorded separately, allowing the pedigree number to serve as a means of identifying the family member on the pedigree). Spouses or partners may be given the same number with a different lowercase alphabetical letter (e.g., I-2a and I-2b). This method of identification is particularly useful for large research pedigrees and for pedigrees that will be published. As new offspring are added to a pedigree drawn using this approach, however, the numbers in each generation may not be in ascending order.

Additional confidentiality concerns arise when it is necessary to release information (with a signed medical record release) to a third party: an extended family member, insurer, or employer. Not all the information recorded in a pedigree may be necessary or appropriate for other individuals or parties to obtain. For example, a pedigree may contain information about pregnancy termination, pregnancies conceived through assisted reproductive techniques, or presymptomatic carrier status that is not relevant to the purpose of the request for information. Genetic counselors need to carefully determine what information to include on a pedigree, or review their contents before sending it to a third party, since it may be appropriate to omit information deemed irrelevant or potentially stigmatizing. One alternative, especially if the pedigree was not specifically included in the request for information, is to provide the clinic chart note and genetic counseling letter without the pedigree, as these documents generally contain a summary of the pertinent family history information. The safest course is to discuss potential areas of concern with clients before releasing pedigrees to third parties.

INTERPRETING THE FAMILY HISTORY AND PEDIGREE ANALYSIS

The pedigree should be an accurate and easily interpretable diagram from which risk information can be derived. DNA analysis in many instances may define risks for patients and clients in absolute terms, but decisions to undertake testing in the first place may still be partly based on an assessment of the family history data. In other instances, where DNA analysis is not feasible, risk information is based solely on the analysis of the pedigree. It is therefore critical for the genetic counselor to carefully evaluate the data and to consider all possible modes of inheritance when providing risk information. Three important considerations in the interpretation of the family history data are the variable expressivity of inherited disorders, reduced penetrance, and the value of a negative family history.

Variable Expressivity

The concept of variable expression of inherited conditions, especially those that are dominantly inherited, should always play a role in determining what questions to ask the client when obtaining a family history. We have already noted the need to ask not only about the presence of a given disorder in relatives but also about the presence of *associated* physical features. For example, the proband in the pedigree in Figure 2.2 was referred for a diagnostic evaluation because of cleft palate and micrognathia. It was also reported that there was a positive family history of cleft palate in two first-degree relatives, the mother and a sibling. The differential diagnosis for these anomalies and family history includes a condition known as Stickler syndrome, an autosomal dominant condition involving cleft palate, micrognathia, myopia, retinal detachment, hypermobile joints, and degenerative arthritis. When obtaining the family history, the genetic counselor inquired about the presence of physical features associated with this condition in other family members, in an effort to compile data that

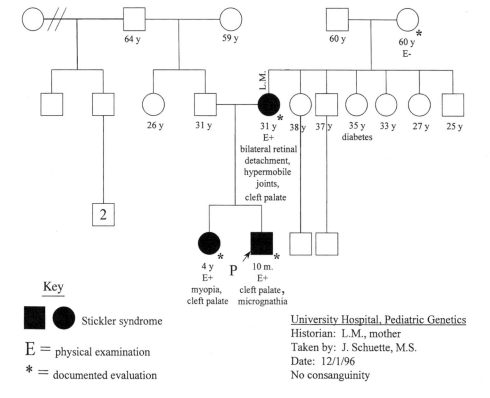

Figure 2.2   *Pedigree showing family history of physical features associated with Stickler syndrome.*

would support or refute the diagnosis that relatives other than the proband were affected. As indicated in the pedigree, the patient's mother reported a history of bilateral retinal detachments and hypermobile joints in addition to the cleft palate. The patient's sister was reported to have early-onset myopia in addition to a cleft palate. This information, along with the diagnostic evaluation, helped confirm the diagnosis of Stickler syndrome in the proband as well as in the mother and sibling.

It is important to remember that disorders may present diversely within a family, and in some instances the sum of varying manifestations among multiple family members will suggest the diagnosis of a particular disorder. Figure 2.3 is a pedigree illustrating such a family. The 15-year-old proband in this family was referred for an evaluation for mild cognitive impairment. The patient's medical history of swallowing difficulties and failure to thrive in infancy, together with a history of

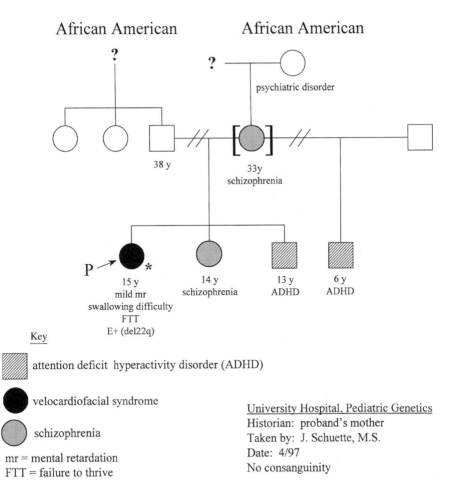

Figure 2.3 *Pedigree (obtained at time of initial evaluation) of a patient later determined to have velocardiofacial syndrome.*

schizophrenia in her mother and sister, suggested a possible diagnosis of velocardiofacial syndrome. Velocardiofacial syndrome includes cleft palate and velopharyngeal incompetence, cardiac defects, specific facial features, mild intellectual impairment, and psychiatric disorders. This diagnosis was confirmed in the patient by the identification of a deletion of chromosome 22 [46,XX,ish del(22) (q11.2)(D22S75-)] by fluorescent in situ hybridization (FISH) spectroscopy. In this case, an analysis of the family history and pedigree was the basis for pursuing a particular diagnosis.

Reduced Penetrance

A small number of autosomal dominant disorders demonstrate reduced penetrance, in which family members known to carry the gene (because they have an affected parent and child or two affected children) have no apparent symptoms. It is important to identify such family members and recommend evaluation and genetic counseling, since the potential implications to their health and medical care must be addressed. The pedigree in Figure 2.4 shows a family first referred to the genetics clinic after the

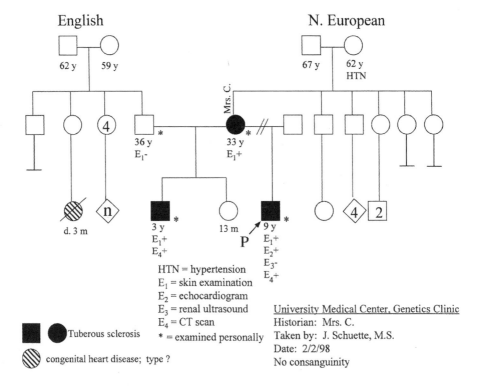

Figure 2.4   Pedigree demonstrating the autosomal dominant inheritance of tuberous sclerosis. The mother, Mrs. C., was determined to be affected at the time of her youngest son's diagnosis after a single manifestation of the condition was identified.

diagnosis of tuberous sclerosis was made in two of the children. Tuberous sclerosis is a neurocutaneous disorder with multisystem involvement; it is inherited in an autosomal dominant fashion. Physical examination of the mother disclosed only a single manifestation of the disorder (a periungual fibroma, a small tumorlike growth around the finger and toenails), and because she had two affected children, a diagnosis of tuberous sclerosis was made. Several important and specific recommendations for the mother's future medical care were then made, including the recommendation for renal ultrasound exams to check for the development of the renal complications associated with this condition.

It is also important for the genetic counselor to bear in mind that the penetrance of some disorders is age dependent, as in the case of Huntington disease. The ages of family members are an important consideration in an assessment of risks.

Value of an Extended Negative History

An extended negative family history provides information that is often as important as a history of a genetic disorder in multiple family members in several generations. For example, the pedigree in Figure 2.5 includes family history information for a client referred for genetic counseling because of a family history of Duchenne muscular dystrophy (DMD), an X-linked recessive disorder. Several of the client's distant relatives were reported to have died from complications of DMD. Of significant value is the presence of multiple unaffected males in the client's pedigree: she has two unaffected brothers and three unaffected maternal uncles. Such a history provides valuable information in arriving at an estimate of the patient's risk for being a carrier (Bayesian analysis would significantly reduce the client's risk from her a priori risk of 1/8) and may play a significant role in her decision to seek carrier testing, linkage analysis, or prenatal diagnosis.

PSYCHOSOCIAL ASPECTS OF OBTAINING A FAMILY HISTORY

Gathering family history information provides the counselor with an excellent opportunity for engaging the client in a relationship that promotes mutual trust. It is an activity that requires participation from all parties—the client, the partner, the counselor, and perhaps extended family members. It can be an "ice breaker," an opportunity to establish an ongoing dialogue and interaction with the family. And, because many intimate family events and details are reported, such as death and loss, abortion, infertility, and consanguinity, the counselor can use these exchanges to set the tone for unqualified acceptance.

During the time spent obtaining family history data, there is an opportunity to make observations about the client and his or her family that may ultimately assist the counselor in coming to know the family. Interactions between couples and family members as information is gathered often provide clues about family dynamics. Does one member of the couple or family provide most or all of the information because he or she apparently has the most knowledge and is the family

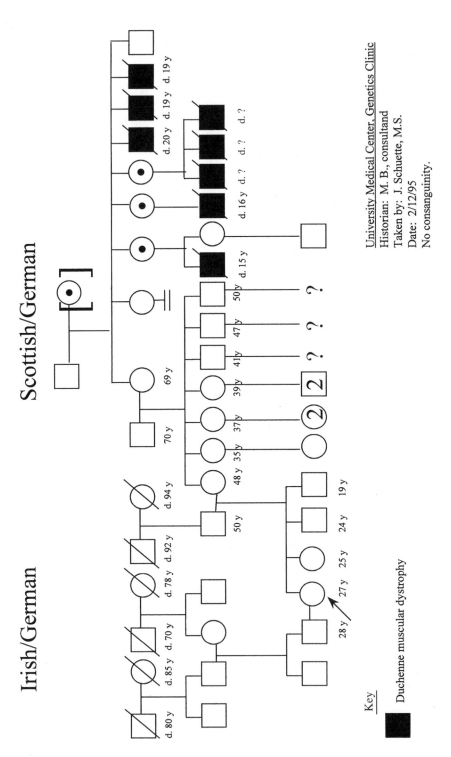

Figure 2.5 Pedigree showing family history of Duchenne muscular dystrophy.

Scottish/German

Irish/German

d. 80 y d. 85 y d. 70 y d. 78 y d. 92 y d. 94 y

d. 20 y d. 19 y d. 19 y

d. 16 y d. ? d. ? d. ?

d. 15 y

69 y

70 y

50 y

48 y 35 y 37 y 39 y 41y 47 y 50 y

? ? ? ? ?

27 y 25 y 24 y 19 y

28 y

Key

■ Duchenne muscular dystrophy

University Medical Center, Genetics Clinic
Historian: M. B., consultand
Taken by: J. Schuette, M.S.
Date: 2/12/95
No consanguinity.

historian? Is he or she the dominant partner, or is there another reason? Does one family member consistently interrupt, correct, or contradict another? Is there tension, or does the exchange between family members appear relaxed and mutually supportive?

Assessments about the level of family stress can sometimes be made during the gathering of family history information. The client and family may appear anxious when identifying members of the family as affected with disorders. A sense of guilt may also be apparent. For example, the counselor obtaining a family history of a patient who may have an X-linked disorder has an opportunity to observe as the client provides information about potentially affected family members and carriers. When there is a positive history, the counselor should watch for clues that might provide an indication about the level of the family's sensitivity to this information.

A family's assumptions about the presence of a disorder in a family member often become apparent during the process of gathering the history. For example, the client may state that the cause of mental retardation in a sibling was "lack of oxygen" during delivery. This may be an explanation that makes sense to the family and has been believed for a long period of time. A sensitive counselor will understand that the suggestion of an alternative explanation for the mental retardation could be disruptive. Families may also have myths that have served to explain why certain members are affected with a disorder while others are not. For instance, some assume that sharing a family resemblance has predictive value in determining who will inherit a particular disorder. In other instances, a family may assume that only the females or only the males are at risk, even though the disorder in question is autosomal dominant. The genetic counselor needs to be cognizant of family assumptions and myths that may present themselves while the history is being recorded, as these may impact on the client's understanding of genetic principles.

In addition to the subjective observations that can be made during the process of gathering the family history, the counselor can benefit from the objective information contained in the pedigree itself. Information about the nature of relationships (e.g., whether there has been a marriage, multiple marriages or partners, separation, or divorce) is valuable in coming to know the client. Ethnic background and religion can provide clues about the ways clients may view their world, family, and the events that have brought them to the genetics clinic.

The patient's experience with genetic disease, loss of close family member(s), infertility, pregnancy losses, and infant death(s) are in "black and white" in the pedigree. When, for example, a history of mental retardation in a sibling (regardless of whether there is a known diagnosis) is obtained from a client being counseled for advanced maternal age, this information is potentially useful. Despite the obvious potential impact on the patient's risks for having a similarly affected child, this personal experience is highly likely to have had an impact (whether negative or positive) on the client's life and is likely, as well, to play a role in the decision to have prenatal testing. Later exploration of this information is often useful for patients for whom the decision of whether to have prenatal testing is difficult.

SUMMARY

There are many methods of drawing pedigrees and describing kinship, but for my own purposes, I still prefer those that I designed myself.
—Galton, 1889

The family history and pedigree are the basis for providing clients referred for genetic evaluation and counseling with a diagnosis, as well as risk assessment, education, and psychosocial support. Accuracy, detail, and relevance are paramount. The genetic counseling student must develop a mastery of this fundamental task, and (unlike Francis Galton) must employ standardized pedigree symbols and nomenclature that enhance the utility of the information obtained. An accurate and complete family history and pedigree lay the foundation for the highest standards of patient care and genetic counseling.

REFERENCES

Adato, A (1995) Living legacy. *Life* 18:60–69.

Bennett RL, Stenhaus KA, Uhrich SB, O'Sullivan C (1993) The need for developing standardized family pedigree nomenclature. *J Genet Couns* 2:261–273.

Bennett RL, Steinhaus KA, Uhrich SB, O'Sullivan CK, Resta RG, Lochner-Doyle D, Markel DS, Vincent V, Hamanishi J (1995a) Recommendations for standardized human pedigree nomenclature. *Am J Hum Genet* 56:745–752.

Bennett RL, Steinhaus KA, Uhrich SB, O'Sullivan CK, Resta RG, Lochner-Doyle D, Markel DS, Vincent V, Hamanishi J (1995b) Recommendations for standardized human pedigree nomenclature. *J Genet Couns* 4:267–279.

Bennett RL, Steinhaus KA, Uhrich SB, O'Sullivan CK, Resta RG, Lochner-Doyle D, Markel DS, Vincent V, Hamanishi J (1995c) Reply to Marazita and Curtis (Letter to editor). *Am J Hum Genet* 57:983–984.

Cowley G (1993) Family matters. The hunt for a breast cancer gene. *Newsweek* December 6, pp. 46–55.

Galton F (1889) *Natural Inheritance.* London: Macmillan, quoted by Resta (1993).

Gelehrter TD (1983) The family history and genetic counseling. Tools for preventing and managing inherited disorders. *Postgrad Med* 73:119–126.

Gordon H (1972) The family history and pedigree chart. *Postgrad Med* 52:123–125.

Gotto AM, Wittels E (1987) The family history in health care evaluation. *Health Values* 11:25–29.

Green P (1983) Skeletons in the closet: Exploring personal family background as a prerequisite for family nursing. *J Adv Nurs* 8:191–200.

Gutmann DH, Aylsworth A, Carey JC, Korf B, Marks J, Pyeritz R, Rubenstein A, Viskochil D (1997) The diagnostic evaluation and multidisciplinary management of neurofibromatosis 1 and neurofibromatosis 2. *JAMA* 278:51–57.

Holsinger D, Larabell S, Walker AP (1992) History and pedigree obtained during follow-up for abnormal maternal serum alpha-fetoprotein identified additional risk in 25% of patients. *Clin Res* 40:28A.

Juberg RD (1972) Making the family history relevant. *JAMA* 220:122–123.

Krause C (1995) *How Healthy Is Your Family Tree?* New York: Simon & Schuster.

Marazita M (1995) Standardized pedigree nomenclature (Letter to editor). *Am J Hum Genet* 57:982–983.

Papazian R (1994) Trace your family tree: Charting your relatives' medical history can save your life. *Am Health* 13:80–84.

Pearson K (ed) (1912) *The Treasury of Human Inheritance* (Parts I and II). London: Dulau and Co. Quoted in Resta (1993).

Resta RG (1993) The crane's foot: The rise of the pedigree in human genetics. *J Genet Couns* 2:235–260.

Schuette J, Spergel S (1989) Prenatal diagnosis counseling: Current practices and policy. Poster at annual meeting of the National Society of Genetic Counselors, Baltimore, MD, Nov. 9–12.

Spence MA, Hodge, SE (1996) Segregation analysis. In: Rimoin DL, Connor JM, Pyeritz RE (eds), *Emery and Rimoin's Principles and Practice of Medical Genetics,* 3rd ed. New York: Churchill Livingstone, pp. 103–109.

Stang PE (1985) The family history and medical genetics. *Physician Assist* 9:33–34, 38–40.

Steinhaus KA, Bennett RL, Uhrich SB, Resta RG, Lochner-Doyle D, Markel DS, Vincent VA (1995) Inconsistencies in pedigree nomenclature in human genetics publications: A need for standardization. *Am J Med Genet* 56:291–295.

Weiss R (1994) Medical family tree is revealing. *Ann Arbor News* June 13, p. D1.

Wolpert C (1994) Genetic history . . . recertification series. *Physician Assist* 18:23–24.

Appendix 2.1

Example History and Pedigree

The fictitious example reproduced here and in Figure 2.6 demonstrates the recording of family history information using recommended symbols and nomenclature. The details of the case (adapted from Bennett et al., 1995a) are as follows:

FICTITIOUS FAMILY HISTORY AND PEDIGREE USING RECOMMENDED PEDIGREE NOMENCLATURE*

The consultand, Mrs. Wang, age 40 years, and her husband, Mr. Wang, age 36 years, were referred for genetic counseling regarding advanced maternal age. Mrs. Wang is 16 weeks pregnant. One of her prior pregnancies ended with an elective termination (TOP) at 18 weeks of a female fetus with trisomy 21.

Mrs. Wang's History

Mrs. Wang had three prior pregnancies with a former husband, the first a TOP, the second a spontaneous abortion (SAB) of a female fetus at 19 wk gestation, and the third a healthy 10-year-old son who was subsequently adopted by Mrs. Wang's 33-year-old sister, Sue.

Sue had three pregnancies, two SABs (the second a male fetus at 20 wk with a neural tube defect and a karyotype of trisomy 18), and a stillborn female at 32 wk.

*Adapted, by permission, from Bennett et al., *Am J Hum Genet* 56: 752. Copyright © 1995 by The University of Chicago Press.

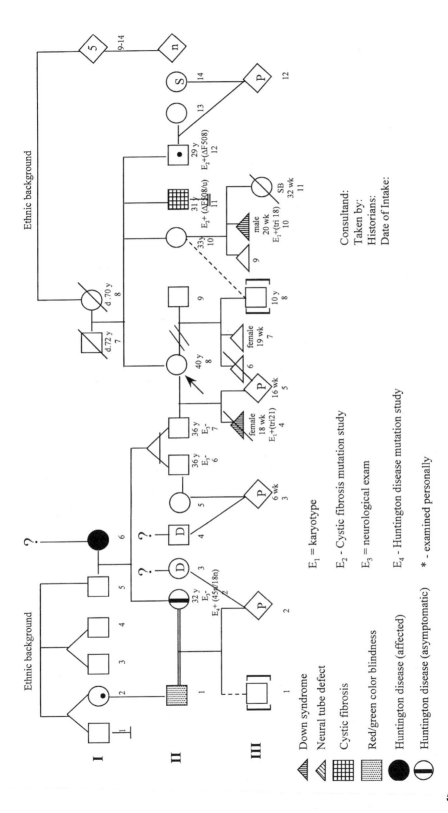

Figure 2.6 Hypothetical pedigree, using recommended nomenclature. (Reprinted, by permission, from Bennett et al., Am J Hum Genet 56:752. Copyright © 1995 by The University of Chicago Press.)

E_1 = karyotype

E_2 - Cystic fibrosis mutation study

E_3 = neurological exam

E_4 - Huntington disease mutation study

* - examined personally

Down syndrome

Neural tube defect

Cystic fibrosis

Red/green color blindness

Huntington disease (affected)

Huntington disease (asymptomatic)

Consultand:
Taken by:
Historians:
Date of Intake:

Ethnic background

53

Mrs. Wang has a 31-year-old brother, Sam, who is affected with cystic fibrosis (CF) and is infertile.

Her other brother, David, age 29 years, is healthy and married. Using gametes from David and his wife, an unrelated surrogate mother has been successfully impregnated.

Mrs. Wang's father died at age 72 years and her mother at age 70 years, both from "natural causes." Mrs. Wang's mother had five healthy full sibs who had many healthy children.

Mr. Wang's History

Mr. Wang had two siblings, an identical twin brother, Lee, whose wife is 6 weeks pregnant by donor insemination (donor's history unknown), and a 32-year-old sister, Audrey.

Audrey is married to Liam, her first cousin (Audrey's father's sister's son), who has red/green color blindness. She is carrying a pregnancy conceived from Liam's sperm and ovum from an unknown donor. Audrey and Liam have an adopted son.

The family history of Audrey's mother, who has Huntington disease, is unknown.

Mr. Wang's father has a set of twin brothers (zygosity unknown) and another brother and sister (Liam's mother) who are also twins.

Information Obtained After the Initial Meeting

Mrs. Wang's brother, Sam, with CF, has one Δ F508 allele, and one allele that cannot be identified.

David carries the Δ F508 allele.

Mr. Wang and his twin brother show no clinical signs of HD on examination by the neurologist.

Audrey has a normal neurological exam (for symptoms of HD) by the neurologist, but her DNA testing shows that she has a CAG expansion of 45 repeats (second allele has 18 repeats), consistent with the presymptomatic diagnosis of HD.

3

Interviewing Techniques

Diane L. Baker

Interviewing can be viewed as purposeful conversation.
—Donner and Sessions, 1995

You are about to leave the staff room and head down the hallway to a clinic room where two parents and their infant await you. This meeting will be your opportunity to learn firsthand about the expectations, concerns, knowledge base, questions, and hopes of these parents. During this clinic visit you will elicit, comprehend, evaluate, clarify, and confirm information about the family, and the basic method of inquiry that will guide your exploration will be interviewing.

Interviewing is both a method for guiding the exchange of information with others and an art—an art that draws on one's critical thinking, personal style, and experience. Interviewing is a two-way interaction between the genetic counselor and the client. The counselor is in the role of host–interviewer, assuming responsibility for the setting and context of the relationship. The client is in the role of guest–interviewee, whose unique experience, circumstances, and worldview the counselor seeks to understand. The genetic counselor invites the client into a working alliance as together they examine the genetic, medical, and psychosocial dimensions of health, disease and at-risk status.

Genetic counselors use interviewing techniques to elicit information throughout a clinic visit as they obtain histories, furnish education, and provide counseling. Some of the specific activities a counselor accomplishes through the use of interviewing skills include:

A Guide to Genetic Counseling, Edited by Diane L. Baker, Jane L. Schuette, and Wendy R. Uhlmann. ISBN 0-471-18541-8 (cloth), 0-471-18867-0 (paper). Copyright © 1998, Wiley-Liss, Inc.

- clarifying goals for the visit
- obtaining family, pregnancy, developmental, and medical histories
- inviting narratives about events and experiences
- exploring knowledge about a condition or an inheritance pattern
- eliciting beliefs, hopes, or concerns about personal genetic circumstances
- inquiring about views on the use of genetic testing and/or research participation
- assessing understanding about information presented
- identifying support systems and unmet needs

Interviewing is something we do throughout our daily lives, whether talking with a friend, clarifying with a mechanic the work needed on our car, or designing a project with a colleague. Therefore, we bring insights and techniques from these personal experiences to our professional interviewing. However, analysis of the structure and purpose of professional interviewing indicates that techniques learned from our personal experiences often need to be modified before being used in interviewing relationships with clients.

This chapter examines the goals of interviewing, inventories the toolbox of interviewing techniques used by genetic counselors, and identifies the categories of topics to be listened for during an interview. We examine the dynamics of the counselor–client relationship for their influence on the interviewing process, and we conclude by looking at some common pitfalls the new genetic counselor may encounter.

GOALS OF INTERVIEWING

The primary goal of interviewing is to elicit and understand the knowledge, experiences, assumptions, and beliefs that are unique to each client. The counselor then builds on and connects this information to the client's specific genetic circumstances and stated goals. Throughout the process, the counselor will listen for the client's overt or subtle invitations to explore paths of her own choosing, such as nonpaternity, previous pregnancy termination, family dynamics, or personal fears.

The genetic counselor applies interviewing skills by tailoring them to the specific tasks at hand. For example, when obtaining intake information from a client, the counselor will use interviewing to facilitate recall. Later, specific forms of questioning can serve to focus the conversation on critical issues or to encourage the client to elaborate on particular points. Later still, the counselor will use interviewing to assess the client's understanding of, and personal identification with, the information provided in fulfillment of the educational goal of the session.

As the visit progresses, the counselor will use interviewing to clarify and interpret client statements, to explore perceived consequences of choices, and to uncover assumptions or misunderstandings. At the conclusion of the visit, interviewing can help to ascertain the client's perception of what happened during the visit and to assess her readiness to participate in a follow-up plan.

GETTING STARTED

Let's consider the counselor's self-awareness about the behaviors associated with attending to body language, engaging in small talk, and maintaining a nonjudgmental demeanor. Our awareness of these dynamics throughout the visit sets the stage for effective interviewing by promoting a trusting relationship with the client.

The genetic counselor can take certain simple, direct actions early in the session to promote the client's comfort. Examples include encouraging clients to remove their outer coats and suggesting that everyone might be more comfortable sitting rather than standing. Encourage couples to sit together so that you may easily talk to them both without having to swivel your chair. Rearrange the furniture in the room if necessary to create a more suitable seating arrangement that encourages interaction. Try not to place any furniture, not even a desk, between you and the client.

Attending to Body Language

The genetic counselor will convey a great deal to the client by the level of comfort, interest, and genuineness she conveys through her face and body. A relaxed, comfortable, and open posture; facial expressions that convey interest and attentiveness; and easy eye contact set a welcoming stage for most clients of western European background. (The reader is referred to Chapter 7 on multiculturalism for a more thorough consideration of other cultural backgrounds.) Maintaining a constant and attentive physical presence throughout the session conveys assurance and calmness. If the counselor frequently looks down at her notes, sits in a manner that conveys discomfort (frequent shifting of positions, lack of eye contact when asking questions, or turning away while the client is responding to questions), or maintains a tight, controlled posture (arms crossed in front, remaining standing while the client is sitting), she will be visibly conveying distance or discomfort to the client. In the same manner, the counselor can look for these clues from clients to indicate their level of comfort or discomfort.

Early in the session, you may wish to position your chair at a slight distance from your client, to give them space. As the session progresses, it may then be appropriate to pull your chair up closer, to reinforce your conversation, or to assist in showing laboratory results or diagrams. Additionally, moving from an erect sitting position to leaning forward or "into" the conversation conveys interest, familiarity, and trust. Look for and respond to these cues, and ask other colleagues and supervisors for feedback on your own body language.

Directed Small Talk

Small talk is used throughout many social interactions to put participants at ease, to establish common ground, and to allow people to relate to strangers. However, small talk in the clinical setting is not the same as small talk in a social setting. In our role as professionals, small talk with clients is focused on topics related to the visit or drawn from information available about the client (e.g., the distance traveled for the appointment, the length of the wait prior to meeting with you, or general acknowledgment of factors about the clinic setting). The counselor is using this directed small talk in several ways: to get to know the client; to communicate awareness of her circumstances; to allow her to know more about the counselor, the clinic, and the visit so that she may relax; and to begin establishing a foundation for a trusting relationship.

Obvious circumstances about the client or family that need to be acknowledged can be expanded upon to continue this directed small talk. For example, if there is an infant to be examined, is the mother breast-feeding? On what schedule? Suggest a plan that will accommodate the baby's feeding during the visit. How in general has the baby been feeding? If there is a young child to be examined, did the parents bring toys or snacks? Reassure parents that they may get them out whenever they think it necessary. Ask the parents to tell you how the staff can best interact with their child. What will help, or should be avoided, when interacting with the child?

Generally, one does not make idle comments or pursue a line of inquiry without a purpose. This is not a time for social small talk, which can mislead the client into thinking that you expect a social rather than a professional relationship or could convey that superficial interactions will be more valued than genuine, personal interactions.

Gathering Information in a Nonjudgmental Manner

It is not at all unusual during an interchange with any particular client to hear of behaviors, child-rearing practices, health beliefs, or personal circumstances that are unfamiliar or possibly disconcerting to the genetic counselor. Topics that frequently arise in a routine genetics clinic that may distress a student include nonpaternity, suicide attempts, recent death in the family, pregnancy as a result of rape or incest, abandonment by a spouse, addiction, experiences of insensitivity by others toward an individual with a birth defect or health condition, and a variety of other powerful personal stories. A professional who has dual responsibilities of inviting clients to share relevant areas of their life and respecting each client's experiences must receive these stories in the manner in which they are offered—as a personal and critical piece of the client's experience. It is important to maintain your professional demeanor in light of revelations that can have a powerful impact as you hear about them. Avoid an "open-mouthed" response. You are obviously not immune from being significantly affected by your client's experiences, but you should strive to receive and understand these experiences in a professional and nonjudgmental manner. When you encounter such a topic, be sure to discuss it later with your supervisor or colleagues, with respect to your immediate handling of the situation and your own personal reaction to it. Surprising or unanticipated information is likely to be elicited frequently, and it is important to learn how to manage your reactions on such occasions.

INTERVIEWING TOOLS

The interviewing techniques that follow are drawn from experience in a pediatric genetics setting and therefore reflect the flavor of that environment. These are general principles, however, and are applicable in other clinical settings. Table 3.1 summarizes these techniques.

Questioning

Questioning is the most direct, useful, and regular method of inquiry available to the genetic counselor. Questioning can be used to introduce a topic or to clarify what has already been said.

> Tell me more about. . . .
>
> When did you first become aware of . . . ?
>
> What do you mean by . . . ?
>
> Could you give me an example of . . . ?
>
> What do you think about . . . ?

Asking direct questions of a client may initially be uncomfortable for students. It represents a significant departure from conventional social interactions, which have fairly well-defined limits on the asking of personal questions. This is exactly what you are now called on to do. As a professional, you now have a responsibility to engage in a "purposeful conversation" that encompasses understanding your client's experiences, values, and beliefs. This understanding will assist you in connecting clients' genetic circumstances to their personal experiences and values. When such questioning is conducted in the context of a relationship that is professional, trusting, and respectful, clients not only respond, but welcome the opportunity to share their story.

You will want to use a variety of questioning techniques as they will yield different results. *Open-ended* questions are broad and framed to invite the client to respond as expansively or specifically as they choose. *Narrow* or *focused* questions can help

---

TABLE 3.1   Interviewing Techniques

Questioning
   Open-ended: invites broad responses
   Focused: guides response toward specific circumstances
   Closed-ended: asks for yes/no answers or for specific details; does not encourage elaboration
Rephrasing: restate your understanding of what the client has said
Reflecting: repeat the last phrase of a client's statement as a question
Redirecting: change the direction or focus of the conversation
Promoting shared language: mirroring some of the client's language and encouraging the use of relevant genetic terminology
Using silence as a part of the counselor–client interaction

---

guide the client toward specific moments or circumstances. *Closed-ended* questions are generally those that have yes or no answers or ask for only a specific piece of information (a date or evidence of a symptom).

When opening a new subject area with a client, you will generally want to start with open-ended requests or questions to create the opportunity to broadly hear what is relevant and to invite the client to share the richness of her experience.

> Tell me what living with a brother with hemophilia has been like for you.
>
> What have you and your partner talked about regarding your hopes and plans for starting a family?
>
> How does the prospect of a shortened life for Amanda affect how you spend a typical day?

Most clients will respond to open-ended questions in the manner you invite, and their reply will take the conversation in the direction that is most important or relevant to them. Other clients may pause, look quizzically at you, and provide only a brief reply, such as:

> He was like any other brother.
>
> We haven't really talked about it.
>
> I don't really think about it.

Evaluate these responses. Is the client intrigued by your question, but not sure whether you really want to hear the answer? If so, you will want to encourage a further response. For many clients this will be the first time they have been invited to tell their story to a professional, and hesitance is understandable. Alternatively, they may be conveying that they don't feel ready to talk about the topic at this time. If the latter is your assessment, you may want to provide a summary of your thoughts, so that either party can return to this topic later, with a statement like the following:

> I asked about this because it may be relevant to your current situation, and I want to understand more about what you think.

Specifically crafted, focused questions may then be employed to bring the discussion closer to the heart of an issue.

> *Counselor:* When did you first realize that your brother had hemophilia?
>
> *Client:* Oh, I guess it was during his first hospitalization.
>
> *Counselor:* Tell me what you remember about that hospitalization.

The intent of the focused questioning here is to create opportunities for the client to identify specific experiences or insights about her experiences with her brother and his hemophilia that may help her relate to her current circumstances or decision-making process.

Focused questioning can also be used as a prompt to continue a discussion the client may have truncated because she felt she had said enough or was unsure whether you wanted to hear more. A simple "Tell me more about that day" (that conversation, that moment at his bedside) can help you to learn more about an experience that was powerful and is important to the client.

Closed-ended questions often have yes or no answers, or are used to pin down a time, a decision, or other specific information. This type of questioning is used frequently in medical history taking ("Do you have . . . ? Have you ever experienced . . . ? How much did the baby weigh?). The strength of closed-ended questions is in their specificity, and their application depends on your purpose. The following closed-ended questions ask only for a yes or no answer.

Did you know at that time that he had hemophilia?

Have you been trying to become pregnant?

Have you told your friends about Amanda's life expectancy?

When working with couples as clients, you will strive to give each person an opportunity to respond to questions. If one partner answers an open-ended question placed to them both, acknowledge the response and then specifically invite the other partner to respond. At times a client will bring along a parent or other family members as a support person. In these situations, you will need to consider exactly who your client is. It may not be appropriate to involve every party who is present as though all have an equal stake in the genetic situation. It may be necessary to acknowledge that different agendas are present in the room and to develop a consensus with the family members present about what will be addressed during the visit. Follow-up visits can then be scheduled for other family members.

The counselor may use questioning to clarify the meaning of something the client has said. For example, a mother whose newborn was diagnosed with Down syndrome, when asked if she knew any individuals with Down syndrome, replied, "Oh, yeah, I know what that means." She had not distinguished between having first- and secondhand knowledge about Down syndrome or other forms of mental retardation. A counselor could clarify matters by focusing such a response with another question (e.g., "Tell me what it means to you."). Depending on the reply the counselor would either proceed with the interview or clarify that Down syndrome is a unique and specific form of mental retardation, and address its significance to the client's situation.

Clarifying questions are very useful with interactions around topics that appear to be uncomfortable for the client or the new counselor. In such circumstances the counselor may be prone to make assumptions rather than ask for more information to elucidate the situation. For example, suppose a couple whose only child had recently died of a congenital heart defect is seen in the genetics clinic for a follow-up visit to discuss the autopsy report. To the counselor's question ("How are you doing?"), the father gives a brief reply ("Okay, . . . I guess."). A new genetic counselor, seeing the flat affect and sad demeanor of both parents, might simply proceed, assuming that it was appropriate for them to be sad. Thus the counselor's own discomfort over the parents' loss might prevent her from responding to what could have been a subtle move to open up the topic further. Instead of moving on, the counselor could have tried to clarify the father's meaning ("Tell me what 'okay' means.").

The most important guidepost when questioning a client is this: don't make assumptions. If you're unclear about an aspect of the client's experience or perception, or if you perceive that a particular issue is significant to the client, ask about it. The following examples, though not all framed as questions, demonstrate how you can respond to a client's narrative in a way that avoids making assumptions about what was said.

> When you said . . . , did you also mean . . . ?
>
> Tell me more about what your nod means.
>
> You just said . . . about yourself, but do you think other people in the family feel the same way?

Then speak about your interpretation of what has been said or is happening during the session, and invite the client to correct or expand on your understanding.

> So, even though you felt the procedure was pretty routine, it sounds like your husband was troubled by watching you go through it.
>
> You've brought up the issue of . . . several times today. Do you think that is influencing how you feel about having the test?
>
> Do I understand you to mean . . . ?

Generally when a mutually respectful relationship has been established with the counselor, a client not only will respond to questions but will welcome them as an opportunity to explore relevant experiences and how they have shaped the person's views. However, if a client provides a general, unfocused reply to your query, assess the reasons for this. Might you have been unclear? Is the client trying to avoid the topic you are leading the conversation toward? If so, why? Interviewing is a human interaction, and at times every counselor will make a wrong turn—for example, by unwittingly acting on an incorrect assumption about the client or misinterpreting certain cues or responses. If there are indications that the counselor was premature or unfounded in pursuing a line of questioning, it is necessary to accept responsibility and correct the direction of the conversation.

Rephrasing

Rephrasing involves stating in your own words what the client has just told you. It is a valuable tool for demonstrating to the client that she is being listened to and for ensuring that you have understood what she intended to convey. It reinforces that the two of you have visited and understood a part of the client's experience. It is especially valuable when you and your client have different language skills or when a client uses a great deal of slang or colloquial speech that is unfamiliar to you.

Suppose your client, the mother of two children, is now pregnant by a new partner about whom she states: "So, I just don't know if he's gonna disappear or what." What she has just said seems to be a clear and fairly frank assessment of the strength of the relationship. The counselor may feel this statement can be taken at face value, or she may choose to rephrase it to ensure that she fully understands the woman's meaning. The counselor may reply with something like, "So, you don't think that he's very committed to the relationship?" And pause (is she nodding in agreement?) and then proceed to inquire about the basis for her statement ("Why do you think that is?"). The counselor will have ensured that she indeed understood what the client was saying and will have laid the groundwork for ongoing clarification of her understanding as they cover topics such as risk assessment, prenatal testing, pregnancy termination, and other areas in which an understanding of the client's personal relationships and beliefs is paramount to providing quality care.

Reflecting

Reflecting involves repeating the last phrase of a client's statement in the form of a question to encourage further exploration of the topic. It is also used to maintain the direction along which the conversation is going. The use of reflection is demonstrated in the following exchanges:

> *Client:* Well, since my nephew was diagnosed with muscular dystrophy, my husband just seems scared about another pregnancy.
> *Counselor:* He seems scared?
> *Client:* Well, I guess we're both a little scared, you know, that it might happen to us.

> *Client:* So, that's my decision—a mastectomy is better than cancer.
> *Counselor:* Better than cancer?
> *Client:* Well, sure. That way I don't have to worry any more.

Reflecting encourages the client to amplify her thoughts and ideas in a way that more clearly identifies the significance of her feelings or observations. This also gives the counselor more information regarding the basis for the client's thoughts and beliefs, so that she can respond without making assumptions. In the first example, reflecting has helped the client reveal more to the counselor (and perhaps to herself) about the fear that both partners are feeling. In the second example, the client's comment about mastectomy may indicate that she believes surgery will permanently protect her from cancer.

Redirecting

Redirecting is used by the counselor to manage the rate of information exchange: to direct the introduction and flow of topics, or to refocus the discussion when the client has gone off on a tangent. The following are examples of statements that can be employed to redirect the counseling interaction:

> That's an important issue, but first I'd like us to go back to. . . .
> We will get to that, but first I think it would be helpful to hear. . . .
> Before moving on, let me ask you a little more about. . . .

How about that other matter you mentioned regarding . . . ?

I'd like to slow down here a bit. Do you think what you just said about . . . is relevant to the decision you're facing today?

That's an important question which is best answered by your cardiologist. So, could we come back to. . . .

Consider the example of a young adult male client referred by a urologist following the diagnosis of congenital bilateral absence of the vas deferens (CBAVD), which increases his risk for carrying mutations in the *CFTR* gene responsible for cystic fibrosis. When asked by the counselor what he understands about the diagnosis of CBAVD, he replies, "Well, I guess I won't ever be a father." Rather than answer within the limits of the counselor's specific question, he has provided his own bottom-line answer. The counselor is at a branch point here. She initially sought to clarify the client's knowledge of the anatomical diagnosis provided by the urologist and to determine whether he understood the specific and chronic nature of the condition. She had planned later to explore his awareness of the implications of this diagnosis—infertility. He, however, has gone straight to the implication of CBAVD and in a way that personalizes it for himself. Nevertheless, the counselor still does not know whether the client specifically understands the underlying mechanism that led to his diagnosis. She may choose to join him at the point he has taken the conversation—exploring his feelings about the implications of infertility and returning later to an exploration of his understanding of CBAVD. Alternatively, she may determine that this client will be comfortable waiting for the more personal discussion he has initiated. In the latter case she could redirect through a statement such as, "You've just stated an important bottom line of CBAVD. However before we talk about that, I'd like to know if anyone has explained to you what the underlying mechanism is that leads to this condition."

Promoting Shared Language

One technique that can enhance client comfort and understanding is an effort by the counselor to mirror the client's language or communication style (to an appropriate degree). The counselor should not abandon her style, but to convey that the client is being heard and understood as an individual, she might say, "Well, what I call the 'nonworking' gene and you call the 'bad' gene are really the same thing," and from this point on use the client's term. Additionally, the counselor might model or invite the client to use alternative or new terms that could be helpful to her in the future. For example, "Neurofibromatosis is a mouthful, isn't it? It's also called 'NF,' which is what you'll hear me use today."

Communication style is a broad area and includes word usage, speaking tempo, and the use of colloquial terms. The counselor's ability to echo the client's communication style, if appropriate and comfortable, can facilitate communication. For instance, if your client has a deliberate manner of speaking, while you are naturally a rapid talker, you might choose to slow down your tempo to be more in step with your client's. Likewise, if certain colloquial terms the client uses are comfortable for you, you might want to include them in your conversation throughout the session.

Silence

It is not at all unusual for silence to be a part of a typical genetic counseling interaction. Those who come into this field strongly guided by western socialization generally feel uncomfortable with more than a few seconds of silence in most interpersonal interactions. This is another learned response that the new genetic counselor should leave at the door. Silence can be an important and constructive component of our interaction with clients. There are moments when the client is simply not ready to put into words certain thoughts or reactions. Give people time. Let them find the words if they seem to be trying to do so. You may want to offer a soft, "It's okay, we're in no hurry." Your patient, calm presence can make the silence comfortable and safe. Many times a client who has been emotional, as when describing a personal experience or reliving the birth of a child, may simply need time to recover from the return to this memory, to cry, to sit in quiet contemplation, or to share the experience with you silently. Do not feel that you need to fill silences. Your supervisors can be very helpful and supportive in this area as you encounter and explore the use of silence in the genetic counseling setting.

LISTENING SKILLS

As one is communicating with a client, what exactly does one listen for? What is it that cues the counselor to explore another topic, select a different approach to a subject, or consider a new strategy? Cues from the client are many, yet sometimes subtle, and can indicate opportunities to transition into a more focused educational or counseling mode. What are these cues like?

The Client's Narrative

A client's personal narrative about the circumstances that led him or her to seek genetic counseling is often rich in content, with overt and subtle indicators about the significance of various aspects of the individual's experiences. A narrative may include references to issues within the extended family, the status of intimate relationships, and medical and reproductive concerns, along with stated goals and specific concerns. These stories are what you listen to and use to guide your decisions about when to introduce a new topic, return to a previous one, explore connections between the client and the genetic information, initiate an educational intervention, explore options, and complete the many other tasks of a genetics clinic visit.

Gaps, Omissions, and Inconsistencies

Recognition by the counselor of missing or discrepant information in a client's narrative can indicate a subject area that might benefit from further discussion because the client is uncomfortable with it, or ambivalent or uninformed about it. A **gap** refers to the recognition by a counselor that an issue typically associated with a specific genetic situation has not been raised. The term **omission** describes a circumstance in which the client neglects to include relevant information in her personal narrative or history. An **inconsistency** is a discrepancy identified by the counselor between what a client says at different points during the session or between what the client says and then does. The ability to recognize gaps, omissions, and inconsistencies in a dynamic client interaction comes with experience.

Consider two parents and their infant daughter who pay a return visit to a genetic counselor in a myelodysplasia clinic. The parents were very knowledgeable about their daughter's myelomeningocele, including the anticipated implications to her development. However, the counselor recognized that the topic of the baby's future sexual functioning had not been raised by the parents and might not have been covered by the multidisciplinary team. Realizing that this is a distant, yet relevant issue for the parents to understand, and that they might be reluctant for a variety of reasons to bring it up, the counselor introduced it ("Has anyone discussed . . . ?"). Her "ear" for disorder-specific issues allowed her to recognize a potential gap in the parents' information. Finding that the topic had not been covered, she provided guidance as to what might be expected and reassured the parents that they could bring up the topic again at any time.

An omission would be exemplified by a female client who gives a pregnancy history that mentions only two live-born children, whereas the counselor knows from the referring obstetrician that this client has had an elective termination. There could be a variety of reasons for this omission, and the counselor will need to determine whether the failure to answer completely is incidental to the purpose of the visit, hence does not require exploration, or is relevant and needs to be addressed.

Consider the following example of an inconsistency. Two parents and their 15-year-old son were seen in the clinic for a workup regarding the possible diagnosis of Marfan syndrome in the youth. Following a physical examination and cardiac evaluation, the parents were informed that there was very strong evidence of the genetic condition. Earlier in the visit, both parents had declared their full support of whatever needed to be done for the sake of their son's health (limiting his activities, or scheduling more frequent medical visits). They conveyed concern and affection for their child. After the diagnosis was made, however, each expressed disbelief when informed that the boy might not be able to continue playing on his school basketball team. The parents spoke of their hopes of eligibility for a college scholarship, described their son's talent at basketball, and appeared to be hard put to believe that he might have a serious health condition.

The counselor recognized that the contrast between the parents' earlier statements of concern and their present expressions of disbelief could be explained by the shock of the reality of the diagnosis. She considered responding in a way that would validate their anxiety over the diagnosis, but ultimately decided on a more direct approach and pointed out to the parents the contrast between their earlier statements and their present reaction to the confirmation of earlier suspicions, and asked why this might be. The couple paused, seemingly unaware of the inconsistency. Then the mother slowly replied that she felt she would be betraying her son if she did not oppose the placement of restrictions on him. She went on to acknowledge that she so strongly identified with his desire to play basketball that resisting the placement of limitations felt like a way of protecting him. The father stood by quietly and nodded at his wife's words. The counselor then explored with the parents the loss they were experiencing around this diagnosis. Later she came back to this interchange in a discussion of how the parents might anticipate and respond to the reactions of friends, family, and teammates who also, in an attempt to protect the youth, might be expected to question the diagnosis.

Recurrent References

If a client repeatedly brings up a concern, an issue, or a question, this area should be given special attention. When a client returns to a topic that appeared to have been well covered or was incidental to the primary circumstances, the counselor will want to ask more about it to understand its significance to the client. It may be that a client has a related concern but is unsure of how to introduce it. Some clients, on the other hand, have difficulty giving up a misperception about a risk, or a mythical explanation they have carried for a long time. Such a situation can be acknowledged—perhaps with the additional support that the client may still continue to question the new or different interpretation of information, even after the visit and follow-up letter.

Analogies

Very often clients use analogies to help convey the essence of their experiences. Analogies are ways to make something more accessible to others by relating the matter of concern to something with which listeners are familiar. For example, a client describing her caretaking of a parent with Huntington disease might say, "It's as though I traded in being a daughter and became hired help." This analogy says a lot about what the client has given up and taken on and can be an invitation to explore this changed relationship. Analogies may convey something more about the meaning or significance an experience holds for the client. They can also present a safer way to reveal a dimension of an experience that seems too risky to state outright. Listening closely to analogies and exploring them can increase the counselor's understanding of significant events and experiences.

Client Closing Statements

Often the client's closing comments reveal a great deal about what he or she is experiencing. Also referred to as "hand-on-the-doorknob" comments, these are statements made as the family and the counselor are about to exit the room or close the visit. Consider the following case. A father indicated several times that he felt the prenatal diagnosis of Down syndrome in his wife's pregnancy might be a mistake. The counselor explained the basis for, and reliability of, the diagnosis, and also supported both parents' hope for the best possible outcome for the unborn child. At the conclusion of the visit, the couple said they were comfortable with their decision to continue the pregnancy. As he was leaving, the husband turned and said to the counselor, "Of course, since we are Asian, they won't be certain at his birth if our son has Down syndrome." The counselor recognized this as a recurring attempt by the father to distance the prenatal diagnostic implications from this pregnancy (some would call this a form of denial). She acknowledged the overlap of features, asked if she could share this concern with the OB staff, and encouraged the couple to stay in touch regarding any other questions that might come up.

Subsequently, during postclinic conference, the genetic counselor discussed with the rest of the genetics staff how they could help prepare the OB service for the possibility that the father may continue to hold this position at the time of delivery. A client's closing statements can often be used to help direct future interactions.

INTERVIEWING OR COUNSELING?

Interviewing is distinct from, but intertwined with, other clinical skills—particularly psychosocial counseling. Although similar techniques are used in both interviewing and counseling (e.g., rephrasing, reflecting, the use of silence), in general it is the counselor's application of these techniques that differentiates interviewing from counseling. The distinction with respect to whether a counselor is interviewing or counseling a client may not always be apparent. In fact, the experienced counselor sometimes moves so seamlessly from one mode to another that she is not aware of the transition. The importance here is to recognize that as a counselor, one has access to many tools to assist in serving clients during the genetic counseling process.

COUNSELOR–CLIENT DYNAMICS

Many circumstances can positively or negatively influence the dynamics of the counselor–client relationship. Understanding and managing these dynamics can enhance the sharing of information and promote the establishment of trust throughout the interviewing relationship. The predominant factor in promoting a productive interaction with clients is to see and treat each client as an individual—not as a male, a female, a blue-collar worker, or a 37-year-old primagravida. It is important to provide adequate attention, time, and skill to uncovering and addressing the issues outlined below. Only in this way can clients adjust their assumptions and expectations, becoming able to experience a truly fruitful interaction with the genetics staff.

Client Communication Style

As soon as she walks into a session, the counselor will begin assessing the client's communication skills and style. This includes fairly obvious information, such as whether English is the client's first language, to more subtle interpretations about the client's word choice, literacy level, and familiarity with medical terminology and the medical model (e.g., the experienced vs. the inexperienced patient). The "experienced" patient can arrive with a variety of past encounters, from having lived with the condition, having cared for someone with the condition, or having had extensive health care interactions for another reason. Even such experienced patients may not be familiar with, or ready for, the more egalitarian client–provider relationship that is a hallmark of the genetics clinic visit. Most clients will benefit from reassurances and additional invitations to interact and participate in the manner we strive to encourage.

Agendas

What happens during a patient visit is strongly influenced by the agenda of the genetics staff, which generally includes progressing through the activities of intake, diagnosis, risk assessment, education, counseling, referral, and follow-up. This agenda must be accomplished within the limited time frame of a scheduled visit and through efficient use of the staff's time and skills. The clinic visit is therefore a process controlled by the professional staff. Into this structure the client arrives with an agenda, either stated or unstated. Quite often it emerges and defines itself as the client comes to understand more fully what can be obtained from a professional relationship with the staff. While great importance is given to allowing the client's agenda to emerge, the success of this endeavor is to a significant degree under the control of the staff (Rapp, 1988).

Assumptions

The genetic counselor and the client also arrive in the clinic room with stated and unstated assumptions or expectations about the interaction they are about to begin. These assumptions are a natural response to something new. They serve as a way to prepare for encountering an unknown or for trying to anticipate the consequences of actions. Anticipating and examining these assumptions can help us to use the interview relationship more successfully.

Generally high on the list of the new genetic counselor is the assumption that the client will clearly and specifically state what she is seeking. However, clients may not be prepared to articulate their reasons for attending the clinic or to identify a goal for their visit. The new counselor tends to assume that clients are ready to forge a trusting relationship and work toward a defined or soon-to-be defined goal. We also tend to assume that clients seek increased insight into their own coping mechanisms around a specific risk or condition and will welcome opportunities to access their own resources as well as the counselor's resources. However, clients generally feel cautious and protective of their personal information and family experiences, and many are not prepared to share this information with a stranger. Indeed, until we have helped the client understand what a relationship with us can provide, he or she is unlikely to take advantage of this resource.

Counselors also tend to assume that the client is there voluntarily. In fact, the client may have scheduled the visit at the strong recommendation of a primary care or specialty physician and may not have really considered her own willingness to be present. Alternatively, she may have been unduly influenced by a parent, partner, or other significant individual. We also generally assume that our clients will be compliant with reasonable medical recommendations appropriate to their situation. Counselors may assume that the cost of the clinic visit is not an issue. However, costs can be quite important to clients, and sometimes are the determining factor as to when and how they avail themselves of genetic services.

Clients may arrive with expectations that include a broad range of presumptions, both accurate and unfounded. They may assume that the counselor will be an advice giver, will know what is "best" to do in a particular circumstance, and will make a recommendation that must be followed. Some simplistically expect that during their visit, the counselor will provide all the information necessary for arriving at a comfortable decision at the end of one session. Clients may assume that they are entering into an interaction that will necessitate, perhaps even demand, that they take a passive role. A client may feel that she is there only to please a referring physician, who insisted on this visit. Alternatively, she may believe that once she has a few facts and a risk figure, her decision will be obvious.

Clients tend to cast themselves and their counselors into the respective roles of novice and expert regarding genetic and medical experience. Some clients may carry this assumption further and presume that the counselor is also more knowledgeable about the choices and options that should be considered. We, in turn, tend to reverse this casting with regard to knowledge about living with or being at risk for a condition—seeing the client as the expert and ourselves as the novice. These may or may not be appropriately founded assumptions.

Responsibilities

The counselor and the client come to their interaction with different sets of responsibilities. The counselor will be conscious of her responsibilities to provide accurate medical, diagnostic, and genetic risk information and to protect patient autonomy. Additionally, she is accountable to the sponsoring health care institution to follow the established procedural, consent, and documentation duties required of each patient visit. The client meanwhile arrives with a set of responsibilities to an existing child, partner, pregnancy, family member, or health professional, as well as conscious or unconscious responsibilities to long-standing patterns of personal coping styles, secrets, or beliefs. The degree of accountability that each feels to these responsibilities will be a factor in the dynamics of the interviewing relationship.

Power

We should also examine the real and perceived power differentials in the counselor–client relationship (Brunger and Lippman, 1995; Kenen and Smith, 1995). The client generally seeks genetic counseling because of a belief that the genetics team possesses knowledge, experience, wisdom, or resources not likely to be available through any other avenue. The client may feel anxiety about meeting the expectations of the counselor, about being found worthy to receive all the information in the possession of the medical genetics staff. The client may incorrectly assume that answers to all his or her questions exist, but are subject to be withheld if the client is not seen as qualified (emotionally or intellectually) to receive them. Therefore, some clients feel a need to prove their worthiness. We cannot forget that as professionals we have control over and access to medical and genetic information, and the client must go through us to obtain it. Some clients assume that professionals are in the habit of holding back information—information that might be painful or difficult to understand, or that the client might resent hearing. They may therefore think it necessary to present a demeanor (whether passive or aggressive) that has the greatest likelihood of resulting in access to information. This particular disposition may be acutely present in clients whose loved ones have had lengthy hospitalizations or are diagnosed with chronic conditions. Such clients, having had experience with the culture of hospitalization, may feel that they cannot risk offending a medical professional.

Identifying with the Client

We all identify with characteristics of our clients or with circumstances of their lives—similarity in age, for example, or in deciding to undertake a second pregnancy—or with specific experiences (e.g., infertility or loss of a parent). These identifications with the client by the counselor can cause us to pay attention to our own needs and, therefore, do not belong in our professional interaction with clients. Likewise, the sharing of our own experiences detracts from attending to the client's affect and narrative and risks shifting the focus of the session away from the client.

SOME COMMON PITFALLS OF THE NEW INTERVIEWER

Anticipating the areas in which new counselors are more likely to commit oversights in client interviewing can facilitate the recognition and correction of such errors. These are not necessarily experiences to be avoided, but, if they occur, they should be explored with supervisors so that resulting insights and understanding can be applied to future interviewing experiences.

Wanting to Be Liked

One very common problem new counselors create for themselves has to do with how they frame their relationship with clients. It is easy to misapply the socially based measure, "Do they like me?" to one's interactions with clients. This measure is not appropriate in a professional's relationship with clients. Professional relationships with clients are not a social alliance. Successful client relationships are evaluated by examining outcome measures of communication, trust, information exchange, client satisfaction, and other standards, not by whether the counselor felt "liked" by the client. Many times a very effective counseling relationship is established despite a counselor's feeling that the client did not personally "like" her. It takes instruction, clinical training, and discussion with supervisors to move beyond one's initial tendency to apply learned social measures of successful interpersonal interactions and to use more appropriate and replicable measures that focus on the outcomes of a professional relationship.

Asking Too Many Closed-Ended Questions

We often have a lot of familiarity with closed questioning from our personal lives. We comfortably ask a friend or family member closed-ended questions (Did THAT really happen?, Do you think this is the right thing for us to do?). But someone who knows us well automatically provides the narrative, embellishment, or details that are necessary for our full understanding of their reply. We often don't use open-ended questions in our personal relationships. Students in genetic counseling often feel that they have asked an open question if they themselves assuredly intended to invite a narrative, even if their actual words were stated in a narrow or closed manner. However, clients often decline to amplify their replies if not specifically invited to do so with open questions.

Reluctance to Control the Agenda

Many new counselors are reluctant to control the agenda of a clinical session or to interrupt a patient who has gone off on a tangent. This reaction from the counselor has implications for the entire session, by potentially allowing the clinic visit time to elapse without completion of all the components of the visit. Some counseling students experience significant discomfort over the need to stop or redirect a client who is providing information that is not relevant to the visit. Remember that it is your responsibility to manage not only the individual components of a visit, but the flow of the overall visit. This controlling of the visit agenda by the counselor, whether through redirecting or other strategies, will ultimately serve the client best by allowing for the most productive use of the time available.

Avoiding Emotionally Charged Issues

Being present with a client as she experiences an emotional reaction is challenging. At such times clients may cry, or express feelings of hopelessness, sadness, or anger, or become temporarily withdrawn. Yet, because of the developmental path of skill acquisition, students are likely to be eliciting and encountering these client reactions before they have completely developed their psychosocial skills. Therefore, students may avoid exploring these issues with clients because they feel unprepared to respond. Remember that your supervisor will interject if he or she feels it is necessary. You should not shy away from interviewing in a manner that may allow the client to access the emotional dimension of her experiences.

CONCLUSIONS

Let's return to the scenario with which this chapter began: the couple waiting for your arrival. Upon entering the room and opening the session, you learn that their infant daughter has recently been diagnosed with cri-du-chat syndrome. It also becomes apparent that the two parents are reacting very differently to the situation. The father is very interactive, even domineering, offering and seeking information about the diagnosis, the child's well-being, and future medical care. The mother is quiet, holding the infant, making eye contact with you, apparently wanting to be engaged in the visit, but choosing, for the moment, a quiet role for herself. How would you direct the interviewing components after opening this session?

There is not a right or wrong way to proceed with this family. There are strategies that can be employed for any one of several issues that appear to be present in the dynamics of the couple. You might proceed by focusing on the mother—inviting her input and acknowledging her needs—or by responding to the father—reassuring him that his agenda is important and will be addressed. Alternatively, you could work to identify common ground for the two parents or focus on the infant and what they have learned about her diagnosis. Any or all of these approaches could be proper starting points for this case. The direction taken will depend on your experience, judgment, and style.

Counselors in training can look forward to exploring and discovering their own style, drawing on the techniques and tools discussed here, and learning from the approaches and strategies of their instructors and supervisors. Your effectiveness as an interviewer will be enhanced by bringing to this role your own insight, attention to detail, and personality. The more you are able to use your own inherent communication skills, the more consistent, reliable, and effective you will be as an interviewer.

Your interviewing skills will serve your clients throughout your career, even as advances continually alter the medical and diagnostic landscape of our field. As you acquire interviewing skills, keep in mind that in general, if you aren't making mistakes you aren't taking risks, and you are depriving yourself of the opportunity to grow. When you do make a mistake, try not to place the blame elsewhere or pretend it didn't happen—take responsibility, learn, and grow.

REFERENCES

Brunger F, Lippman A (1995) Resistance and adherence to the norms of genetic counseling. *J Gene Couns* 4(3):151–167.

Donner S, Sessions P (eds) (1995) *Garrett's Interviewing: Its Principles and Methods,* 4th ed., p. 5. Milwaukee, WI: Families International, Inc.

Kenen RH, Smith ACM (1995) Genetic counseling for the next 25 years: Models for the future. *J Gene Couns* 4(2):115–124.

Rapp R (1988) Chromosomes and communication: The discourse of genetic counseling. *Med Anthropol Q* 2(2):143–157.

4

The Medical Genetics Evaluation

Elizabeth Petty

A HISTORICAL PERSPECTIVE

The medical genetics evaluation is a critically important component of a comprehensive genetics clinic visit. Although hereditary diseases and genetic disorders have been described since biblical times and depicted in art throughout the ages, it was not until the last half of the twentieth century that the specialty of medical genetics blossomed and grew into a widely recognized clinical subspecialty. In the mid- to late 1800s and early in the twentieth century, many genetic syndromes were first being recognized, delineated, and named by a wide variety of medical specialists. Many syndromes continue to bear the name(s) of their original describers (Hecht and Hecht, 1991; Jones, 1997).

The medical genetics evaluation has evolved in response to the rapid explosion of new knowledge regarding genetic diseases. Earlier in the twentieth century clinic visits and medical genetics literature were focused on delineating the characteristic features of specific syndromes, describing new syndromes, and determining their inheritance pattern. With the discovery of human chromosome abnormalities in the late 1950s, geneticists began to use cytogenetic testing as a means to identify the chromosomal basis of specific syndromes. In the early 1960s, increased knowledge regarding the biochemical basis of metabolic diseases, the development of diagnostic biochemical assays, and, ultimately, improved management of metabolic diseases sparked the development of newborn screening programs and the emergence of biochemical ge-

A Guide to Genetic Counseling, Edited by Diane L. Baker, Jane L. Schuette, and Wendy R. Uhlmann.
ISBN 0-471-18541-8 (cloth), 0-471-18867-0 (paper). Copyright © 1998, Wiley-Liss, Inc.

netics clinics to identify and manage individuals with inborn errors of metabolism. The recognition of restriction fragment length polymorphisms (RFLPs) and Southern blotting ushered in the era of molecular genetics in the 1970s. In the late 1980s DNA diagnostics became more widely available and more regularly utilized. The advent of the polymerase chain reaction (PCR) in 1985 and the implementation of the Human Genome Project in 1990 revolutionized the field of molecular and clinical genetics (Epstein, 1995; Winter, 1995). The availability of DNA diagnostic tests for monogenic disorders skyrocketed during the 1990s and continues to grow. It is anticipated that the use of DNA diagnostic testing for more complex and, perhaps, polygenic and/or multifactorial traits will become available in the near future.

The overall biomedical technology explosion has increased the need and demand for highly specialized medical and molecular genetic services. The clinical geneticist now has a variety of auxiliary tests available to help confirm diagnoses, establish carrier status, make predictive diagnoses in asymptomatic individuals, and provide prenatal diagnoses for interested individuals. Indeed, current clinical genetics services are provided through departments or divisions of pediatrics, internal medicine, obstetrics and gynecology, oncology, and pathology, demonstrating today's broader scope of available medical genetic services. These clinical services have grown and developed as tremendous new knowledge in medical genetics has been actively sought and rapidly acquired. Genetic discoveries, the resulting technology, and the subsequent media exposure of such advances have also changed, and likely will continue to change, the character of, public desire for, and the availability of services provided by a genetics clinics. Finally, the impact of an ever changing health care reimbursement system is having, and will continue to have, a major role in determining how medical genetics evaluations are conducted and clinical services are provided.

Despite extensive growth of clinical genetic services over the latter half of the twentieth century, it was not until 1991 that medical genetics was recognized as a bona fide medical specialty by the American Board of Medical Specialists. The American College of Medical Genetics, also established in 1991, was formally recognized by the American Medical Association five years later, in 1996, when it was admitted to the AMA's House of Delegates. Clinical geneticists have always believed in the importance and uniqueness of this specialty, in their role in providing patients with the most accurate diagnostic and prognostic information available, and in offering the most up-to-date strategies for management.

THE MAKING OF A CLINICAL MEDICAL GENETICIST

To provide precise information about genetic disorders to patients, families, and other health care providers, it is essential that a comprehensive medical genetics evaluation be performed by a trained clinician. Most often the medical genetics evaluation is performed by a formally trained clinical geneticist (Graham et al., 1992). Practicing clinical geneticists are, for the most part, physicians who had their initial primary medical training in another area of medicine (usually pediatrics but sometimes internal medicine, obstetrics and gynecology, pathology, or other specialties) and subsequently obtained at least two years of additional subspecialty training in clinical medical genetics. Traditionally, clinical genetics training was available through specialized fellowship programs, accredited by the American Board of Medical Genetics (ABMG), for M.D.'s and D.O.'s. The ABMG began certifying medical geneticists and genetic counselors in 1981. Since 1997, however, accreditation for M.D. and D.O. clinical genetics training programs has been granted by the Accreditation Council for Graduate Medical Education (ACGME), in conjunction with the ABMG. Occasionally, other professionals such as formally trained Ph.D. geneticists and dentists with interests in genetic syndromes have assumed active primary roles as certified medical geneticists. After completion of an accredited training program, physicians are eligible to sit for board examinations in clinical genetics, which are given every three years. Certification is time-limited and must be renewed. Thus, board-certified clinical geneticists are individuals who, after completing specialized training, have passed the ABMG examination in clinical genetics.

Depending on the individual clinical geneticist's interests, and previous medical training, his or her area of practice may focus on genetic disorders of specific types or on specific age groups. For instance, a board-certified clinical geneticist with a background in obstetrics and gynecology may limit practice to prenatal and perinatal genetics, whereas an individual with an internal medicine background may have a practice limited to adult-onset disorders. There are many other subspecializations within the broader realm of clinical genetics, including neurogenetics, biochemical genetics, and cancer genetics. There are also geneticists who study dysmorphology (literally, the study of abnormal forms) and syndrome delineation. These individuals may further specialize in particular diseases or syndromes, sometimes becoming the "world's expert" on diagnosing and managing one particular disorder.

Most clinical geneticists work closely with genetic counselors and laboratory-based geneticists in providing and delivering comprehensive clinical genetic services. In fact, during a clinic session the roles of the counselor and the clinical geneticists are often closely intertwined, to ensure full optimization of care for patients and their families. Thus, the medical genetics evaluation is intimately connected to and reliant on genetic counseling services. A good medical genetics evaluation depends on appropriate case preparation, including family history and medical records review, patient education and counseling, and case follow-up. Thus, the formal medical genetics physical evaluation is only one part of any full clinical genetics service.

PURPOSES OF THE MEDICAL GENETICS EXAMINATION

There are several purposes of the medical genetics evaluation (Table 4.1), which may vary considerably depending on the particular disorder under consideration or the unique concerns of the patient or patient's family. Most often, a complete physical examination and clinical evaluation is performed to help establish or confirm a particular diagnosis for an individual or for several individuals within a family. The medical genetics evaluation not only addresses the disorder under consideration, but questions such as the following:

TABLE 4.1 Purposes of a Medical Genetics Evaluation

New Patients

Establish or confirm a specific diagnosis
Enable accurate, individualized counseling
Determine precise recurrence risks
Obtain necessary diagnostic tests
Provide specific education and support
Initiate appropriate referrals
Plan for focused medical management and follow-up

Established Patients (follow-up care)

Assess new medical problems and related concerns
Determine compliance with recommended management
Keep patients informed about new diagnostic and management strategies
Provide ongoing, age-appropriate education
Help coordinate necessary referrals and evaluations
Evaluate other at-risk family members

Why did it occur?

When did it likely happen?

Who else may be affected?

What are the chances that it may occur again?

What future problems should be anticipated?

Can these problems be avoided?

How can one's present and future health and psychological well-being be optimized, given this condition?

How should it be managed?

Establishing a diagnosis is important for providing accurate patient education, genetic counseling, and recurrence risk information. In addition, patients and their primary health care team need to be educated with respect to the particular diagnosis and provided with anticipatory guidelines regarding potential problems as well as state-of-the-art therapeutic or management options. This helps provide focused medical management, or referral to other subspecialists, for management of unique problems related to the particular condition. A better sense about an individual's prognosis may be based on the medical genetics evaluation, and individualized counseling can be directed to particular patient concerns. The ability to recognize a specific genetic disorder segregating in a family may provide interested individuals with an opportunity for specific family planning, including the use of prenatal testing to plan for or avoid the birth of an affected child, as desired. During follow-up visits for patients with a known genetic condition, a focused medical genetics evaluation is important in identifying and managing new problems, addressing any new patient concerns or questions, and providing anticipatory guidance. The regular follow-up medical genetics examination can also monitor the extent of systemic involvement in patients with an established diagnosis.

Another important role of the medical genetics evaluation is the assessment of other at-risk family members for the condition that has been identified in the proband. Although at present there are often no cures, there frequently are medical management strategies that can improve an individual's medical condition and, subsequently, that person's daily life. In addition, anticipation and watchful evaluation for potential problems will enable early detection of complications, improved medical management, and decreased morbidity. Providing appropriate psychosocial support and education may enable affected individuals to manage their lives more effectively.

A clinical geneticist should take the time to do a thoughtful and comprehensive evaluation of a patient, even when a particular diagnosis at first glance seems quite likely. Sometimes this evaluation can be accomplished in one visit; at other times a series of visits in a tiered fashion, may be necessary, to accommodate additional genetic tests and specialized medical examinations. As important as it is to give the patient a precise diagnosis, it is even more important not to issue an incorrect diagnosis based on a hasty evaluation or incomplete review of the patient's medical history, family history, or medical records. A misdiagnosis could negatively impact on various aspects of a patient's life including, but not limited to, continued health care management, sense of self and well-being, and recommendations for other family members. It is estimated that approximately one-third to one-half of patients presenting to a genetics clinic for a diagnosis leave the clinic without one (Aase, 1990, p. 267). Inasmuch as some syndromes or conditions become recognizable as the patient ages, regular reevaluation of "undiagnosed" patients is a common and important practice in medical genetics clinic. Additionally, the rapid growth of genetic knowledge and resulting diagnostic technology may facilitate making a diagnosis in some patients at a later time.

THE MEDICAL GENETICS EVALUATION: BASIC COMPONENTS OF A MEDICAL HISTORY

Today's medical genetics evaluation generally encompasses elements from both the general medical physical examination, used routinely by all physicians, and the specialized dysmorphology examinations used to define and characterize syndromes.

A good medical history of a patient, along with a focused and detailed family history, will often provide significant clues pointing to a particular diagnosis, even prior to the physical examination. Frequently portions of the medical history are provided by the referring physician or other health care provider. Knowledge about the patient's medical history is critically important in identifying particular problems, unique concerns, and valuable clues to an individual's diagnosis. The sources of information for the medical history may be the patient, the patient's family members, or the patient's physician. When obtaining a medical history from a patient or a family member, it is important to determine the reliability of the informant's knowledge about diagnoses and medical problems. The general components of the medical history, utilized in virtually all medical practices, are presented in Table 4.2.

In addition to talking with the patient or family member, it is essential to obtain pertinent medical records for careful review prior to diagnosing the patient. Having this background information will enable a more focused and clinically relevant diagnostic evaluation. This information is also useful for preparing appropriate educational materials, obtaining specific diagnostic test requisition forms, and conducting appropriate background research. The history obtained from the medical record is then reviewed and updated with the patient at the time of the clinic visit. This sequence helps to ensure the accuracy of the information obtained through the record review.

It is important to review pertinent medical records of other individuals in the family who have similar symptoms or other related medical problems to verify the histories provided by patients, family members, and health care providers. Failure to verify this information with objective data could lead to an erroneous diagnosis despite the best intentions and most accurate recollections of all parties involved.

The Chief Complaint

The first part of the medical history is the determination and documentation of the patient's "chief complaint." This simply means identifying and recording the major concern(s) or question(s) that initiated the genetics evaluation and need to be specifically addressed during the clinic visit. For children and older patients unable to express their concerns, the "chief complaint" is generally stated by a parent or legal guardian. The referring physician may have provided a specific indication or reason for the referral of a particular patient. However, it is extremely important to understand from the patient's (or family's) perspective each individual's own sense of the problems at hand and his or her understanding of the purpose of the consultation at the genetics clinic. Therefore, the chief complaint is often recorded in the patient's own words and elaborated on as needed by describing the concerns of the referring physician or of another family member. In addition, it is important to understand from patients what questions they have regarding their current health. It is often helpful to ask patients to prioritize their concerns, for sometimes there are several.

The History of Present Illness

After the chief complaint, it is necessary to understand and record in detail, as precisely and concisely as possible, the patient's present illness or present problems. It is important to define the onset of particular problems, to determine what has already been done to help diagnose and manage them, and to understand how these problems affect the patient in both a medical and a psychosocial sense of well-being. When obtaining the history of a patient's present illness or condition, it is necessary to consider information obtained from the patient, the patient's family members, and health care professionals, with objective information as presented in the medical record, including any previous studies or laboratory tests, to get a full and accurate sense of the constellation of problems.

TABLE 4.2   Basic Components of a Medical History

Identify sources of information and reliability of those sources
    Patient and/or patient family member
    Other health care provider
    Medical records
Identify chief complaint (CC)
    "What is your primary reason for this visit?" (record patient's own words when possible)
Determine main questions and concerns of patient, family members, and health care providers
when possible
    "What do you hope to learn or receive as a result of this visit?"
History of present illness (HPI)
    Description of patients problems related to CC
    Identify nature of problems, duration of symptoms, changes in quality of symptoms, previ-
    ous medical management; problem-oriented approach
Past medical history (some portions may be part of the HPI, depending on the particular case)
    Prenatal history
        Parents' ages
        Prenatal exposures (timing during gestation)
            Maternal illnesses
                Fevers
                Rashes
                Systemic disorders (lupus, hyperglycemia)
            Medications
                Prescribed and over the counter
            Alcohol, tobacco, recreational drugs
            Environmental
        Maternal immunization history
        Maternal complications
    Neonatal history
        Birth history (delivery type, complications, gestational age)
        Initial newborn examination (Apgar scores, length, weight, head circumference)
        Nursery course
    Infant and childhood developmental history
        Social, motor, adaptive milestones
        Hearing and speech assessments
        Vision assessments
    Childhood illnesses
    Surgeries
    Trauma
    Other hospitalizations
    Current medications
    Other nontraditional treatments
    Medication allergies
Family history (may be part of the HPI for inherited conditions)
Social history
    Habits
    Diet
    Education
    Living situation
    Employment
    Support systems
    Religious beliefs
Review of systems

It is important to assess how much the patient (or family member) knows or perceives about the condition. Oftentimes, patients have been given a tentative diagnosis by a local health care provider without much further education or information. Other patients make "self-diagnoses" based on family history, the medical problems of people they know, or something they have read in a magazine or newspaper. Many patients come to clinic after having researched their condition in libraries or on the Internet. It is important to ascertain what resources or preconceived ideas patients already have regarding their symptoms or diagnosis. They also may have received inaccurate information or misunderstood what they have been told about the disease from family members, friends, or even physicians. It is likewise important to know whether the patient has already initiated contact with support groups regarding the diagnosis and what materials and resources are available. A patient may have unrealistic hopes or expectations regarding the outcome of the genetics clinic visit, and it is best to clarify such matters immediately. Therefore, it is essential to determine each patient's own sense of his or her symptoms and knowledge about the disease condition early in the course of the evaluation.

The Past Medical History

Review of the past medical history of a patient is important in understanding present condition and problems. It includes a review of hospitalizations, major medical illnesses, and any surgeries. Further details about injuries, major trauma, and frequencies and types of different illness are also important to understand. For neonates, infants, and children, it is absolutely essential that a detailed history regarding the mother's pregnancy and prenatal circumstances be elicited. The subsequent neonatal and newborn course should also be fully explored. For instance, in a child presenting with short stature it is important to determine whether the growth retardation is of prenatal or postnatal origin. This will point the clinician down different diagnostic paths in the continued evaluation of the child. It is equally important to delineate and understand the child's development from the neonatal period onward. Often through key questions about history alone, a general sense of a child's developmental progress can be elicited relatively quickly. Abbreviated general guides to average milestones in childhood development are presented in several textbooks (Behrman, 1992; Johnson, 1993). It is important to determine whether the child's development shows continued progress, is static, or is regressing. A child with Down syndrome will generally make slow and steady developmental progress on a time course lagging behind average children, whereas a child with delays associated with a metabolic disorder, such as X-linked childhood adrenoleukodystrophy, may have normal early development followed by a period of regression. Many children seen for evaluation of developmental delay will ultimately require formal developmental testing, including vision and hearing tests, to document the extent and cause of their delays as precisely as possible.

In addition to standard questions regarding development, it is worthwhile to ask about any unusual behaviors or special talents. The hand-wringing behavior of developmentally delayed girls can be a very helpful observation when the diagnosis of Rett syndrome is under consideration. The hypotonia and failure-to-thrive history of a neonate who later, as a young child, has a rapid onset of abnormal weight gain and unusual food-seeking behaviors would lead a clinician to consideration of Prader–Willi syndrome in the differential diagnosis. Thus, important clues leading to a differential diagnosis may be gained by asking specific and thoughtful questions about prenatal and past medical history, developmental milestones, and unique behaviors.

In the pediatric genetics clinic, it is not uncommon for a mother to suspect that something she did or did not do during the pregnancy or prior to conception caused the child's specific problems or condition. By asking questions about the prenatal history, including maternal illness, medication use, and exposures to any environmental, ingested, or inhaled substances, it is often possible to determine whether a mother assigns blame for a child's condition to herself, her partner, or perhaps an obstetrician. It is also appropriate and worthwhile to directly ask parents or family members what they feel has caused the problems they or their children have.

It is important to determine a patient's compliance with past medical recommendations, including their participation in routine health care and preventive medicine strategies, and their overall general health behavior patterns, including diet and exercise. These may be useful in guiding future recommendations for continued care. It is important to explore whether patients have sought alternative forms of health care treatment. It is not uncommon for individuals who have rare syndromes to explore alternative health care options in seeking answers and treatments for their condition. Individuals from various cultural backgrounds may also seek alternative medical care that deviates from the standard mainstream western medicine practiced in the United States. It is important to understand the various medications, both prescribed and over the counter, that patients are taking to help alleviate any symptoms they are having. In addition, it is critically important to know of any specific drug allergies a patient has, in the event that it is necessary to prescribe a particular medication.

The Social History

A social support system may have a significant influence on a patient's overall health, as well as on the resources available for continued health management. The social history documents an individual's living situation, support system, key relationships, education, employment, activities, and religious beliefs. Exploration of a patient's social history may help identify specific concerns about environmental exposures, provide information about cognitive functioning based on a history of the person's education, or provide information about an individual's physical limitations in the home or work environment.

The Family History

As already noted, the family history is a critical component of a medical history. In genetics, this is especially important and therefore Chapter 2 was devoted entirely to this subject.

The Review of Systems

Generally the last, but very important, component of a medical history is known as the "review of systems" (ROS), during which the physician or health care provider will ask the patient specific questions about all parts of the body and their functions.

This determines whether there may be other related symptoms that were not identified either in the history of the present illness or in past medical history. In addition, this detailed review of systems helps to determine other interrelated illness or problems that may influence the patient's sense of well-being. A detailed review of systems as part of the genetics evaluation may also provide additional clues to an underlying genetic diagnosis. Generally the review of systems is a series of questions eliciting problems with all systems of the body, including eyesight, dental care, hearing, bowel and bladder function, and sexual function. In addition to specific questions, patients are asked more about their own sense of their health—whether they feel they are generally in excellent, good, fair, or poor health.

Special Considerations in the Medical History

It is from the medical history that the physician obtains significant information about the patient, including the patient's concerns about the disease or condition, and develops a tentative differential list regarding possible underlying diagnoses. This enables the physician to do a medical examination that focuses on the particular concerns of the patient as well as the areas of concern that have been made apparent through the medical history.

The sequencing of the elements in the medical history will vary somewhat depending on the case. For example, an evaluation of a newborn or young child with multiple congenital anomalies would place critical information about the pregnancy history, prenatal history, and neonatal history in the history of present illness rather than in the past medical history. Similarly, the case of a child presenting with regression of cognitive or motor skills would call for the inclusion of information about childhood development as well as neonatal and prenatal histories in the history of present illness. Therefore, although all the basic components of a medical history are always documented, their order and emphasis tend to vary depending on the circumstances.

Medical genetics evaluations often involve reviewing medical records of other family members, including pathology reports from biopsies and autopsy reports of family members who reportedly had similar conditions or features. This is especially important when an asymptomatic person comes to the clinic because of a family history of a particular condition. For example, if the patient reports a family history of gynecological tumors, it is important to review the medical records of affected individuals to determine whether the tumors were ovarian in origin, which may indicate an increased risk for a *BRCA1 or BRCA2* mutation, or whether cervical or uterine cancer was present, which is not associated with a significant increased risk for a *BRCA1* mutation. Similarly, in the instance of an asymptomatic adult at 50% risk for having inherited a mutated gene causing Huntington disease from a deceased affected parent, it is worthwhile to try to obtain as much information as possible about the affected parent prior to doing specific genetic counseling and/or genetic testing. This step is important because several other neurological conditions have symptoms that may overlap or mimic Huntington disease. Therefore, to provide the most accurate and precise genetic counseling and the most appropriate evaluations for a patient, it is important to critically evaluate medical records, not only of the patient, but often of additional family members as well.

THE PHYSICAL EXAMINATION

The General Examination

In a medical genetics evaluation, the physical examination is based on a standard physical examination (Bates, 1983) (Table 4.3) but contains some additional elements, depending on the particular case under consideration. The basic components of a physical examination include ascertainment of an individual's vital signs (heart rate, blood pressure, and respiratory rate) and an overall assessment of the individual's general health and nutritional status. It is important to record parameters of growth, including height, weight, and head circumference, and to evaluate the individual's general body habitus. The physical examination then often proceeds in a systematic fashion, covering the head and neck, thorax, heart, lungs, abdomen, breasts, genitals, musculoskeletal system, and neurological system including mental status, followed by an evaluation of the skin. The order of the examination may vary depending on the patient's chief complaint and age. In pediatric examinations, it is often best to do the most "threatening" parts (in children this includes looking in the ears and throat) near the end of the examination and to begin by simply observing the child, then earning trust and cooperation by playfully engaging young patients in the parts of the examination.

The Dysmorphology Examination

There are several special components of a physical examination in genetics that are used in specific situations (Table 4.4). In an individual who presents with major and/or minor congenital anomalies, a dysmorphology examination is conducted in which key measurements are taken and compared to published normalized standards. As the term implies, the dysmorphology examination is specifically focused on recognizing "abnormal forms." This examination identifies aberrant size, shape, place-

---

TABLE 4.3   Basic Components of a Complete General Physical Examination

Vital signs (heart rate, blood pressure, and respiratory rate
Assessment of growth, body habitus, general proportions
Overall general health
Examination by system or structure
    Head and neck
    Thorax
    Cardiac
    Pulmonary
    Breast
    Abdominal
    Genitourinary
    Pelvic
    Rectal
    Musculoskeletal
    Neurological, mental status, developmental
    Cutaneous

TABLE 4.4   Special Components of Focused Genetics Examinations

Dysmorphology Examination

Recognition of major and minor anomalies
Pattern recognition of syndrome
Key measurements
Dermatoglyphics
Photographic/video documentation

Assessment for Specific Conditions (Two Examples)

Neurocutaneous disorders
    Careful cutaneous examination for specific signs of syndromes
    Detailed neurological examination
    Ophthalmologic examination
    Specialized neuroimaging studies
Connective tissue disorders
    Detailed examination of joints
    Examination of body proportions
    Careful cutaneous examination
    Specialized examinations of other organs (eyes, heart)

ment, or structural anomalies of certain features, such as the eyes and ears (Friedman, 1990; Hall, 1993; Gorlin, 1990; Saksena, 1989). It is important to specifically look for, characterize, precisely describe, and document major or minor anomalies that may be present. The specificity of the dysmorphology examination is illustrated by examples of some of the terms used to describe alternations in the structures of the head and face listed in Table 4.5. Figure 4.1 identifies some common facial measurements taken during a dysmorphology examination.

Major anomalies or malformations are defined as those that create significant medical problems for the patient or require surgical or medical management. Minor anomalies are described as features that vary from those seen in all but a small percentage of the normal population but in and of themselves do not cause increased morbidity. Most often, a diagnosis is not based on a single major anomaly or minor anomaly; rather, a constellation of major and minor anomalies will point to a specific syndrome diagnosis or known association. A major anomaly, such as a unilateral cleft lip, has a defined set of implications as to etiology, diagnosis, and prognosis when it is the only such defect present in an individual, but there is a significantly different set that applies when one person has a median cleft lip and palate associated with polydactyly (extra digits), a congenital heart defect, and low-set ears. However, to simply say that an individual has polydactyly without defining the number of extra digits, the extent of extra digits (a small nubbin or an additional well-formed finger), and the precise location (pre- or postaxial, referring to the radial/thumb or ulnar/little finger) of the extra finger would not be helpful in narrowing down the differential diagnosis. It is also important to consider an individual's ethnicity and familial features, which will influence physical appearance.

TABLE 4.5   Examples of Terms Used in a Dysmorphology Examination of the Head and Face

Head

Acrocephaly: "tower skull," also known as oxycephaly
Brachycephaly: decreased anterior/posterior diameter of skull
Craniosynostosis: premature ossification of the skull; obliteration of sutures
Frontal bossing: prominent, protruding forehead
Macrocephaly: abnormally large head circumference
Microcephaly: abnormally small head circumference
Trigonocephaly: triangular head; sharp angulation ventrally; craniostenosis of metopic suture

Orbital (Eye) Region

Canthus: the angle at either end of the fissure between the eyelids (inner, outer)
Epicanthal folds: redundant fold of skin that obscures the inner canthus
Hypertelorism: abnormally long distance between paired organs, such as eyes
Hypotelorism: abnormally short distance between paired organs, such as eyes
Microphthalmia: small eyes
Palpebral fissures (slant): gives some idea of the growth rate of the brain (above the eye) versus the facial area (below the eye); up-slanting palpebral fissures can be seen with microcephaly, and down-slanting fissures with maxillary hypoplasia
Ptosis: drooping of upper eyelid from paralysis
Synophrys: eyebrows grown together in midline

Perioral and Mandibular Region

Macrognathia: large jaw
Micrognathia: small jaw
Macrostomia: large mouth
Microstomia: small mouth
Philtrum: the region of the face between nares and upper lip, which usually has normal vertical grooves

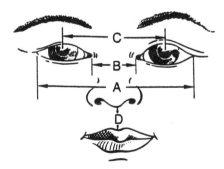

Figure 4.1 *The measurements of body proportions is an important part of a dysmorphology evaluation. For instance, measurement of the spacing of the eyes can be useful in helping to determine whether the eyes are widely set (hypertelorism), as seen in Waardenburg syndrome, or closely set (hypotelorism), as seen in holoprosencephaly. Measurements of various distances between the eyes, using a clear ruler or blunt calipers, to determine the outer canthal distance (A), inner canthal distance (B), and interpupillary distances (C) will help the geneticist determine whether abnormal placement is present, possibly providing a clue in the differential diagnosis. Reproduced with permission from* Birth Defects, 10 *[Suppl 13]: 1974, Feingold and Bossert, March of Dimes Birth Defects Foundation.*

When considering dysmorphic features it is important to keep in mind how any abnormalities identified might be related (Barness-Gilbert et al., 1989; Saksena et al., 1989; Cohen, 1989, 1990; Friedman, 1992; Sharony et al., 1993). A **syndrome** is generally recognized and defined when a well-characterized constellation of major and minor anomalies is observed together, in a predictable fashion, presumably as a result of a single underlying etiology, which may be monogenic, chromosomal, mitochondrial, or teratogenic in origin. For example, Down syndrome is associated with a predictable constellation of major and minor anomalies that create a recognizable phenotype, although not all the characteristic anomalies are present in any one affected individual. It is relatively common for patients with a specific syndrome to exclaim "I saw pictures of people who look like me!" when they first see or read patient newsletters or educational brochures about their particular diagnosis. Recently, when a mother saw a textbook photograph of a baby with Carpenter syndrome (acrocephalopolysyndactyly type II—a craniosynostosis syndrome) during counseling about her 31-year-old son's recent diagnosis of the syndrome, she remarked that she had never before seen "a baby's picture that looked exactly like my son's newborn baby picture."

An **association** is a group of anomalies that occur more frequently together than would be expected by chance alone but do not have a predictable pattern of recognition or known unified underlying etiology. For example, the CHARGE association is named from the acronym formed by the associated findings: **c**oloboma, **h**eart disease, **a**tresia, choana, **r**etarded growth, **g**enital anomalies, and **e**ar deformities or deafness. A **sequence** is a group of related anomalies that generally stem from a single initial anomaly that affects the development of other tissues or structures. Potter's sequence is recognized by a constellation of physical findings in which the newborn manifests flattened abnormal facial features and deformations of the hands and feet. These features, along with poor lung development, are secondary to decreased amounts of amniotic fluid (oligohydramnios), which is often due to major renal abnormalities associated with decreased fetal urine output. The term **field defect**, often used to describe related malformations in a particular region, sometimes is used interchangeably with sequence.

When considering dysmorphic features, it is important to keep in mind the various ways in which structures and tissues can become abnormal (Figure 4.2). A structure may be visibly abnormal due to a deformation. A **deformation** (Figure 4.2**D**) is caused by an abnormal external force on the fetus during in utero development that results in abnormal growth or formation of the fetal structure. For instance, when not enough amniotic fluid is present (oligohydramnios), the face of the fetus may be flattened because it is compressed against the uterine wall, and yet there is full development of the face and facial features. In another type of abnormality, known as a **disruption** (Figure 4.2**C**), the normal growth process of a fetal structure is interrupted, and growth is arrested. Disruption is seen in the condition of amniotic bands where a digit or extremity that is growing normally ceases to grow properly, or at all, because of constriction of amniotic bands surrounding that extremity. The result may be missing fingers, toes, or hands and feet. Generally, disruptions and deformations are relatively isolated and are not associated with multiple congenital anomalies. A **malformation** (Figure 4.2**B**) signifies that fetal growth and development were prevented from proceeding normally by an underlying genetic, epigenetic, or environmental factor that altered development. Another type of generalized anomaly is due

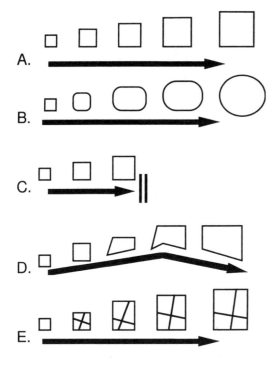

Figure 4.2 Graphic representations of abnormal processes in human development. **(A)** Normal fetal growth and development (small square becomes a large square) during gestation, as designated by the arrow. **(B)** A **malformation**: over time, instead of the formation of the normal anticipated shape, a different shape of the structure is created (here a circle develops instead of a square). **(C)** A **disruption**: full, normal development of the structure is halted prematurely. **(D)** A **deformation**: an abnormal extrinsic force some time in the course of development (depicted by the bent arrow) transforms and distorts the normal shape of the resulting structure. **(E)** **Dysplasia**: the overall resulting shape or structure may be roughly maintained but the underlying architecture of the tissue making up the shape is different.

to **dysplasia** (Figure 4.2**E**), where the intrinsic cellular architecture of a tissue is not maintained normally throughout growth and development. Many of the skeletal syndromes of short stature are due to dysplasia in developing bone and cartilage.

It is important to determine the type of abnormality that a patient has as this has implications for the diagnosis and management of the patient's condition. A relatively common congenital anomaly, unilateral cleft lip and cleft palate, is most often due to an isolated malformation in the structural development of the tissues forming the lip and palate. However, it is important to note that a cleft lip and palate, or other facial clefts, may be caused by amniotic bands, in which case the clefting would be considered a disruption of normal growth, rather than a feature of a genetic syndrome. These different causes of facial clefts are associated with different recurrence risks for the individual and his or her family.

When evaluating malformations, deformations, disruptions, and dysplasias, it is useful to keep in mind the timing of development of particular organs, tissues, and fetal structures. This may help determine the time and etiology of the particular abnormality. Basic charts of fetal development can be found in standard medical genetics textbooks and are worth having readily available in the genetics clinic. A general rule of thumb is that most internal organs are formed during the first 4 to 6 weeks of fetal development, digits and facial features are defined within the first 8 weeks, and neurological development occurs throughout gestation.

Another component of the genetics evaluation is a detailed cutaneous or skin examination to look for any characteristic lesions that are associated with neurocutaneous disorders of other syndromes (Spitz, 1996). For instance in Gorlin syndrome, which is characterized by multiple basal cell carcinomas as well as other birth defects, individuals will have a pattern of small pits in the palms of their hands and soles of their feet. These pits may be missed unless they are specifically looked for because the syndrome is being considered. Similarly, in a condition such as tuberous sclerosis, specific cutaneous lesions are seen, including shagreen patches, ash leaf spots, and hypopigmented confetti lesions. These characteristic findings are best identified with the aid of a special source of ultraviolet light known as a Wood's lamp.

Assessment of body proportions is important to the medical genetics evaluation, especially in the consideration of certain connective tissue disorders, such as Marfan syndrome. Individuals with Marfan syndrome generally have long arms and legs and arachnodactyly (long fingers and toes), hence arm span tends to be significantly greater than height, and the lower portion of the body longer than the upper portion (Hall, 1989; Pope, 1995; Belshton, 1993). Therefore measurements of upper to lower segment ratios, arm span, length of fingers compared to length of the palm, and total hand length are all important for the diagnostic evaluation of Marfan syndrome. Considerations of various short stature syndromes call for measurements to determine whether there is proportional or disproportionately short stature. Many of the skeletal dysplasias are classified based on the specific area of shortened bone growth. In considering skeletal and connective tissue disorders, it is also important to carefully evaluate joints for contractures or increased laxity, characteristics that may provide important clues to the diagnosis.

A unique part of the clinical genetics evaluation is the viewing of photographs of the patient at various ages. Some syndromes typically evolve or change with age, and therefore a review of such a series of photographs may be useful. Although it is important to physically evaluate other affected family member, if this is not possible, inspection of their photographs may be of great utility.

A component of the "classic" genetics examination, no longer routinely conducted, is dermatoglyphics ("skin carvings" in Greek), or the analysis of finger patterns or fingerprints. The term was coined by anatomist Harold Cummins in 1926. Ten years later he published information about the dermatoglyphic patterns in individuals with trisomy 21, then known as mongolism. Now that more sophisticated genetic testing is available, dermatoglyphics is less utilized as a specific diagnostic tool. Since, however, dermatoglyphic patterns are very easy to evaluate and readily available by noninvasive means, there still are some advantages to looking at them before a patient's diagnostic testing is begun. There is a wealth of older genetic literature describing different dermatoglyphic patterns in various conditions.

Types of Medical Genetics Evaluation

The broad and different reasons for medical genetics referrals necessitate specialized evaluations and individualized considerations within the medical genetics examination. Medical genetics evaluations vary with the age of the patient, the nature of the referral, and the need for subsequent evaluations. Prenatal evaluations, which are not dealt with in this chapter, are rooted in the same general concepts as medical genetics evaluations in terms of the critical importance of a family history, medical history, and specific diagnostic testing (Evans, 1992).

One of the major focuses of pediatric genetics clinics is the evaluation of multiple congenital anomalies in the infant, with or without developmental delay. Timely evaluation of newborns with congenital anomalies, birth defects, or ambiguous genitalia is obviously critical if one is to provide specific information and counseling to the often surprised and distraught family, as well as specific information regarding prognosis and medical management for the health care team. Learning problems by early school age are another common cause of concern for families, teachers, and health care workers, often resulting in referral of a child to a pediatric genetics clinic. Syndromes associated with learning problems and minimal dysmorphic features include fragile X syndrome and sex chromosome abnormalities (e.g., aneuploidy).

The area of inborn errors of metabolism or biochemical disease is a specialized field in medical genetics (Scriver, 1995). Many genetic centers employ biochemical geneticists whose area of expertise is the diagnosis and management of these specific disorders. A handful of these conditions, including phenylketonuria, galactosemia, homocystinuria, and maple syrup urine disease, are screened for at birth in many states. Since states differ as to which conditions are included in their newborn screening programs, it is important to know what type of newborn screening is performed in the state in which your patient was born.

Neurocutaneous disorders, or germatodermatoses, are conditions that are often associated with a variety of features involving the neurological system, skin, skeletal system, and potentially other organ systems. Individuals having neurofibromatosis or tuberous sclerosis are often referred to a medical genetics center not only for initial evaluation but for continued management and follow-up. These multisystemic disorders are relatively rare in the population, and many primary health care physicians have neither the time nor the specialized educational background to fully manage affected individuals in their practices. Similarly, individuals with connective tissue disorders such as skeletal dysplasia, Marfan syndrome, or one of the many forms of Ehlers–Danlos syndrome are often referred to genetics clinics for continued management and coordinated follow-up with the primary care provider.

Adult medical genetics clinics evaluate, diagnose, and follow individuals with adult-onset disorders including neurodegenerative disorders, cancer predisposition syndromes, and psychiatric conditions; perhaps in the future more complex polygenic and multifactorial conditions will be added to this list. With the advent and utility of DNA diagnostics for many of these conditions, the area of presymptomatic or predictive genetic testing is a widely growing field that deserves special attention in terms of genetic counseling, consideration of genetic testing, and cost-effective application of available genetic tests. Growth in the field of presymptomatic or predisposition genetic testing is anticipated as more gene discoveries are made.

DOCUMENTATION OF THE MEDICAL GENETICS EVALUATION

As with any medical evaluation, precise and concise documentation of relevant findings, including both the presence and absence of significant findings, is extremely important to a patient's care and in conveying accurate information to other health care providers. A pedigree is obviously a special form of documentation that the medical genetics team can supply to other health care providers. Documentation of the medical history and record review should also be clearly presented. The physical findings should be well described and may be facilitated by the use of forms designed to efficiently record specific measurements and diagram specific features or cutaneous lesions (Figure 4.3). Additionally, photographs and videotapes of

---

NAME: _____ **DATE:** ___
REG #: _____
DOB: _____ **AGE:** _____

VITAL SIGNS

BP: RA _____ LA _____
RL _____ LL _____
HR: RR ___ LR ___ RDP ___ LDP ___
RR: _____
HEIGHT: ____in _____cm (%)
WEIGHT: ____lbs _____kg (%)

HEAD - NECK

General description:

HEAD CIRCUM.: _____cm (%)
Cranium/Sutures:
Hair distribution:
Eyes: IC: _____cm (%)
OC: _____cm (%)
IP: _____cm (%)
RPF: _____cm (%)
LPF: _____cm (%)
Ears: shape:
placement:
R _____cm (%)
L _____cm (%)
Philtrum: _____cm (%)
Lips:
Palate and mouth
Teeth:
Chin:
Neck:

TRUNK - ABDOMEN

General description:
sternum:
spine:
Breast: Tanner Stage: _____
Heart:
Lungs:
Chest circumference: _____cm
Inter nipple distance: _____cm
IND_____ / CC_____ = _____ (%)
Abdomen:
Umbilicus:
Genitalia:
penile length: _____cm (%)
testicular vol.: _____cm (%)
Tanner Stage: _____

SKIN

General description:

Freckles - Iris:_____Axillary:_____ Inguinal:_____
CAL spots: _____ Size range:_____cm
Neurofibromas: _____ Size range:_____cm
Hypopigmented lesions:
Other lesions:

EXTREMITIES - PROPORTIONS

General description:
laxity:
contratures:
digital abnormalities:

LOWER segment: _____cm
UPPER segment: _____cm (HT-LS)
US/LS ratio: ____/____ = % (SD)
ARM SPAN: _____ cm
WRIST SIGN: ____ THUMB SIGN: ____
HAND: R _____cm (%)
L _____cm (%)
MIDDLE FINGER: R _____cm (%)
L _____cm (%)
PALM : R _____cm (%)
L _____cm (%)
FOOT: R _____cm (%)
L _____cm (%)
Finger to Hand: ____/____ (%)
Hand to Height ____/____ (%)
Foot to Height ____/____ (%)

NERVOUS SYSTEM

Mental status:

Cranial nerves:

Muscle bulk and tone:

Sensation:
Motor strength: Deep tendon reflexes:

DIAGRAM TO DOCUMENT FINDINGS

R FRONT L	L BACK R
RIGHT HAND	LEFT HAND

Figure 4.3 Example of a form that may be utilized to facilitate documentation of the medical genetics examination.

patients often provide a wealth of information that is otherwise too time-consuming, cumbersome, and difficult to convey in concise and accurate words (Hall, 1996).

In addition to having a plethora of names to identify hundreds of different rare syndromes and disorders, the clinical geneticist has been well equipped with a special vocabulary of descriptive words and terms. These have Greek and Latin roots and are used to describe in a concise manner the specific nuances and physical findings of a genetics examination. For genetic counselors just entering the field, as well as for many physicians in other specialties, these words sometimes seem more confusing than useful. As is the case in any medical specialty, it is important for all those who provide clinical genetics services to make a concentrated effort to fully familiarize themselves with the "language of the specialty," to ensure effective communication with other geneticists and to permit absorption and appreciation of the medical genetics literature. It is equally important for clinical geneticists to communicate information to patients and referring physicians by means of terminology that can be readily understood. A physician or counselor who speaks to a patient accurately but in unfamiliar terminology ultimately does that person a disservice. Table 4.5 provides some common terms used in a dysmorphology examination of the head and face and demonstrates both the specificity and uniqueness of this terminology.

DIAGNOSTIC STUDIES

Utility of Diagnostic Studies in the Medical Genetics Evaluation

A variety of diagnostic studies may be required in concert with a clinical evaluation and medical history to make a diagnosis in a patient or to provide information about prognosis and medical management. These tests include specific genetic testing (Table 4.6) as well as more conventional laboratory studies. In a newborn boy who has excessive bleeding after a circumcision, for example, a precise DNA-based mutational analysis may reveal the molecular basis for his bleeding disorder and confirm a diagnosis of hemophilia A. However, more routine hematology laboratory tests, such as clotting factor studies, may be useful in making the initial diagnosis and are critical to the patient's immediate medical management.

Application of Diagnostic Genetic Tests

Diagnostic genetic tests can be divided into three primary but overlapping categories: cytogenetic studies [routine karyotypes, high resolution karyotypes, and molecular fluorescent in situ hybridization (FISH) studies], biochemical tests, and DNA-based diagnostic tests.

Cytogenetic studies can be conducted on virtually any cells but are most often performed on peripheral blood leukocytes (white blood cells) collected from a whole-blood specimen. Cytogenetic studies are also conducted on bone marrow samples, amniocytes, and chorionic villus and skin biopsy samples. Tissue samples from abnormal fetuses or other products of conception can also be studied. Most cytogenetic laboratories prefer receiving blood samples in sodium heparin tubes, since this

TABLE 4.6   General Overview of Diagnostic Genetic Tests
___

Cytogenetic Studies

General types of test:
  Karyotypes
  Fluorescent in situ hybridization (FISH)
Samples utilized vary depending on specific test
  Blood in sodium herapin tubes
  Amniocytes
  Chorionic villus samples
  Fibroblasts
  Products of conception

Biochemical Assays of Metabolic Disease

General types of test:
  Metabolic screening tests
    Metabolite spot tests
    Amino acids
    Organic acids
  Specific biochemical or enzyme assays
Samples needed vary depending on specific test
  Urine on ice or 24-hour sample in special container
  Plasma collected from blood drawn in lithium heparin tubes placed on ice
  Red blood cell pellets
  White blood cell pellets
  Fibroblasts
  Organ or tissue biopsy samples, which may require immediate dissection and/or
  placement on dry ice

DNA-Based Diagnostic Tests

General types of test:
  Direct mutational analysis
  Indirect analysis–linkage analysis
  Functional assays
Samples needed vary depending on specific test
  Blood in EDTA or buffered sodium citrate tubes at room temperature
  Cheek swabs for PCR-based tests
  Other nucleated cells
  Functional assays may require special processing and shipment of cells on dry ice
___

medium does not inhibit cell growth. Other tissue samples are best submitted in sterile tissue culture media. Any sample that is handled and shipped at room temperature is easily utilized for cytogenetic studies.

Biochemical studies, depending on the specific tests, may be conducted on samples from urine, plasma, or red or white blood cells, as well as prenatal samples of amniotic fluid or placental tissue. For biochemical tests it is important that specimens be collected in the appropriate tube. Lithium heparin tubes for plasma isolated from whole blood can be utilized for many of the plasma-based screening tests, including those for quantitative plasma amino acids. For biochemical tests it is often critical to keep blood and urine samples on ice to prevent any breakdown of the metabolites being analyzed.

In addition, for biochemical tests, it is important to know precisely what medications or dietary supplements the individual has taken and when the person last ate.

DNA-based testing has grown rapidly over the past decade. Indirect DNA testing, or linkage analysis, is still used when precise mutation testing is not yet available. For DNA diagnostic linkage studies, it is important that genetic heterogeneity be minimal or nonexistent, and that samples be available from multiple appropriate family members. Patients undergoing such studies need to understand the ambiguities that may be associated with linkage studies, including the potential for recombination. As more disease genes are cloned, the availability of specific mutation screening for many disorders will become increasingly possible. DNA testing can be done on any nucleated cell specimen from which DNA can be obtained, even from pathology specimens of deceased family members. Blood samples for DNA testing are usually sent in EDTA tubes or buffered sodium citrate tubes to ensure high quality DNA for testing. Some mutational studies are based on functional assays of mutations requiring special specimen handling. For many of these tests it is important to have intact RNA as the sample source.

Prior to ordering any genetic tests, it is extremely important to determine exactly what type of sample is to be obtained and how it should be sent. Failure to obtain, handle, or ship specimens properly could jeopardize the results, especially for tests involving sensitive biochemical assays.

Other more routine laboratory studies may be required in the medical genetics evaluation to help reach a diagnosis or manage patient symptoms. These include blood counts, clotting factors, liver function tests, kidney function tests, acid/base status, and measurements of other breakdown products of metabolism. Various diagnostic imaging studies are also of great importance. Radiographic studies, including skeletal surveys, may help in the diagnosis of skeletal dysplasias or the recognition of bony congenital anomalies that may point to a specific syndrome diagnosis. Specialized imaging studies such as CT scans, MRI scans, and echocardiograms, may be required. For example, since aortic root dilatation and rupture are serious complications in Marfan syndrome, an echocardiogram documenting aortic root size is useful for confirmation of the diagnosis and for management. Once the diagnosis has been made, routine echocardiograms or, in some cases, other imaging studies such as transesophageal echocardiograms or serial CT scans of the aorta, are recommended on a regular basis to monitor the patient for signs of progression of aortic root dilatation, necessitating more aggressive medical management. Specialized imaging studies in the prenatal period such as detailed ultrasound and fetal echocardiography exams are routinely used for the detection of congenital anomalies. Therefore, imaging studies in patients can be quite useful for diagnosis and routine follow-up and management, alike.

TOOLS AND RESOURCES OF THE CLINICAL GENETICIST

The usual "tools" of the physician, including stethoscope, reflex hammer, otoscope, ophthalmoscope, and tuning fork, are supplemented by various devices used by the clinical geneticist. Obviously, the utility of this equipment will vary depending on the type of examination done. A few basic additional pieces of equipment that are generally useful are as follows: a long (2.5-meter), flexible measuring tape (for measuring arm spans and other extremity proportions), a short clear ruler (for measuring proportions and size of facial structures such as palpebral fissure length, inner canthi, outer canthi, and interpupillary distance, a jointed ruler with degree markings for measuring the angles of either contracted or hyperextended joints, and large and small blunt calipers for measuring distances spanning irregular body parts, such as various proportions of the calvarium or skull. For measuring the sizes of long bones, especially in squirming children, it is helpful to have a long, durable piece of string that can be marked and later measured, and a calculator to help quickly determine ratios between body segments. In addition to measuring devices, it is sometimes helpful to have a magnifying lens to closely examine cutaneous lesions. A Wood's lamp is useful for examining fair-skinned individuals for hypopigmented lesions such as the ash leaf spots associated with tuberous sclerosis. Cameras, still and video, are also essential tools in the genetics clinic. Photographs can provide documentation of specific findings for the medical record and can be shared with consulting specialists. Other tools or aids may be useful in certain cases and need to be individualized. For instance, the use of red dye disclosing tablets (the type used in dental offices) can be useful in the genetics clinic to demonstrate the enamel pits in the teeth of patients with tuberous sclerosis. For skin biopsies, a 3–4 mm punch biopsy kit and local anesthesia need to be available in the clinic.

Resources for the clinical geneticist include several well-written textbooks regarding syndromes and metabolic disorders, a handful of which are listed in the References (Baraitser & Winter, 1996; Gorlin et al., 1990; Evans, 1992; McKusick, 1992; Pope and Smith, 1995; Scriver et al., 1995; Spitz, 1996; Emery et al., 1997; Jones, 1997). Guides to normal measurements are found in many texts but are also well addressed in the pocket-sized *Handbook of Normal Physical Measurements,* which should be readily available to any geneticist (Hall et al., 1989). Computerized dysmorphology and skeletal databases (e.g., POSSUM, OSSUM, and the London Dysmorphology Database) are available for purchase and are used by clinical geneticists to help fully explore a differential diagnosis (Evans, 1995; Winter and Baraitser, 1987; Stromme, 1991). The regularly updated On-line Mendelian Inheritance in Man (OMIM) is used by many clinical geneticists as an important initial resource. This database is connected through the National Center for Biotechnology Information (NBCI) to other useful genetic databases and MEDLINE searches. The HELIX database to identify DNA diagnostic testing laboratories for various conditions is also a useful resource that all geneticists should be familiar with. Also useful for clinical geneticists is familiarity with some of the major Internet resources, which are discussed in Chapter 15.

CLINICAL CASE CONFERENCES AND OUTSIDE EXPERT CONSULTANTS

As in any medical specialty that diagnoses and manages a wide variety of relatively rare conditions, there is great utility and benefit in conducting clinical case conferences. These conferences may be used to review cases prior to and following clinic visits, for discussion of a differential diagnosis, and for determination of appropriate management. Given the rapid growth in genetic technology, such conferences are increasingly important and can serve to keep one abreast of the most current diagnostic options and management strategies for patients. Some conditions are so rare that a clinical geneticist may have difficulty in recognizing them when first meeting one of the few patients affected. Fortunately the clinical genetics community is peppered with experts in virtually every disease who are generally open and willing to provide curbside consultations and expert advice. Formal consultation with an outside expert should be actively sought when needed, however.

Other medical specialists that may be utilized for consultation about the medical genetics evaluations include neurologists, cardiologists, oncologists, orthopedists and other surgery subspecialists, ophthalmologists, developmental pediatricians, physical medicine and rehabilitation specialists, pain management physicians, audiologists, plastic surgeons, psychiatrists and psychologists, social workers, pathologists, dermatologists, and radiologists.

SUMMARY

The field of medical genetics is rapidly evolving, reflecting our ever increasing knowledge about the human genome. The broad scope and complexity of disorders seen in the medical genetics clinic necessitates a broad range of expertise among the clinic staff to provide appropriate diagnosis and management for patients. Therefore, the genetics clinic is best served by a true team approach, where genetic counselors, medical geneticists, laboratory geneticists, and other health care providers interact with patients and families to provide comprehensive care and appropriate evaluation. Because of the relative rarity of many genetic conditions, it is important to have access to a library of current medical texts and journals, access to computerized databases, and availability to other medical specialists to fully evaluate patients. The interaction of all these health care professionals together with the genetics team, often is necessary to provide patients with the most appropriate management for multisystemic disorders.

REFERENCES

Aase JM (1990) *Diagnostic Dysmorphology.* New York: Plenum Medical Book Company.

Baraitser M, Winter RM (1996) *Color Atlas of Congenital Malformation Syndromes.* London: Mosby-Wolfe.

Barness-Gilbert E, Opitz E, Barness LA (1989) The pathologist's perspective of genetic disease. Malformations and dysmorphology. *Pediatr Clin North Am* 36(1):163–187.

Bates B (1983) *A Guide to Physical Examination,* 3rd ed. Philadelphia: Lippincott.

Behrman RE (1992) *Nelson Textbook of Pediatrics,* 14th ed. Philadelphia: Saunders, pp. 15–43.

Belshton P (1993) *McKusick's Heritable Disorders of Connective Tissue,* 5th ed. St. Louis: Mosby-Year Book, Inc.

Cohen, MM Jr (1989) Syndromology: An updated conceptual overview. VI. Molecular and biochemical aspects of dysmorphology. *Int J Oral Maxillofac Surg* 18(6):339–346.

Cohen MM Jr (1990) Syndromology: An updated conceptual overview. IX. Facial dysmorphology. *Int Oral Maxillofac Surg* 19(2):81–88.

Emery AEH, Rimoin DL, Connor JM, Pyeritz RE (eds) (1997) *Principles and Practice of Medical Genetics,* Vols. 1 and 2, 3rd ed. New York: Churchill Livingston.

Epstein CJ (1995) The new dysmorphology: Application of insights from basic developmental biology to the understanding of human birth defects. *Proc Natl Acad Sci USA* 92(19):8566–8573.

Evans CD (1995) Computer systems in dysmorphology. *Clin Dysmorphol* 4(3):185–201.

Evans MI (1992) *Reproductive Risks and Prenatal Diagnosis.* Norwalk, CT: Appleton & Lange.

Friedman JM (1990) A practical approach to dysmorphology. *Pediatr Ann* 19(2):95–101.

Friedman JM (1992) The use of dysmorphology in birth defects epidemiology. *Teratology* 45(2):187–193.

Gorlin RJ, Cohen MM, Levin LS (1990) Oxford Monographs on Medical Genetics No. 19, *Syndromes of the Head and Neck,* 3rd ed. New York: Oxford University Press.

Graham JM Jr, Curry CJ, Hoyme HE, Stevenson RE, Hall JG (1992) Fellowships and career development in dysmorphology and clinical genetics. *Pediatr Clin North Am* 39(2):349–362.

Hall BD (1993) The state of the art of dysmorphology. *Am J Dis Child* 147(11):1184–1189.

Hall BD (1996) Photographic analysis: A quantitative approach to the evaluation of dysmorphology (Editorial; comment). *J Pediatr* 129(1):3–4.

Hall JD, Froster-Iskenius UG, Allanson JE (1989) *Handbook of Normal Physical Measurements.* New York: Oxford Medical Publications.

Hecht F, Hecht B (1991) Creativity in medical genetics and dysmorphology. *Am J Med Genet* 40(1):115–116.

Johnson KB (ed) (1993) *The Harriet Lane Handbook, A Manual for Pediatric House Officers,* 13th ed. St. Louis: Mosby, pp. 131–142.

Jones KL (ed) (1997) *Smith's Recognizable Patterns of Human Malformations,* 5th ed. Philadelphia: Saunders.

McKusick, VA (1992) *Mendelian Inheritance in Man, Catalogs of Autosomal Dominant, Autosomal Recessive, and X-Linked Phenotypes,* Vols. 1 and 2, 10th ed. Baltimore: Johns Hopkins University Press.

Pope FM, Smith R (1995) *Color Atlas of Inherited Connective Tissue Disorders.* London: Mosby-Wolfe.

Saksena SS, Bader P, Bixler D (1989) Facial dysmorphology, roentgenographic measurements, and clinical genetics. *J Craniofac Genet Dev Biol* 9(1):29–43.

Scriver CR, Beaudet AL, Sly WS, Valle D (1995) *The Metabolic and Molecular Bases of Inherited Disease,* Vols. 1–3, 7th ed. New York: McGraw-Hill.

Sharony R, Hixon H, Pepkowitz SH, Carlson DE, Platt LD, Graham JM Jr (1993) Experience with a fetal dysmorphology/pathology service in an academic medical center. *Birth Defects Orig Artic Ser* 29(1):219–233.

Spitz J (1996) *Genodermatoses.* Baltimore: Williams & Wilkins.

Stromme P (1991) The diagnosis of syndromes by use of a dysmorphology database. *Acta Paediatr Scand* 80(1):106–109.

Winter RM (1995) Recent molecular advances in dysmorphology. *Hum Mol Genet* 4(Spec No):1699–1704.

Winter RM, Baraitser M (1987) The London Dysmorphology Database (letter). *J Med Genet* 24(8):509–510.

5

Patient Education

Ann C. M. Smith

It is often said that knowledge is power. For individuals affected by or at risk for a genetic condition, having adequate knowledge or access to knowledge about a specific diagnosis, its etiology, and its management implications imparts the power to respond to their particular life situation. Patients who are struggling with basic questions, uncertainties, and/or confusing information may have difficulty in understanding their situation and in seeking help.

Genetic counselors are often the health care professionals who first begin the discussions of the underlying genetic basis for a condition and provide insight about implications for the individual and other family members regarding current health concerns and future reproductive planning. *What* is said and *how* information is presented during the early genetic sessions can have a significant impact on the ability of a person or a family to process, understand, and assimilate such new knowledge. However, the very nature of genetic counseling often places the counselor in the role of delivering "bad news," or information that families want but are also apprehensive of hearing. The challenge is to communicate genetic information in a supportive, culturally sensitive climate that encourages individual autonomy in decision making.

Communication is the cornerstone of genetic counseling, as outlined in the profession's early definition: "Genetic counseling is a *communication* process which deals with the human problems associated with the occurrence or risk of occurrence of a genetic disorder in a family" (ASHG Ad Hoc Committee, 1975). As defined by Webster, "communication" is the act or process of imparting knowledge or interchange of thoughts, opinions, or information. A variety of media are used to communicate—the spoken word, writing, graphics, pictures, and the Internet.

A Guide to Genetic Counseling, Edited by Diane L. Baker, Jane L. Schuette, and Wendy R. Uhlmann.
ISBN 0-471-18541-8 (cloth), 0-471-18867-0 (paper). Copyright © 1998, Wiley-Liss, Inc.

Inherent in communicating about genetics are a number of barriers that can interfere with a client's understanding of the genetic information. By its very nature, genetics information is often complex and highly technical, with a lexicon or vocabulary of its own. Individuals and families dealing with genetic disease must assimilate large amounts of new information that is often complicated and abstract. Many late-onset genetic conditions, such as cancer and Huntington disease, evoke a variety of feelings of fear, anxiety, or depression that may hinder the genetic counseling process itself. Similarly, the birth of a child with unexpected birth defects often precipitates a cascade of psychosocial reactions (e.g., denial, anger, grief) that affect not only the parents' readiness to hear but how much they can hear. Culture, beliefs, and traditions also impact a person's perception and understanding of biology and genetics, and may conflict with long-held personal beliefs.

This chapter focuses on how the genetic counselor serves as an instructor during the genetic counseling process, highlighting the essential tools used to communicate technical information in a way that ultimately fosters informed decision making. An undertanding of the ways in which adults learn and the use of effective tools for communication can enhance the education process. When coupled with compassion, empathy, and sensitivity to ethnocultural values, the genetics encounter moves from a simple educational contact to the multifaceted practice of genetic counseling.

ADULT LEARNERS

In communicating about genetics, genetic counselors takes on the role of instructor, sharing their expertise in most circumstances with adult patients or parents. The methods of educating adults differ from those used in teaching children. In contrast to younger students, who are motivated by external pressures from parents and teachers, adult learners are motivated by internal factors such as self-esteem, recognition, and better quality of life (Knowles, 1973, 1980). Adults are generally motivated to learn and seek out information for specific reasons; they are often goal-oriented and pragmatic, seeking information to solve problems, build new skill sets, pursue job advancement, or make decisions (Cross, 1981). Recognition and understanding of different learning styles is critical for counselors to be successful instructors. Learning styles—typical ways of feeling, behaving, and processing information in a learning situation—change over a person's life span (Hyman and Rosoff, 1984; Kuznar et al., 1991).

Instructors working with adults should consider what the learner's motivation is for seeking information and what the learner already knows. Failure to connect the knowledge or information to the daily needs and life of the learner does not permit a bridge for common understanding.

Four characteristics of a skilled and motivating instructor that help to enhance adult learning are defined by Wlodkowski (1993): expertise, empathy, enthusiasm, and clarity. These have particular relevance to genetic counselors and can be learned and improved on through continued practice and effort. Just as musicians perfect their singing repertoire or orchestral talents through dedicated training and practice, genetic counselors can also perfect these four skill areas.

The practical definition of *expertise* according to Wlodkowski (1993) involves three essential elements: we know something beneficial; we know it well; and we are prepared to convey it through an instructional process. As instructors, genetic counselors are uniquely trained to integrate knowledge from the fields of medicine, human genetics, and psychology. This skill permits them to communicate their knowledge about genetic conditions in terms that can be more clearly understood and are likely to have practical applications for the family.

There is no substitute for knowing a topic inside and out. However, experts in many fields often make the classic mistake of believing that if they know a lot about a subject, they can automatically teach it effectively. We should ask ourselves if we understand WHAT we are going to educate families about. First, can we explain it in our own words? Is there more than one example—a story, fact, research finding, or analogy—that would be useful in explaining a particular genetic concept? Are there any visual aids (pictures, diagrams, etc.) that are effective in demonstrating the concept? Do we know what we don't know and feel comfortable admitting it? Counselees don't expect us to know everything—but they desire an honest explanation of what is known. Human genetics is rapidly changing, requiring counselors to continually keep abreast of new information of relevance to their patients. Given the rarity of many genetic conditions, maintaining up-to-date knowledge of every possible condition is virtually impossible; however, having the skills and ability to access the most current information is critical and essential!

Information should be conveyed using an instructional process preceded by intensive preparation and organization of materials. A well-prepared counselor can display a relaxed familiarity with a specific genetic condition or area of information. Preparation permits us to spend most of our time looking at the learner/couple and talking with them—not to them! Rather than lecturing to the client, the counselor seeks to facilitate a two-way, give-and-take communication flow. For novice genetic counselors this may feel awkward, given the amount of information that needs to be gathered and shared. Yet the ability to do several things at once (think, observe, ask questions, construct a pedigree, give feedback, etc.) must be mastered. Student genetic counselors may tend to keep their eyes glued to intake sheets, notes, and/or templates for pedigree construction, rather than relating to the patient.

We are all guilty of trying to teach as we have been taught—and in the historical pedagogical model, it is the teacher who decides what will be taught and delivers the lecture; the student is expected to be ready to listen, absorb, and learn what the instructor lectures about. A parallel can be drawn between this model of learning and the biomedical model of physician–patient interaction. For both, the relationship is asymmetrical, with the teacher/physician taking a dominant role, controlling the interaction by asking questions and initiating topics for discussion. In the context of the biomedical model, emphasis is placed on communicating technical information rather than on how it is understood or acted upon.

Educators have suggested at least three better ways to instruct adults: (1) by treating the adult student as a partner in the learning process, (2) by building and placing value on previous learning and life experiences, and (3) by promoting personal direction and control of learning. Malcolm Knowles, considered by many to be the grandfather of adult education, has written extensively about the adult as a self-directed learner and offers the androgogical or adult model of learning (Knowles, 1973, 1980). Contrary to the traditional or pedagogical model in which the learner is dependent, with little experience of value, Knowles's model is characterized by a learner who is self-directed and whose experience becomes a resource to be used, valued, and accepted. The "readiness" to learn stems from a need to know or do something rather than a requirement for advancement. Genetic counseling practice adheres to the principles of Knowles's model of adult learning, viewing the counselee's life experience, attitudes, and needs as important to the genetic counseling process.

Kenen and Smith (1995) suggest two models stemming from the sociological literature with major relevance to this discussion: the mutual participation model and the life history or narrative model. The **mutual participation model** (Szasz and Hollender, 1956) is premised on the ideas that the counselor and counselee have approximately equal power as adults, are mutually interdependent, and engage in activity that will be satisfying to both. Under this model, the counselor assists the patient/counselee to help himself or herself, thereby enhancing rather than narrowing autonomy. Counselor and counselee work as equal partners in seeking solutions, with free give and take, permitting clarification of information about genetic facts as well as the psychological, social, and cultural determinants. The interaction is of one adult to another.

Under the **life history/narrative** model described by Mischler (1986), counselor and counselee possess dual roles: the counselor is both interviewer and listener, and the counselee is both respondent and narrator. The main focus for the counselor is to listen (actively!) to the stories the counselee/narrator is telling. The interaction recognizes and values the client's beliefs and cognitive perceptions and fosters client power. Under this model, the counselor is interested in *themal* coherence (i.e., how utterances express a client's recurrent assumptions, beliefs, and goals or cognitive world) (Mischler, 1986). The counselor—as interviewer and listener—actively enters the counselee's storytelling by the form and intent of questions asked, assessments made, acknowledgments, and silences.

TENETS OF HEALTH EDUCATION AND PROMOTION: APPLICATION TO GENETIC COUNSELING

As a complement to the tools and techniques used by the genetic counselor cum instructor, some discussion of contemporary models of health behavior and their relevance to health promotion programs is warranted. Geneticists and others in medical practice have been rather slow to recognize and utilize the modern principles and practices of learning and behavior theory. However, similar to genetic counseling, health promotion programs seek to improve health, reduce disease risks, and improve the well-being and self-sufficiency of individuals, families, and communities, as well as organizations. Utilizing a multilevel interactive approach, such programs go beyond traditional educational activities by including advocacy, policy development, economic supports, organizational and environmental change efforts, and multimethod programs. Formal awareness and application of the health promotion literature and health behavior models provides genetic counselors with new avenues for clinical practice and research.

Contemporary models of health behavior stem from cognitive behavioral theories, which are premised on two key concepts: (1) that behavior is mediated through what we know and think (cognition), and (2) that knowledge, while needed, is not alone sufficient to change behavior; factors such as perception, motivation, skills, and personal and social environmental factors also have important roles (Glanz and Rimer, 1995). Behavior is influenced at multiple levels, as shown in Table 5.1, including intrapersonal (or individual) factors, interpersonal factors, institutional or organizational factors, community factors, and public policy factors. Several theories describing the interrelationship between these factors and health behaviors have emerged and warrant a brief discussion.

The review of contemporary behavior theories about health promotion that follows is not intended to be exhaustive but rather to heighten awareness of a body of literature of particular relevance to the genetic counseling field. The tenets of health education are certain to have widespread application for genetic counselors involved in the development of patient education materials and service delivery programs. The concepts derived from contemporary behavior theories and health promotion programs are consistent with the underpinnings of the field of genetic counseling, which include accessibility of services, attitudes toward health care, perception and beliefs about susceptibility (threat of illness), knowledge about disease, social structure, self-efficacy, decision making, and autonomy. Health education programs improve health, identify and reduce disease risks, manage chronic illness, and improve the well-being and self-sufficiency of individuals, families, organizations, and communities. While it is easy to see the relevance of these concepts to the planning and development of wide-scale genetic testing and screening efforts (e.g., Tay–Sachs disease, cystic fibrosis), these concepts also have applicability to individual genetic counseling encounters, a matter we explore shortly.

TABLE 5.1   Factors Influencing Health Behavior

Factor Level	Definition
Intrapersonal	Individual characteristics, such as knowledge, attitudes, beliefs, personality traits, self-concept, skills, past experience, motivation
Interpersonal	Interpersonal processes and primary groups (i.e., family, friends, peers) that provide social identity, support, and role definition
Institutional	Rules, regulations, policies, and informal structures that may constrain or promote recommended behaviors
Community	Social networks and norms, or standards, which exist formally or informally among individuals, groups, or organizations (e.g., schools, worksites, health care settings)
Public policy	Local, state, and federal policies and laws that regulate or support healthy actions and practices for disease prevention, early detection, control, and management

Source: U.S. National Institutes of Health (1995) *Theory at a Glance: A Guide for Health Promotion Practice,* NIH publication 95-3896, p. 16.

Health Belief Model

The first and most widely recognized model that adapts theory from the behavioral sciences to health problems is the health belief model (HBM), which attempts to explain and predict health-related behavior in the context of certain belief patterns. Based on the assumption that the degree of "fear" or "threat" of a disease or condition is a powerful motivating force to behavior, the HBM was introduced in the 1950s by psychologists in the United States to explain the public's use of preventive health services available at that time, such as flu vaccines and chest X-ray screening for tuberculosis (Glanz and Rimer, 1995). The model takes into account a person's perception of the susceptibility and severity of a disease condition and the practical and psychological costs and inconvenience (perceived barriers) to taking a "health" action. A person's readiness to act (cues to action) reflects the accompanying appraisal of a recommended behavior for preventing or managing the health problem within the context of the perceived threat (Green and Kreuter, 1991; Glanz and Rimer, 1995). The concept of *self-efficacy* was added to the HBM in 1988, to reflect confidence in a person's ability to take action (e.g., change habitual unhealthy behaviors such as smoking or overeating).

Stages of Change Model

The stages of change model was introduced by Prochaska and DiClemente (1983; Prochaska et al., 1992) based on work with smoking cessation and drug and alcohol treatment programs. This model views behavior change as a circular process, with individuals at varying levels of readiness to change or attempt to change toward healthy behaviors. It is based on the assumption that people do not change chronic and habitual behaviors all at once, but continuously through a series of stages including the following (Green and Kreuter, 1991; Glanz and Rimer, 1995):

precontemplation stage (people have no expressed interest in or are not thinking about change)

contemplation stage (serious thought is given to changing a behavior)

decision stage (the plan to change is made)

action stage (the first 6 months after an overt change in behavior has been implemented)

maintenance stage (the period from 6 months after the change until the unwanted behavior has been completely terminated)

Consumer Information Processing Theory

Originating from studies of human problem solving and information processing, the consumer information processing theory focuses on the process by which individuals, as consumers, acquire and use information in their decision making. Although not originally developed for the study of health behaviors, this model can be used to examine why people use or fail to use health information; hence it can be instructional in the design and development of successful informational and intervention strategies (Glanz and Rimer, 1995). The theory recognizes that there are limitations in the amount of information individuals can acquire, process, and remember; hence, choosing the most important and useful points to communicate is critical. Providing information in small, clear, and concise "chunks" that can be handled according to usable decision rules (heuristics) permit individuals to make choices faster and more easily. The information should be convenient to use and tailored to the audience with respect to amount, format, readability, and processibility. For people to use health information, the material must be available; in addition, it must be viewed as useful or new, and it must be communicated in "user-friendly" fashion.

Social Learning Theory

Complex in its design, social learning theory (SLT) incorporates concepts and processes from cognitive, behavioralistic, and emotional models of behavior change. Synonymous with the social-cognitive theory put forth by Albert Bandura in 1970s, the SLT assumes that individuals exist within an environment in which personal factors, environmental influences, and behavior continually interact. The concept of reciprocal determinism is central to SLT, and this represents a major departure from traditional operant conditioning theory, which tends to view all behavior as a one-way product of the environment (Green and Kreutner, 1991). In SLT, behavior and environment are "reciprocal" systems. While environment shapes, maintains, and even constrains behavior, the process is bidirectional, with individuals able to create or change (self-regulate) their environment and actions. Learning takes place through direct experience, by means of the indirect or vicarious experience of observing others (modeling), or via the storing and processing of complex information. Self-efficacy, or the confidence in one's ability to successfully take action, is the most important aspect of the sense of self (Bandura, 1986). Not only does self-efficacy influence behavior, but it may enhance coping and alleviate anxiety.

Theory of Reasoned Action

The concept of *behavioral intention* is central to the theory of reasoned action and represents the final step in the process before any action takes place. Attitudes toward the behavior and perception of social norms favorable to the behavior influence one's behavioral intention. A person's existing skills—skills already possessed, which don't need to be learned—are closely tied to self-efficacy and behavioral intention. For example, an experienced mother who successfully breast-fed her first child possesses a high self-efficacy about breast-feeding and the self-confidence and skills necessary to support her behavioral intent to breast-feed her second child. The model also adds a strong cultural component to predicting behavior (Shumaker et al., 1990).

THE GENETIC COUNSELING SESSION: A VEHICLE FOR PATIENT EDUCATION

By its relatively structured organization and logical progression, the typical genetic counseling visit serves as a vehicle for patient education and is divided into seven major components or phases: (1) information gathering, (2) diagnosis, (3) risk assessment, (4) information giving, (5) psychological assessment and counseling, (6) help with decision making, and (7) ongoing client support (Walker, 1996). During each of these phases, educational opportunities exist to present new information, correct misconceptions, reinforce information, or lay the foundation for future patient education and counseling. The genetic counselor must be alert to these opportunities and take advantage of teachable moments as they present themselves.

A **teachable moment** is defined as "the point(s) at which an individual, couple or family is most able to comprehend and absorb the information being given" (Andrews et al., 1994). Ideally, genetic counseling is most effective when provided after the initial shock and denial of receiving a distressing genetic diagnosis have diminished, and the client is able to listen and fully comprehend the information presented. However, this ideal is often not achievable, as a result of scheduling, insurance, or managed care requirements. Follow-up counseling over a series of appointments to reinforce earlier discussions has tremendous merit but has become less feasible in today's era of cost containment.

The subsections that follow present several of the major components of genetic counseling, with examples of teachable moments that might typically occur.

Information Gathering

During the information gathering phase, it is important to establish and understand the patient's perception of the problem or reason for coming before setting the agenda for the remainder of the visit (Riccardi and Kurtz, 1983; Walker, 1996). In the context of patient education, learning is tied to clearly understood expectations. Hence, clarification of counselor and counselee expectations, problem identification, and contracting with the patient about the organizational plan and content of the visit provide the foundation for a logically sequenced and well-organized session. Like a teacher following a logical course outline or instructional plan, the counselor has the responsibility for providing overall structure and ensuring smooth transitions, thereby setting the stage for the educational aspects of the genetic counseling session. Jumping from topic to topic can be extremely confusing and disconcerting for the counselee. A logical sequence with smooth transitions is beneficial; however, the counselor must maintain flexibility and be ready to diverge from the established action plan when such modifications are seen to be prudent. This dynamic is what makes genetic counseling a process tailored to the counselee's needs—rather than a standard, rote instruction.

Obtaining the family pedigree can provide an instructional opportunity. The pedigree, which serves as a visual generation-to-generation tool on which to build later discussions of inheritance mechanisms, also provides an opportunity to correct prior misconceptions related to proneness in the family (Kenen and Smith, 1995). The concept of **proneness** refers to lay beliefs about inheritance, and the nature and extent to which people feel prone to a genetic disease or feel that other family members are subject to have or acquire the condition (Richards, 1993). Each counselee's concept of proneness merits further exploration, since it may interfere with that person's understanding of general genetic principles. Green et al. (1993) found that many individuals' views of proneness differ substantially from standard principles of Mendelian inheritance, especially among families with a history of a genetic condition. Often the phrases "in the family," "inherited," and "genetic" are used idiosyncratically (e.g., "Well, of course, I know that hereditary things can't run in the family."). The recognition of such teachable moments permits the counselor not only to identify potential misconceptions or family "myths" but also to lay the foundation for future discussions. Just introducing certain genetic concepts (e.g., dominant inheritance and variable expressivity), albeit briefly and simply, serves as a prelude to the more detailed discussion of inheritance during the *information giving* phase.

Diagnosis

During the physical examination, opportunities exist to educate the counselees about the diagnostic process. It is often helpful to explain that a genetic exam includes the usual general review of systems, including height, weight, and head circumference measurements, complemented by a more detailed assessment and series of measurements for normative comparison, checking for minor findings of the head, face, neck, trunk, skin, and extremities. Care should be taken to ensure that the patient or parents understand that no one finding is diagnostic but, rather, that geneticists look for patterns of findings that suggest a specific diagnosis or syndrome. Technical descriptors or terms (e.g., hemihypertrophy, brachycephaly, microcephaly) can sound frightening, and the counselor can assist the family by providing explanations. Pointing out pertinent findings as well as normal variations to parents can also make the physical examination more understandable, hence less threatening. Furthermore, such introductory remarks open the door for future discussions of physical features of relevance to a specific diagnosis.

In cases of presymptomatic or carrier testing, pretest information must be provided that addresses the risks, benefits, efficacy, and alternatives to testing; it is also essential to furnish information about severity and about decisions that will likely become necessary if test results are positive (Andrews et al., 1994). Once counselees have been identified as carriers, follow-up counseling and education may occur in two stages: (1) support counseling (with the confirmation of carrier status) and (2) more detailed genetic information and counseling (after the person has had an opportunity to evaluate the significance of the new information). Once a diagnosis has been established, the counselor is able to provide information about the etiology and recurrence risk and to facilitate resource support; he or she can also instruct counselees with respect to the natural history and disease manifestations of the condition, management and/or treatment options, and reproductive alternatives.

Information Giving

During the *information giving* phase, the focus is on providing factual data and information in a culturally sensitive and individualized fashion tailored to the client/counselees. Often this centers on helping families to comprehend the medical facts related to diagnosis, natural history, and management and/or treatment options; causation and recurrence risks; available reproductive options; and available support resources. The counselor must now bring together all the information elicited and observations made during other phases of the genetics visit, capitalizing on any groundwork laid during teachable moments. The counselor weaves the information together with counseling and support in a fashion that is easily understandable and relevant to the needs and concerns of the counselees. It is important for the counselor to:

- Distill complex information into clear, understandable concepts.
- Restrict the use of complex terminology and medical jargon, being sure to provide clarification where technical language is unavoidable.
- Refrain from equating the patient/client with the diagnosis; use of words like diabetic and hemophiliac can demean the patient and detract from his or her individuality.
- Provide information, beginning with the general then moving on to more specific information. For example, it is fairly common when reviewing basic genetic concepts to begin with a brief explanation of chromosomes (in all cells of our body) and then move from a normal karyotype to a pair of chromosomes (usually the pair in question if the gene has been linked) to a specific band area to genes to DNA).
- Promote a two-way mutual exchange of information, especially with respect to the patient's understanding of the problem and diagnosis, expectations and requests for the session, and participation in the choice and discussion of treatment and/or action plans.
- Encourage counselees to restate information in their own words.
- Correct and clarify potential misunderstandings or incorrect information.
- Offer ample opportunities for questions and answers.
- Be attentive to both verbal and nonverbal cues, providing adequate reassurance and messages of support.

In this way, the counselor goes beyond simply educating counselees about the genetic and medical facts, using a decision support model, recognizing that before making decisions, counselees must "process and personalize facts effectively and cognitively" (Walker, 1996).

Communication of Risk

One of the major focuses of genetic counseling is risk assessment and the provision of recurrence risk information to the client in an understandable fashion. The term **risk** refers to the probability that an event will happen. For the public, this term has an inherent negative connotation, imparting a sense of danger (e.g., risk of being struck by lightning, hit by a car, or dying of cancer). Everyday words like "chance" and "likelihood" are more neutral. Families may also have preconceived beliefs about their risk status (proneness).

Several factors should be considered when one is providing risk information to clients/counselees. While there is no one correct way to present risk information, the presentation that is made must be balanced, accurate, and tailored to the client. Individuals vary in how they understand risks numerically (i.e., **objective** risk estimate) and in their interpretations (i.e., **subjective** risk estimate). While studies have found no numerator bias in the manner in which a risk is framed (e.g., 4/100 or 4/1000) (Chase et al., 1986), almost a third of women with less than a college education failed to recognize that 1/1000 is equivalent to < 1%. For many, the fraction 1/800 *sounds* higher than 1/400 because of the larger denominator. People have difficulties interpreting abstract probabilities and evaluating what they mean; they frequently perceive chance in a binary fashion—all or nothing. The numeric equivalent that individuals attach to nonnumeric phrases of probability (e.g., *often, rarely, never*) is also highly subjective; and the use of such nonnumeric phrases by the counselor introduces a potential for bias.

Ideally, counselors should provide risk information in several different ways. Rather than refining the risk to the fraction of a percent, it is more important to consider how to present recurrence risks in a manner that will have meaning for the family. Thus it is important to frame risks by providing the likelihood of a negative outcome balanced by the likelihood of a positive outcome. A 1% risk for complications may be perceived as much higher than its corollary—the 99% chance of no complications. The manner in which risks are presented or framed also impacts on the client's subjective perception. In prenatal counseling, major information to be provided pertains to risk; procedure-related risks, the likelihood of detecting a fetal abnormality using available techniques, the patient's individual risks based on age or other factors, and background population risks. Walker (1996) suggests that it is more helpful to present prenatal risks to the counselee in terms of *how* a specific risk factor (e.g., age, triple marker result, exposure) *changes* the woman's chances from what they would have been without that factor. Stating numeric data both as fractional risks (1 in 4 chance) and as percentages (25%) can also improve client understanding.

Ultimately, it is the counselee's perception of risk rather than the actual risk that is meaningful. It is important, therefore, to explore the counselee's own impression or perception of risk figures compared to the actual or objective risk given. Families view risks differently, depending on their experience with the condition in question and their view of any attendant problems. For example, the perceived burden of the outcome associated with 1% risk to have a child with Down syndrome may be higher for a couple who already has a child with Down syndrome than for a couple at risk due to advanced maternal age alone. (I recall a family in which one parent had neurofibromatosis and the other had a dominant connective tissue disorder; for this couple, the 25% risk of having a child with both conditions was "much lower" than the 100% they had originally anticipated.)

An individual's estimate of the likelihood that a given outcome will occur is affected by the ease with which actual or "concrete" examples of the instance can be called to mind—a concept referred to as the **availability heuristic** (Tversky and Kahneman, 1973). Actual or dramatic instances of certain outcomes will increase the perception that the outcome is likely to occur. How many times have the parents asked, "If it was one in a million before, how could the risk be only 1% now?" Families who have experienced a rare event may perceive their future risk as 100% (since it has happened already). Similarly, a woman undergoing prenatal diagnosis who knows someone who has undergone the procedure is likely to be influenced by that personal encounter—or concrete instance—in her perception of the procedure, its risks, and the likelihood of the test leading to a "good" outcome (birth of a normal baby). If the friend's experience involved complications (e.g., multiple taps, miscarriage), this concrete instance may affect the counselee's perception of the risks of the procedure for herself. Women undergoing amniocentesis for advanced maternal age tend to overestimate failure not only in terms of perceived probabilities of procedural complications but with respect to the chances that the test will reveal a problem with the fetus (Adler et al., 1990). Similarly, it has been my experience that parents who have had a child with a rare de novo chromosome abnormality or sporadic malformation syndrome perceive their chances of having a second affected child as higher than the actual recurrence risk.

Explaining Mendelian inheritance can, at first glance, appear relatively straightforward, but confounding issues such as genetic heterogeneity, variable expressivity, nonpenetrance, gonadal mosaicism, new mutation, sex limitation, imprinting, phenocopies, and genocopies can introduce great confusion. Add to this the nontraditional forms of inheritance (i.e., uniparental disomy, mitochondrial inheritance, and trinucleotide repeats), and the counselor is truly challenged to make the relevant concepts understandable.

The use of metaphors, such as tossing a coin, can provide concrete examples of probability. However, it should also be remembered that in some cultures this activity is seen as a form of gambling, hence may have negative connotations. In the case of Mendelian inheritance, tactile or visual examples illustrating patterns of inheritance can prove extremely beneficial. Use of standard diagrams illustrating dominant, recessive, and X-linked inheritance not only help the counselor explain the different possible genotypes but can also be given to the parent or counselee to take home as added reinforcement of the concepts. Encouraging counselees to restate what has been said in their own words is also effective in ascertaining their level of understanding of the concepts and information discussed. Often, asking them how they might explain the inheritance to a relative or friend encourages them to restate what they have learned and offers the counselor an opportunity for clarification.

Ongoing Client Support

In supportive counseling, the counselor tries to facilitate each counselee's ability to cope and deal with information presented. By encouraging counselees to discuss what the information means for them (e.g., exploring their feelings about raising a child with a disability, reactions to learning a diagnosis, personal values or previous experiences with handicapping conditions), the counselor seeks to foster independent decision making.

Information and resources constitute one of the most critical needs of parents and/or families of a child or family member newly diagnosed with a birth defect or a chronic medical or genetic condition. Parents of a child with a newly diagnosed birth defect or genetic condition often have no idea how to identify resources. Many parents express a strong desire for parent-to-parent referrals, yet one study showed that fewer than 20% of physicians made such referrals (Sharp et al., 1992). For some families the need to seek as much information as possible and network with other families can be intense and immediate; others may feel somewhat threatened by this. Ideally, the counselor knows what the agency or group can offer families and should be able to provide primary contact information. While families vary in how much additional information they want and when they want to seek out peer family support, they generally gladly accept the ready accessibility of such information.

During the 1980s, in the midst of the self-help era, many voluntary genetic support groups were established to provide information, peer support, and advocacy for their respective populations, both locally and nationally. Given the rarity of certain genetic conditions, not every condition will have an established support group; thus the counselor is challenged to identify other resources in the community or on a broader scale that may have relevance to the family. Families and patients affected by the specific disorder are an extremely practical and rich source of information. Ideally, an organized parent support group is best. Some genetic conditions are so rare, however, that families may have to fall back on networking with other parents who have had the same experience. This option, nevertheless, can be extremely beneficial.

Many counselors maintain personal files of resource information. However, maintaining up-to-date information in the "age of information" becomes almost impossible. Two major resource groups that maintain and provide such information for families are the Alliance of Genetic Support Groups and the National Organization for Rare Disorders, Inc. (see Chapter 15 for further information).

ADDITIONAL ASPECTS OF GENETIC COUNSELING

Consumer Preferences

In a study that examined parental experiences and preferences related to communicating medical bad news (e.g., a diagnosis of Down syndrome), a difference was found between what parents *desired* and what they actually *experienced* from physicians who convey bad news. Parents preferred significantly more communication of information and feelings by their physician. They expressed strong preferences for the physician to show caring (97%), to allow parents to talk (95%), and to allow parents to show their own feelings (93%) (Sharp et al., 1992). Families often feel lost, inadequate, fearful, and/or even angry at learning "bad news." Genetic counselors seek to develop a climate of support that encourages parents to express their feelings and emotions, simultaneously promoting a mutual exchange of information between counselee and counselor.

There is a general paucity of outcome-based research about genetic service delivery and expectations about services from the consumer's perspective. Middelton and Smith (1997) recently examined consumer expectations about and satisfaction with genetics services among individuals receiving pediatric or prenatal genetic services in the Mid-Atlantic region. Clients were asked to prospectively indicate the level of "importance" of discussing a series of issues prior to their initial genetics visit. After the visit they were asked about the "level" to which these same issues were discussed during the counseling session. Table 5.2 summarizes pediatric and prenatal clients' expectations at pretest based on a ranking of the five most and least important issues to be discussed. In general, pediatric clients desire information that assists them in learning about the impact of the genetic condition on themselves and/or their child, specifically as it relates to medical treatment, management, and diagnosis. Prenatal clients, on the other hand, appear to place primacy on the administrative details of prenatal testing (e.g., how they will learn about results; test accuracy; procedural risks). Relationship issues were among the least important issues for both groups. While limited, these results are worthy of further study because they offer a consumer-based perspective about expectations for the content of genetic information to be discussed. At the very least, such findings stress the importance of understanding what the client's expectations are, and how they relate to patient satisfaction and improved genetic service delivery.

Duration of Interaction

It has been shown that the amount of time a professional spends with a patient greatly affects patient information seeking behaviors. In one study of 106 rehabilitative medicine patients, physician–patient interactions lasting 19 minutes or more were associated with increased patient desire for information and participation in medical

TABLE 5.2 Consumer Perspectives of the Most and Least Important Issues for Discussion Prior to Pediatric or Prenatal Genetics Visit

Pediatric Genetics Patients	Prenatal Genetics Patients
MOST Important Issues to Discuss or Learn About	*MOST Important Issues to Discuss or Learn About*
Available medical treatment/management	How I will learn test results
What is wrong with me/my child	Accuracy of prenatal testing
Learning coping skills	Risks of prenatal testing
If the condition can be cured	Finding out chance of genetic condition occurring in me, my child, or other family member
Chance of the condition occurring in me or my child	What to expect when having prenatal testing
LEAST Important Issues	*LEAST Important Issues*
My relationship with my partner or spouse	Talk about my/my partner's pregnancies
Plans for future pregnancies	Pregnancy termination options
Availability of prenatal testing	Discuss my feelings about pregnancy
My or my partner's/spouse's pregnancies	Decide whether I should have another child
Alternative reproductive options	Discuss reproductive options

Source: Based on Middelton and Smith (1997).

decisions (Beisecker and Beisecker, 1990). While patients sought information on a wide range of medical topics, they failed to engage in many information seeking behaviors when their visits with physicians were shorter than 19 minutes.

Recall

Experience has shown that patients, like students, forget almost half of what they are told. In genetic counseling, when the information being communicated is technical and new, comprehending and remembering a series of new terms, genetic concepts, and risk information can present a formidable task. Recall can be significantly improved, however, when information is organized and categorized verbally (Riccardi and Kurtz, 1983). Studies of physician–patient communication have shown that patients recall best what they are told first and what they consider important (Ley, 1972). The most effective means of increasing long-term patient recall is to couple patient restatement with ample opportunities for feedback from the physician or professional (Kupst et al., 1975).

Riccardi and Kurtz list several factors that influence patient recall, compliance, and satisfaction. Important information should be "categorized" or highlighted verbally (e.g., "I am going to tell you what we think your son has and what tests we want to perform to confirm this diagnosis."). Then the counselor might continue, "First, we feel that both his delay in development and the birth defects may be due to an abnormality of his chromosomes. To determine this, we need to study his chromosomes by getting a blood sample." Important information should be given first ("It's important for you to remember that . . ."). Encouraging patients to restate the "message" in their own terms is extremely beneficial. Potential misunderstandings or incorrect information should be addressed immediately. Providing patients with ample opportunities to ask questions is critical and has been shown to improve understanding and recall. Try to restrict the use of technical and medical jargon, relying instead on everyday language to increase comprehensibility. Provide an appropriate level of reassurance and messages of support, and be attentive to both verbal and nonverbal cues.

BARRIERS TO CLIENT UNDERSTANDING

There are both internal and external barriers to client understanding of genetic information. Internal barriers include a client's personal and cultural beliefs, and his or her education level. Economical constraints comprise an important external barrier in today's health care marketplace.

Ethnocultural Barriers to Health and Genetic Services

The U.S. population is becoming increasingly multicultural, with an ever increasing influx of new immigrants, many of whom do not speak English as their first language. Since the early 1990s, 600,000 legal immigrants have been added to the U.S. population each year, the majority arriving from Latin American and Pacific Asian countries. Ethnocultural differences in the perception of health and disease, and disability and disease burden, as well as interpretation of risk factors, pose formidable barriers to those attempting to deliver adequate and quality health and genetic services to this diverse population.

Many new U.S. immigrants are unaware of the availability of genetic services and have little prior experience with this type of medical interaction. In some cultures, individuals seek help from a healer or "professional" to be told what to do. Faced with the traditional genetic counseling process, which emphasizes and respects patient autonomy, many are bewildered or confused; they may view the nondirective approach in genetics as unauthoritative or indecisive, calling into question the counselor's expertise and professional competence (Walker, 1996).

Finding genetics instructional materials for the non-English-speaking patient is difficult. Simple translation of English text into another language (Spanish, French, etc.) is generally not satisfactory. One useful resource compiled by genetic counselors, the *Catalog of Multilingual Patient Education Materials on Genetic and Related Maternal/Child Health Topics* (Center for Human and Molecular Genetics, 1993), provides a listing of over 200 genetics and health-related topics covered in 32 languages.

Scientific Literacy

Educators increasingly are aware of the public's need to have a basic understanding of genetics, disease risk, and health choices within the context of broader scientific and biological literacy. According to the Biological Sciences Curriculum Study (1997), about 95% of American high school students take biology in ninth or tenth grade; moreover, human genetics is now included in virtually all high school biology curricula, with a focus primarily on Mendelian genetics (BSCS, 1997). Yet, despite the increased exposure to human genetics and issues related to genetic testing among high school students, the scientific literacy of the general public continues to lag far behind the technological trends and innovations occurring in human biology, genetics, and medicine. Only 7% of Americans are considered scientifically literate, and only 24% understand the relation of DNA to inheritance (Andrews et al., 1994). Genetic concepts are often not well understood and are influenced by a person's ethnocultural and educational background as well as his or her life experiences.

A genetic counselor's approach may be influenced by whether a given patient has any background knowledge on which to build. Many counselees lack a frame of reference for standard genetic terminology, such as gene, chromosome, risk, inheritance patterns, and carrier, representing the basic building blocks for explaining genetic mechanisms. In many instances, the genetic counselor must resort to a minicourse in biology and inheritance before beginning to provide individualized risk assessment information.

It is estimated that 40–50 million people, about one-quarter of the U.S. population, read at a very basic level of proficiency, at or below the eighth grade (Weiss and Coyne, 1997). This deficiency has important ramifications with respect to counselees' ability to access genetic information. While the use of written materials such as pamphlets and booklets can be beneficial for patient education, it is important to evaluate the readability as well as comprehensibility of such materials by the intended audience. Moreover, clients who are unable to read may be embarrassed to reveal this limitation.

Economic

In an era of cost containment and managed health care, genetic counselors are being seriously challenged to develop genetic service delivery models that provide the best genetic care to a growing population of potential clients both efficiently and cost-effectively. These economic pressures come at a time when genetic advances and the population in need of genetic services also are growing continuously. As Walker (1996) writes, "The luxury of the team approach and repeated, lengthy counseling sessions is rapidly becoming a thing of the past" (p. 597). Managed care providers often restrict patients to a single genetics visit by a sole genetics provider. Not only does the new health care economics impact referral and access to genetic services, but the ability to provide long-term genetics follow-up and continuity of genetics care for rare genetic conditions may no longer be feasible.

APPLICATION OF INSTRUCTIONAL AIDS FOR PATIENT EDUCATION

Use of Instructional Aids

The use of instructional aids developed to assist the counselor in explaining concepts is extremely beneficial. The value of colored diagrams, figures, and other materials cannot be overstated. Many graphics already exist, and graphic programs permit the counselor to quickly create useful instructional aids. Colored markers with interchangeable caps work terrifically to illustrate explanations of chromosome translocations, deletions, or duplication. Colored pipe cleaners, colored paper clips, and beads on a string are all three-dimensional aids that can provide an easy-to-understand visual analogy of genes as units strung together on a chromosome or of linked DNA bases.

Experience supports the benefit of developing a personal *genetic counseling resource book*. A three-ring binder can be divided into sections including, for example:

Hospital/center-specific information for easy reference (billing office; major consult contacts, etc.)

Clinical/research protocols in use at your institution for selected conditions

Recurrence risk tables compiled from published data and/or center-specific data

Diagrams illustrating patterns of inheritance (e.g., Mendelian, multifactorial)

Diagrams showing chromosome structure and actual karyotypes (normal and abnormal)

Syndrome-specific information (local support group information; pictures of children with common conditions such as Down syndrome, etc.)

Growth curves (normative and syndrome specific)

Prenatal diagnosis maternal age curves, MSAFP risk tables, etc.

The comprehensive third edition of *Counseling Aids for Geneticists,* published by the Greenwood Genetic Center in 1995 or similar materials developed by the counselor or other groups are essential in the counselor's armamentarium of counseling tools. It is also extremely helpful to have a series of predrawn tear-off tablets showing two parents, their genotypes, and the four possible genotypes for the offspring. Counselors can also create on the spot diagrams that illustrate specific genetic circumstances—a skill to be mastered, especially when drawing upside down for a client's benefit. These materials are convenient and can be given to the client to take home.

The Internet has opened up a virtual library of information accessible to professionals and the public alike. The computer itself becomes a major tool in retrieving and developing educational materials for patient education. For example, computer graphics programs can be used to create individualized translocation diagrams, ideograms for chromosomes can be obtained from commercial Web sites (see, e.g., Figure 5.1).

Patient Letters   The genetic counseling summary letter serves as a valuable and important educational tool for families referred for genetic services. By providing a narrative summary of the information from the genetic visit, the summary letter documents the pertinent aspects of the patient's medical and family history, mechanisms of inheritance, and recurrence risks provided. Moreover, the counselees can refer back to the letter for recall of salient points and/or to share the information with other family members. Diagrams to explain molecular and cytogenic results can also be appended to the letter, along with a copy of the laboratory results (see Chapter 15 for further information).

Audiovisual Materials   Visual aids can further assist the client in understanding complex genetics information. For example, a videotape reviewing prenatal diagnos-

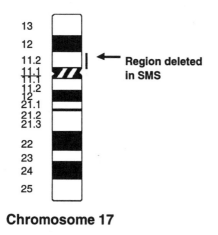

Diagram of del 17p11.2 causing Smith-Magenis syndrome (SMS) created in under 5 minutes using standard chromosome ideograms (http://www.selu.com/~bio/cyto/human/) of chromosome 17, then addition of arrow, text and deletion bar. Further customization is possible. Printed out this provides a simple graphic illustration of the deletion 17p11.2 that causes SMS.

Figure 5.1   Sample instructional aid *(created in five minutes using Microsoft Word and WWW access).*

tic procedures (amniocentesis/chorionic villus sampling) that incorporates the common information to be discussed with all clients provides a means of providing standardized information that not only saves time but can be used for quality assurance. Coupled with client-specific individualized counseling, these materials can enhance the process, providing information both visually as well as in narrative fashion. In response to ever increasing numbers of patients presenting for prenatal diagnosis, the state of California developed a videotape to ensure adequate review of the procedural risks, limitations, and benefits of testing. While such tools can streamline and ensure the quality of standardized pretest information, they are not intended to take the place of genetic counseling, especially with respect to its functions of individualized risk assessment, counseling, and support.

Brochures/Pamphlets Written materials serve as aids to information transfer and are almost universal to health promotion and patient education programs in genetics. They clarify and reinforce the information presented, augmenting the genetic counseling process. Disease-specific brochures are available from individual support groups or through the information Resource Referral Center of the Alliance of Genetic Support Groups (see Chapter 15).

DESIGNING INFORMATIONAL MATERIALS

In response to needs of their clients, genetic counselors have always been involved in the development of a variety of educational materials to facilitate the learning process. Writing about genetics often requires the use of potentially confusing technical terminology and concepts. When drafting such materials, consider writing style, vocabulary, layout, typography, and use of graphics and color. Use the active voice. Write in short, simple sentences, using one- or two-syllable words. Provide clarification through the use of examples. Avoid the use of jargon, technical terms, and unexplained abbreviations and acronyms. In brochures, leave plenty of white space on the printed page, and break up the narrative with subheadings and captions, and by highlighting important terms (bold/italic). Important points should be summarized.

Knowledge about the target audience is crucial to producing informational materials on genetic conditions; care should be taken in the development of materials intended for ethnic minorities and for patients and their families. Printed materials should be simply written, reinforced with graphics, and pretested with the target audience. Just as language skills differ, people will react differently to graphics, illustrations, and/or analogies. Printed materials should not be simply translated directly from English into other languages.

The importance of formative evaluation in the instructional design of instruments (e.g., questionnaires), media, and both written and visual materials cannot be overstated. To be effective, instructional materials should be developed professionally, and there should be systematic pretesting during draft stages. A general three-step review process consists of reading level assessment, content analysis, and review by content specialists (U.S. Department of Health and Human Services, 1992). Readability analysis can be performed using one of many computer software applications. Formal pretesting of materials with the intended audience during development is also beneficial to ensure that the final materials are understandable, relevant, attention-getting, attractive, credible, and/or acceptable to the target audience. Any sensitive and/or controversial elements can be identified and revised as necessary.

FUTURE DIRECTIONS

Many Americans are likely to be faced with difficult and complex decisions about the use of genetic technology to determine personal health risks as well as risks for their unborn offspring. As genetic testing becomes available for more conditions, the traditional format of face-to-face genetic counseling sessions lasting one to two hours may not be feasible because of time constraints and cost containment pressures arising from managed care. New models to deliver required information to facilitate patient education and informed decision making must be investigated with attention to quality outcomes and support of patient autonomy.

With advancing computer and video capability, new ways of educating the patient about treatment options and outcomes are now feasible. Kasper and colleagues (1992) published their experience with shared decision-making (SDM) programs for the improvement of health care in the areas of treatment and management of a variety of medical problems, including breast cancer, mild hypertension, prostate cancer, and low back pain. Designed for use in a provider's setting, these programs supply information about a given medical condition along with descriptions of the benefits and harms associated with related treatment options. Early SDMs used interactive videodisc technology accessed through a computer touch screen; however, CD-ROM and digitized video technology has improved significantly, permitting an expansion in the use of interactive educational media. Not only can the patient and medical staff input relevant medical data (e.g., age, sex, symptoms, status, medical history), but patient-specific probabilities or outcomes of treatment alternatives can be presented in a standardized fashion. The utility of interactive videodisc (IVD) and/or CD-ROM permits personal video "vignettes" for the learner to view, thereby providing a vicarious understanding of possible outcomes in a more frank and personal manner. Ideally, SDM modules can be viewed during the initial visit, accompanied by written materials for reinforcement of information. Such SDMs are intended to *augment* rather than "circumvent" the existing physician–patient relationship by improving patient satisfaction, increasing professionalism among providers, and promoting "informed" treatment choices on the basis of individual values and needs.

With the advent and potential for genetic screening programs for such conditions as cystic fibrosis and breast cancer, researchers, including genetic counselors, have taken a role in the design of prescreening educational materials. They also are involved in the evaluation of the effectiveness of these materials as educational interventions. New models utilizing technologies such as videotape, interactive CD-ROM, and/or World Wide Web access warrant further exploration. It is critical for counselors to continue to assist in the development and evaluation of the benefits, risks, and costs of new intervention models, and to assist in determining the proper roles for such technologies compared to traditional counselor/patient strategies. The chapter appendix describes the design and development of multimedia educational materials for cystic fibrosis carrier screening as an example of an alternative intervention and model of instructional design methodology.

SUMMARY

John Naisbett, author of *Megatrends,* cautions that "We are drowning in information but starved for knowledge." Individuals and families affected by genetic disease face a plethora of high tech information, from which they seek to gain true knowledge about their genetic circumstances. *What* is said and *how* information is communicated can have a significant impact on their ability to process the information, and on their understanding and assimilation into personal life circumstances. To place more power into the hands of individuals and their families affected by genetic disease is to be sure that they have adequate knowledge—not just information—about their genetic circumstance. In the role of instructor, the genetic counselor seeks to effectively "communicate" such highly technical genetic information in a way that is compassionate, empathic, and sensitive to the ethnocultural values of the client. In this way, the genetics encounter moves from basic patient education to the multifaceted practice of genetic counseling.

REFERENCES

American Society of Human Genetics, Ad Hoc Committee on Genetic Counseling (1975) *Am J Hum Genet* 27:240–242.

Adler NE, Keyes S, Kegeles S, Golbus MN (1990) Psychological responses to prenatal diagnosis: Anxiety in anticipation of amniocentesis (unpublished), cited in Adler NE, Keyes S, Robertson P (eds) (1990) *Psychological Issues in New Reproductive Technologies.*

Andrews LA, Fullarton JE, Holtzman NA, Motulsky AG (eds) (1994) *Assessing Genetic Risks: Implications for Health and Social Policy.* Committee on Assessing Genetic Risks, Division of Health Sciences Policy, Institute of Medicine. Washington, DC: National Academy Press.

Bandura A (1986) *Social Foundations of Thought and Action.* Englewood Cliffs: Prentice-Hall.

Beisecker AE, Beisecker TD (1990) Patient information-seeking behaviors when communicating with doctors. *Med Care* 28(1):19–28.

BSCS (1997) Genes, Environment & Human Behavior. Biological Sciences Curriculum Study. Colorado Springs, CO.

Center for Human and Molecular Genetics (1993) *Catalog of Multilingual Patient Education Materials on Genetic and Related Maternal/Child Health Topics.* Newark: New Jersey Medical School.

Chase GA, Faden RR, Holtzman NA, Chawalow AJ, Leonard CO, Lopes C, Quaid K (1986) Assessment of risk by pregnant women: Implications for genetic counseling and education. *Soc Biol* 33(2):57–64.

Cross, KP (1981) *Adults as Learners: Increasing Participation and Facilitating Learning.* San Francisco: Jossey-Bass.

Gardiner RJM, Sutherland GR (1989) *Chromosome Abnormalities and Genetic Counseling.* New York: Oxford University Press.

Glanz K, Rimer RK (1995) *Theory at a Glance: A Guide for Health Promotion Practice.* U.S. Department of Health and Human Services, Public Health Service. National Institutes of Health Publication 95-3896.

Green LW, Kreuter MW (eds) (1991) *Health Promotion Planning: An Educational and Environmental Approach,* 2nd ed. Mountain View: Mayfield Publishing.

Green J, Murton F, Statham H (1993) Psychological issues raised by a familial ovarian cancer register. *J Med Genet* 30:575–579.

Greenwood Genetic Center (1995) *Counseling Aids for Geneticists,* 3rd ed. Greenwood: Jacobs Press.

Holsinger D, Larabell S, Walker AP (1992) History and pedigree obtained during follow-up for abnormal maternal serum alpha fetoprotein identified addition risk in 25% of patients. *Clin Res* 40:28A.

Hyman R, Rossoff B (1984) Matching learning and teaching styles: The jug and what's in it? *Theory into Practice* 23:35–43.

Kasper JF, Mulley AG, Weenberg JE (1992) Developing shared decision-making programs to improve the quality of health care. *Qual Rev Bull; J Qual Improv* 18(6):183–190.

Kenen RK, Smith ACM (1995) Genetic counseling models for the next 25 years: Models for the future. *J Genet Counseling* 4(2):115–124.

Knowles M (1973) *The Adult Learner: A Neglected Species,* 1st ed. Houston, TX: Gulf Publishing.

Knowles MS (1980) *The Modern Practice of Adult Education: From Pedagogy to Andragogy.* Chicago: Follett.

Kupst MJ et al. (1975) Evaluation of methods to improve communication in the physician-patient relationship. *Am J Orthopsychiatr* 45:420.

Kuznar E et al. (1991) Learning style preferences: A comparison of younger and older adult females. *J Nutr Elderly* 10(3).

Ley P (1972) Primacy, rated importance and recall of medical information. *J Health Soc Behav* 13:31.

Ley P (1973) A method of increasing patient recall. *Psychol Med* 3:217.

Magyari T, Smith ACM, Wholey K (1994) Effectiveness of multimedia decision-support materials vs traditional genetic counseling for CF carrier screening. *Am J Hum Genet* 55(3):A21, supplement.

Middelton LA, Reinhart GR, Pugh EW, Irwin AM, Smith ACM (1997) Consumer Expectations and Satisfaction with Genetic Services. 1997 American College of Medical Genetics Conference, Ft. Lauderdale, FL. (Abstract).

Mishler EG (1986) *Research Interviewing.* Harvard University Press. Cambridge, MA.

Mittman I (1990a) A Model Perinatal Genetics Program. *Birth Defects Orig Artic Ser* 26:93–100.

Mittman I (1990b) Immigration and the provision of genetic counseling services. *Birth Defects Orig Artic Ser* 26:139–146.

Mittman I (1997) personal communication; *Genetic Counseling to Communities of Color.* unpublished data.

Mittman I, Fenolio KR, Lee ES, et al. (1988) Perinatal genetic services tailored for a multiethnic, low income patient population at a county hospital. *Am J Hum Genet* 43:A241.

Peters JA, Stopfer JE (1996) Role of the genetic counselor in familial cancer. *Oncology* 10(2):159–166.

Prochaska JO, DiClemente C (1983) Stages and processes of self-change in smoking: Toward an integrative model of change. *J Consult Clin Psychol* 5:390–395.

Prochaska JO, DiClemente CC, Norcross JC (1992) In search of how people change: Applications to addictive behaviors. *Am Psychol* 47:1102–1114.

Riccardi VM, Kurtz SM (1983) *Communication and Counseling in Health Care.* Springfield, IL: CC Thomas.

Richards MPM (1993) The new genetics: Some issues for social scientists. *Soc Health Illness* 15:567–586.

Schneider KA (1994) *Counseling About Cancer: Strategies for Genetic Counselors.* Boston: Dana Farber Cancer Institute.

Sharp MC, Strauss RP, Lorch SC (1992) Communicating medical bad news: Parents' experiences and preferences. *J Pediatr* 121:539–546.

Shumaker SA, Schron EB, Ockene JK (1990) *The Handbook of Health Behavior Change.* New York: Springer.

Smith ACM, Magyari T, and Hernandez M (1993): Use of focus group methodology in the development of interactive educational materials for cystic fibrosis (CF) carrier screening. *Amer. J. Hum. Genet.* 53(3):1504, Suppl.

Szasz T, Hollender M (1956) The basic models of the doctor-patient relationship. *Arch Int Med* 97:585–592.

Tversky and Kahneman (1973) Availability: A heuristic for judging frequency and probability. *Cognitive Pscyhol* 5:2-7-232.

U.S. Bureau of Census (1996) *Statistical Abstract of the United States,* 116th ed. In: *The National Data Book.* Washington, DC: Bureau of the Census, Department of Commerce.

U.S. Department of Health and Human Services (1992) Publication 92-1493, U.S. National Institutes of Health.

Walker, AP (1996) Historical perspective and philosophical perspective of genetic counseling. In: Emery AEH, Rimoin DL, Connor JM, Pyertiz RE (eds) (1996) *Principles and Practice of Medical Genetics,* 3rd ed. New York: Churchill Livingston.

Wang V (1993) *Handbook of Cross-Cultural Genetic Counseling.* Wallingford, PA: NSGC Special Projects Award, National Society of Genetic Counselors.

Weiss BD, Coyne C (1997) Communicating with patients who cannot read. *NEJM* 337(4):272–274.

Wiel, Mittman I (1993) A teaching framework for cross-cultural genetic counseling; *J Gene Couns* 2(3):159–169.

Wlodkowski RJ (1993) *Enhancing Adult Motivation to Learn.* San Francisco: Jossey-Bass.

Appendix 5.1

Design and Development Model: A Case Example

Building on Kasper's experience, genetic counselors Trish Magyari and Ann C. M. Smith (1994) at Macro International, Inc. (Calverton, MD), were involved in the design, development, and evaluation of an integrated package of multimedia decision support materials on cystic fibrosis (CF) carrier screening. The project "Cystic Fibrosis Carrier Testing Educational Materials" was funded by Phase 1 and 2 SBIR (Small Business and Investigational Research) grants from NIDDK (2R44DK44794-02) covering a 30-month period. The process used to produce these materials serves as a model for others and is described below. The aim of these materials was to provide the necessary prescreening education about CF carrier testing in a way that empowers patients to make an informed decision based on personal values. The materials were not intended as a replacement for human input or genetic counseling.

The integrated package of materials "CF Carrier Testing: The Choice Is Yours" consists of a linear video, multimedia instructional program using CD-ROM, and supporting print materials (clinician's manual and take-home patient brochure). Materials were developed for a multicultural population of reproductive-age adults (pregnant or nonpregnant), both with and without a family history of CF, and are written at the low to average reading level. Figure 5-2 outlines the 2-year design and development process. A series of focus groups to inform the design process was held with different groups, including English- and Spanish-speaking women of reproductive age, individuals who had previously undergone CF carrier testing, parents of individuals with CF, and genetic counselors (Smith et al., 1993). The focus groups, coupled with critical review by a technical expert panel and content specialists (genetic counselors), guided the final design decisions on the 10 key learning concepts for core material, language to be used (English only), content areas to be covered, storyboards and scripts, user interface, graphical images, and so on. Pre- and post-

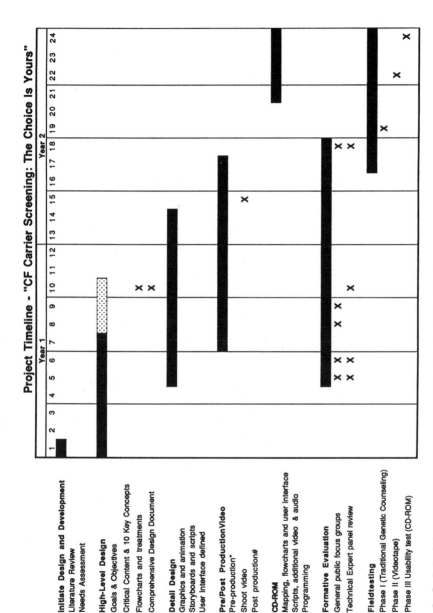

Figure 5.2 Project timeline: "CF Carrier Screening: The Choice Is Yours." *Source: Macro International Inc., Calverton, MD.*

production of the videotape was achieved with consultation and services of a professional video production company.

Both the videotape and the CD-ROM provide factual information about CF carrier screening that empowers informed decision making. The 14-minute linear videotape is designed for use by individuals or couples and uses full motion video, animation, graphics, and text reinforced by voiceover narration. The interactive CD-ROM includes the same core video footage, as well as several optional interactive modules, which provide a more in-depth understanding of the material surrounding CF carrier testing and/or living with CF. All users are exposed to the core content in the introduction and overview followed by a closing summary. Designed to assist the client in understanding the risks, benefits, and limitations of the test, the core content includes the following topic areas:

- *Living with CF:* overview of CF and what this diagnosis means for the individual and family members; users hear from parents and individuals with CF about what it is like to have CF in the family.
- *Genetics of CF:* discussion of the inheritance of CF using animation.
- *The CF Carrier Test:* overview of how the CF test is performed and the risks, benefits, and limitations of this diagnostic tool.
- *Reasons For and Against Testing:* short vignette narratives and text.

Unique to the CD-ROM are two modules: the "Personal Risk Calculation" module, which allows personal determination of one's risk based on ethnicity and/or family history of CF, and the "Personal Decision Making" module, which utilizes a series of decision support vignettes showing three sets of individuals grappling with issues raised by CF carrier testing. Embedded throughout the CD-ROM are opportunities for individual users to test their knowledge with feedback information. Intended to model decision making, these modules permit users to interactively organize and select reasons for and against CF carrier testing, based on their personal values.

The complete package of multimedia materials was formally field tested for three purposes: (1) to assess the impact of the multimedia materials, (2) to evaluate the effectiveness of the two experimental methods (linear videotape and CD-ROM) in providing pretest education on CF carrier testing versus the traditional genetic counseling (control) method, and (3) to examine the acceptability of and satisfaction with the developed materials. Field testing was conducted in three consecutive phases (phase 1, traditional counseling; phase II, videotape; and phase III, CD-ROM) at four U.S. genetic counseling sites offering prenatal CF carrier testing. Subjects who consented to participate in the field test completed questionnaires both before and after the instructional intervention, as illustrated in Figure 5.3. Because of subject recruitment difficulties and site limitations, phase III of the field test (CD-ROM) was limited to an alpha test and a usability test.

Results of the field test demonstrated that all subjects, regardless of instructional intervention used, gained knowledge about CF carrier testing; average knowledge

Subject Recruitment

Pre-test Questionnaire
- Demographic information
- 12-item Knowledge instrument
- Measure of "Intent" to have/not have CF test

Educational Intervention about CF carrier testing
Phase 1: Traditional Counseling (Control)
Phase II: Videotape
Phase III: CD-ROM

Individualized Genetic Counseling appointment (as scheduled)

Post-Test Questionnaire
- 12-item Knowledge instrument
- Measure of "Intent" to have/not have CF test
- Satisfaction measure

Figure 5.3   *Field test methodology used.*
Source: Macro International Inc., Calverton, MD.

gain was 5.5 correct responses (Magyari et al., 1994). A significant difference in knowledge gain was observed at posttest for both the video and CD-ROM compared to traditional counseling ($F = 2.378$; $p = .024$], with the greatest change from the CD-ROM. Demographic variables and personal knowledge of someone who has CF had no significant effect on posttest knowledge gain. Both multimedia approaches initiated a reexamination of the reasons for and against having the test, with subjects increasing their reasons for not having the test more than the control group. There was no significant difference between groups regarding the subjects' intent to have CF carrier testing; however, subjects viewing the video or CD-ROM tended to become less favorable toward the test. Subjects were equally satisfied with the educational materials and traditional counseling methods and found all such aids helpful in making their decision about CF carrier testing.

Both multimedia approaches (videotape and CD-ROM) were received favorably and were as successful as the traditional (control) method at imparting knowledge about CF carrier testing. It must be emphasized that these materials were never designed to replace genetic counseling, but to serve as an adjunct to provide prescreening education to potentially large populations in a fashion that empowers informed decision making.

6

Psychosocial Counseling

Luba Djurdjinovic

In my earliest days as a genetic counselor I struggled to understand what genetic counseling was all about. The cognitive challenge of appreciating the medical and genetic variables in a case was very exciting, but these technical elements took a different shape when I sat with a patient. It was when I recognized that my empathic attunement with patients and families brought the science to life that I really understood what genetic counseling was all about. My wish to more fully "know" the meaning of genetic concerns for individual counselees has led me to provide a genetic counseling experience that is both a scientific and a psychological practice.

The counseling aspect of genetic counseling has a recognizable structure though which a number of psychological perspectives are embraced. At this time, there is a limited body of literature describing the experiences and processes of the genetic counseling session. Interest in describing the psychosocial dynamics has been accelerated by the expansion of services into adult-onset conditions and predisposition testing (Djurdjinovic, 1997; Eunpu, 1997; Lerman, 1997; Matloff, 1997). In this chapter I weave my clinical experience into the existing literature.

Almost all genetic counseling sessions contain competing priorities between the medical and psychological considerations of a genetic condition. How these priorities are set, and to what extent they are addressed, are matters that vary with clinical setting and clinical experience. The increasing occurrence of one-time visits has a limited effect on what can realistically be undertaken, but psychological interventions can still occur. Understandably, their scope will be limited. I would like to demonstrate that whatever the cognitive aspects are in a case, the genetic counseling experience for the counselee can carry a therapeutic value.

A Guide to Genetic Counseling, Edited by Diane L. Baker, Jane L. Schuette, and Wendy R. Uhlmann. ISBN 0-471-18541-8 (cloth), 0-471-18867-0 (paper).   Copyright © 1998, Wiley-Liss, Inc.

Genetic counseling sessions involve a context that is complex in verbal and non-verbal communication between counselee and genetic counselor. The psychological content of this communication forms a stratum at which perspectives, experiences, beliefs, emotions, and expectations of counselor and patient overlap. A family therapist once likened the dynamics in such an encounter to a dance (Gratz, 1997). The dance is initiated through an articulated, or sometimes assumed, understanding of the individuals in the session and what each wants to gain. The dance is set against a backdrop of the community of each participant. The counselee brings into the room—actually or, more often, figuratively—family members and their socioethnic communities. The genetic counselor also brings a socioethnic perspective and also represents an institutional medical structure. The willingness of all parties to join in the dance will have its moments of unfamiliarity, awkwardness, and adjustment, as each person communicates information and gradually comes to a mutual acceptance of style.

The interweaving of a sociocultural stratum with the psychological allows the genetic counselor to consider the counselee in his or her fullest complexity. Understanding this complexity is as cognitively challenging and rewarding as our discussions about the genetic issues. The counselor's challenge is to provide for patients an experience that offers opportunities to consider their concerns and questions in the fullest manner. It thus requires us to be cognizant of, and facile with, psychological dynamics.

THE GENETIC COUNSELOR AS PSYCHOTHERAPIST

The counselor is engaged in a complex psychodynamic process that involves the lessening of denial, the relief of guilt, lifting of depression, the articulation of anger, and gradual, rational planning for the future. Everyone who has provided genetic counseling has witnessed these processes. Yet little is known about the formal dynamics or results of genetic counseling.
—Hecht and Holmes, 1972

Professional uncertainty about the appropriateness of, or our ability to address, psychological concerns has delayed analysis of what we do and why. In a seminal paper, Seymour Kessler provided a perspective that legitimizes genetic counseling practice to include active psychosocial discussions. He describes genetic counseling as "a kind of psychotherapeutic encounter . . . that is concerned with human behavior" (Kessler, 1979). It is critical that the genetic counselor appreciate that a counselee who is having an emotional response to information being presented may not be able to hear, understand, or assimilate the material (Frazer, 1976).

Little has been documented about what patients expect from genetic counseling. Some literature and patient stories provide evidence that patients have a wish for a psychologically oriented experience about the genetic counseling session (Sorenson et al., 1981; Michie et al., 1996). Many genetic counselors will agree that a classic question raised by the counselee is "Why did this happen?" This question frequently directs conversation to a didactic explanation, yet it may represent a more existential concern ("Why did this happen to me?") (Levine, 1979). So, how should we consider the wishes and needs of the counselee? Or, as family therapy theorist Virginia Satir asks, "What is the message about the message?" (quoted in Kessler (1979).

PSYCHOSOCIAL ASSESSMENT AND THE STRUCTURE OF A SESSION

There is no prescribed process or set of identifiable tools common to all counselors. However, the tool most central to all genetic counseling sessions is the family history. It is through the construction of this familial diagram that medical and social information is revealed. Family history construction is also the foundation of a more extensive psychosocial process known as the genogram (McGoldrick and Gerson, 1985). The genogram stresses relationships within families and intergenerational beliefs and attitudes. The pedigree in its simplest form describes to us the patient and the family. The family is presented intergenerationally with a description of ethnicity. The family history provides information about relationships that exist as well as relationships that, as the result of divorce, death, or adoption, no longer exist. The counselee's ability to provide information can suggest the level of connectedness in a family (Eunpu, 1997).

Counselees provide information about themselves in a number of ways. For example, counselees' available social support and their ability to avail themselves of it is sometimes revealed by whether a given patient attends a session alone or with a partner or other family member. Clients provide hints at the burden they may be carrying by statements such as "It's really up to me," "I don't know," and "It's my fault." Counselees give direct information about themselves that a genetic counselor can utilize to form a psychosocial answer to the question "Who is this person?" It also helps the counselor to address two important questions: "Can I understand this patient, and what can I offer beyond didactic information?"

It is by eliciting information to the following questions that we come to understand a counselee and begin to elucidate the meaning of that person's concern.

1. How do you "know" the client?
   a. Review of chart or intake form
   b. Demographic information
   c. Pedigree construction process
   d. Presenting concern
   e. Ethnocultural community
2. What self-disclosures does the patient make, and how?
   a. Social support
   b. Narrative surrounding this issue
   c. Attitudes and beliefs
3. What variables may influence the genetic counseling session?
   a. Level of anxiety or distress
   b. Depression

   c. Issues of education
   d. Language
   e. Timing
   f. Evidence of a working relationship between counselor and counselee
4. How do you as the genetic counselor feel about this client and his or her situation?

Early studies of patient experience of genetic counseling reveal that psychosocial concerns were not always addressed (Sorenson et al., 1981). This was partly due to patients' assumption that psychosocial issues had to be set aside in genetic counseling discussions. And yet, patient satisfaction is related to the degree to which such concerns can be aired (Sorenson et al., 1981). The clinical setting and the issue of time can result in competing needs: The counselor's need to present the medical/genetic information versus the patient's need to seek answers to existential questions ("Why our baby?").

Having a process that encourages counselor and patient to delineate the expectations for the session serves the goal of mutual understanding and offers the potential for the best outcome. It is also important to periodically review with the patient whether the counseling goals are continuing to fit, and if not, what needs to be readjusted. This is most simply accomplished by asking counselees what they hope to learn, what they are feeling and thinking, and what questions have emerged.

THE PATIENT'S STORY

A powerful tool in assessment is asking the counselee to tell us what brings him or her to genetic counseling. Our wish to know these circumstances and experience more fully begins to form a relational context in the genetic counseling session. It is the patient telling a personal story and our attuned listening that allows for assessment of concerns and emotional issues. The counselee's narrative is increasingly valued in the psychological community as practitioners seek to establish the centrality of the patient's voice in the theoretical construction (Miller, 1976; Gilligan, 1982; Belenky et al., 1986; Kleinman, 1988; Rosenwald and Ochberg, 1992). The story can tell us which ethnocultural perspective is identified and how religious affiliation and beliefs are described by the counselee (Kenen and Smith, 1995). Through the narrative we can begin to appreciate the meaning of the issue that has brought the person to the clinic. Paying close attention to the narrative keeps us from making assumptions about what brings people to see us. Furthermore, the narrative "knits together . . . recollected past with a wished-for future" (Brock, 1995). The meaning of an event for the patient can be intergenerationally constructed ("It skips generations"), and it is often attached to perceptions of risk ("I know I'm a carrier"). This meaning from the client's perspective provides access to the emotional content of the experience. The appreciation of the significance of a concern for the counselee will "contribute to the provision of effective care" (Kleinman, 1988).

THE WORKING RELATIONSHIP

The first step of delineating mutual expectations and goals initiates the working relationship. This relationship is sometimes referred to as the "working alliance" or "therapeutic alliance." The concept has its origin in early psychodynamic theory, which has been generalized for other counseling perspectives. Bordin (1979) deconstructed the analytic definitions and offered one in which the alliance grows out of a mutual agreement on goals and tasks. Further, these goals and tasks are set within a relational milieu of the client and counselor that includes understanding and empathy. I believe it is this relational context of genetic counseling that offers the patient and counselor a sense of satisfaction. To date, there are no formal studies to confirm this. A number of psychological scales have been used to evaluate working alliances in various counseling settings. Some of the measures have included patient's perception of the counselor's willingness to help, level of agreement between counselor and patient clinical goals, and level of mutual respect (Tichnenor and Hill, 1989).

The working relationship is central to the genetic counseling process. It begins to emerge through the knowledge we acquire in our initial assessment and in the first exchanges between the counselor and patient: "What brings you to our office today? . . . What do you hope we will discuss today? . . . We would like to use some of our time today to review your family medical history." The genetic counselor's ability to facilitate a balanced discussion of everyone's agenda further shapes this working relationship. Forming this alliance is only one part of the process; maintaining its integrity can be a more challenging aspect. The emerging alliance is maintained through active commitment to facilitating the following characteristics of the working relationship.

Confidentiality and Boundaries

The working relationship is strongly influenced by issues of confidentiality and respect for physical and emotional boundaries. The ability to share personal and family information requires an understood agreement that the information will not be disclosed beyond the clinical genetics experience without permission from the counselee. The understanding includes a clinical setting that limits what can be overheard. It is only when this parameter is in place that we can expect and invite patients to disclose difficult issues and emotions. Similarly, note taking during the session can be alarming if content and purpose are not disclosed in advance.

Sometimes the issue of confidentiality is not related to persons in the room but involves family members seen earlier or scheduled for future visits. It can also involve issues of paternity and previous pregnancy termination, matters commonly kept secret within families. Therefore, anxiety about disclosure of these events may result in less than satisfying discussions.

Emotional and physical boundaries make patient and counselor alike feel "safe" in an intimate discussion. The need for boundaries is central to all forms of connection. Boundaries can be defined as all the safeguards that do not impinge on patient autonomy, self-expression, confidentiality, and physical safety. The degree of boundary enforcement is influenced by the nature of the experience, previous experience (e.g., sexual abuse) in which boundaries have not been respected, and cultural perspectives. The emergence of a working relationship will require that the boundaries one carries into the session be flexible. In forming connections, for example, it is sometimes necessary to be able to drop or shift a boundary. It is important to understand when the counselee needs a more rigid boundary and to appreciate how engaging in the clinical genetics process poses particular challenges to an individual.

Physical contact with the counselee, other than a handshake, should be limited to the clinical exam. Only when permission is given should there be physical contact with a patient. We may choose to comfort a patient during a moment of intense grief or despair if a supportive gesture is invited, but appropriate occasions for such contact are not common. Unconsidered physical comforting of the patient such as hugging and touching is well understood in psychological circles as disruptive, as well as disrespectful to the patient. Physical contact between counselor and patient can be counter to the cultural practices of the patient, as well. For instance, what might seem like an innocent touch of a child's head can be experienced as threatening by parents of Asian background.

On a final note, it is also important to remember that the issues of boundaries are central to the counselor as well. Unless the counselor can feel safe, the process she directs will be fragmented and may not serve the needs of the patient. It is appropriate for the counselor to create and, at times, to define the boundaries that will allow for her full participation.

Patient Autonomy

Seymour Kessler has defined nondirectiveness in the practice of genetic counseling as "procedures aimed at promoting the autonomy and self-directedness of the client" (quoted in Whipperman-Bendor, 1997). He has tirelessly reminded the genetic counseling community of the origins of the nondirective concept and has helped us disentangle the issues of nondirectiveness and autonomy (Kessler, 1979, 1992a, 1997). The issue of autonomy focuses on the potential for coercion to occur. This concern has led to the embrace of the language of nondirectivenss. The difficulty here is that the term has become, over time, "an empty slogan with no concept behind it" (Wolff and Jung, 1995). Adherence to this generalized appreciation of what is meant by nondirectiveness "prevents critical discussion of both the psychological and the societal implications of genetic knowledge" (Wolff and Jung, 1995).

Counselees' questions of "What should I do?" or "What would you do in my situation?" often lead the counselor to retreat into the arms of nondirectiveness with statements about the professional's necessarily value-neutral position. If, as counselors, we truly see our patients as their own moral agents, then the question "What should I do?" could naturally lead into discussion about the question. "I would like to understand your question better," we might say, adding that, "this question is asked frequently for different reasons." For instance, the question may be an attempt to weigh the reasonable options available or to see more information about how those choices are made by others in similar circumstances. In general, the genetic counselor's professional skills in facilitating the decision are being solicited, not the decision. The traditions of some cultures, however, may produce the expectation that the counselor will make the decision for the client.

Patient autonomy is an outcome of a counseling experience in which enhanced decision-making climate is promoted through information sharing, empathic attunement, reduction of extreme emotional states, and professional guidance.

Empathy

The working relationship grows within an experience of empathy. Though essential for understanding a person or situation, empathy can be difficult to define. Many authors have attempted to capture the complex nature of this specific form of connection and understanding:

> "the capacity to understand what another person is experiencing from within the other person's frame of reference" (Bellet and Maloney, 1991)
> "the accepting, confirming and understanding human echo" (Jordan, 1991a, b)
> "comprehending the momentary psychological state of another person" (Jordan, 1991a, b).

Empathy is thought of as having both affective and cognitive functions. The counselor's emotional attunement framed within a cognitive understanding results in the fullest appreciation of the counselee at that moment in the session (Surrey, 1991). It is also important to remember that empathic attunement can be achieved through spoken and unspoken content (Basch, 1980; Kohut, 1984; Wolf, 1988; Jordan et al., 1991).

It is critical that a counselor be able to differentiate feeling sympathy from feeling empathy. The counselor's sympathetic experience of a patient involves an emotional response based in the counselor's personal life. When we feel moved to tears by the despair of our counselee, for example, it is important to differentiate between a personal identification with the counselee (remembering a similar experience) and having, in that moment, deeply appreciated her pain. The sympathetic response on the part of the genetic counselor is real. Yet, it needs to be put aside and attention refocused on the responses of the counselee. It is through attending to the patient and appreciating her dilemma that empathy can be achieved.

Empathic attunement in a genetic counseling session evolves through three phases of information analysis. The first phase attempts to integrate the content of the patient's story, identifying the patient's affective state, and appreciating our own response to the story. The second phase requires a careful consideration of the significance of the patient's story and the messages within. The final phase involves a decision about how to effectively acknowledge what has been discerned (Basch, 1980).

It is through an empathic stance that a patient begins to feel understood. With this experience a patient is ready to consider the potential challenges of genetic information. Being understood provides a platform for the formation of self-empathy and, in so doing, enhances the patient's ability to personally appreciate the complexity of presented information, choices, and the clinical course of a condition. This is believed to be central to the ability to make decisions or take actions (Jordan, 1991b; Jordan et al., 1991; Miller, 1991a; Surrey, 1991).

Empathic attunement is a learned skill. The learning occurs within our own life experiences and relationships. For many counseling professionals, empathy is explored and developed during case supervision. Essential to this skill is "a well-differentiated sense of self in addition to an appreciation of, and sensitivity to, the differences as well as the sameness of another person" (Jordan et al., 1991). An essential extension of this skill is regaining an empathic connection after one is disrupted. The act of refocusing on the counselee and reestablishing an empathic connection is termed "raising one's empathy" (Wolf, 1988).

To illustrate how a working relationship forms through empathic connections in a genetic counseling session, I have selected a case that involves a one-hour session and demonstrates the unfolding of an uncertain agenda.

A family was referred for genetic counseling after the birth of a son with coronal craniosynostosis. At the time the appointment was scheduled, the mother of the child explained that the pediatric surgeon had recommended genetic counseling. The parents arrived for the appointment with their infant son and their 6-year-old unaffected son. After I settled the parents and the children into the consultation room and made appropriate introductions, I opened the session, "I understand your son was born with a form of craniosynostosis and you were referred by Dr. W. How can we help you?" The father responded, "We don't know, the doctor thought we should come here." This response was somewhat unexpected. It was an awkward beginning, since nothing was identified as a common goal. I proceeded to briefly describe the kinds of issue we might discuss in relation to the infant's diagnosis. The mother interrupted, "They gave us a booklet about craniosynostosis. . . . The doctors told us that we have a 4% chance that this will happen again." (The mother reached into her purse and handed me the booklet.) I experienced the mother's reply as having a tone of anger and defensiveness. I will admit that I was a bit put off by the mother's response. I reminded myself that the mother was telling me something important here and it did not necessarily mean that they had all the necessary facts or that the appointment was unnecessary. I asked the parents to tell me how they felt about this risk. This was the question that truly opened the session.

The parents did not hesitate when I invited them to tell me their story. As the parents talked, they looked at each other frequently. At one point I observed that the mother looked very teary. In that moment I began to more fully appreciate the experience for the mother. I said to her, "The feelings are still with you after all these months." She nodded and went on. I believe this early empathic connection formed the path to what unfolded later in the session. As an aside, the relating of the parents' experience was frequently interrupted by the older son. He made many requests for items in the mother's purse and requests to me for drawing paper. Interruptions were also initiated by the mother, who frequently turned to her sons and spoke to them. I understood these interruptions as a tool to diffuse parental anxieties about the content of the session and what still might be said. I made every effort to respect the defense that I understood to be stimulated by anxiety. I also experienced my own anxiety when I considered the remaining time in the session, the still unidentified goal for the session, and the need to further evaluate the son, given the heterogeneous etiology of the condition. I was committed to introducing my concerns about investigating the genetic etiology; yet I felt uncertain as to how to proceed.

I asked the parents to tell me how they have explained the baby's condition to family and friends. The mother quickly responded with a clear description and added, "It's no one's fault." I asked the couple if they had looked for possible causes. The mother quickly responded, "I thought I was responsible but the doctor helped me understand I was not. The baby and I have a flatness here (pointing to the midface). I was stuck to my mother's spine in the uterus, the doctor had to turn me and pull, then I was born. I am worried because we are planning to start trying again." This statement offered me a potential direction. I guessed that a possible unspoken agenda for this couple could be prenatal diagnostic options. I asked the couple if they were interested in discussing a future pregnancy and what options might be available to them. The mother forcefully stated, "We are Christians and accept what is given." I nodded in acknowledgment. This was followed by the mother turning to her husband and telling him that she had thought of ultrasound studies for reassurance. She turned to me and said, "We have not discussed it, but is it possible?" I understood this to be the concern that had led to the appointment. After my discussion about ultrasound, she shared that she very much wants another child but is afraid. This was followed by, "My older son is not from this marriage and the baby was conceived before we got married. I would love to tell people I am pregnant without a reaction of shock." The mother had told me something she felt burdened by, and I accepted her need to share this.

As I was approaching the final 10 minutes of the session, I realized that these issues could not be further addressed at this time, I needed to make a transition to closure of the session, and find a way to encourage the couple to initiate an evaluation process that would offer diagnosis and risk assessment specific to their circumstances. It was my sense that the mother carried guilt from more than one source, one of which may be her own belief that the craniosynostosis is familial.

I offered the following statement to bring the session to a close: "I can see that your plan to have another child carries a lot of meaning for you. What you have told me today allows me to understand what has made this decision important and difficult. As I think about our conversation, I have the sense that you are still trying to understand why this happened and how this may affect your future children." The father said, "It was just a fluky thing," but the mother said, "Yes, (pause) I'm not sure." The session concluded with my offering a formal genetic evaluation process as a way to assist in answering some of their questions. The couple and the children left, with the mother thanking me "for listening." This was for her the unstated, but perhaps, the primary goal of this session.

It is not unusual for counselees to be unable to articulate their questions or agendas for a session. This may be partly because they do not understand the reason for the referral by their medical provider or, as in this case, because of guilt and a possible belief by the mother that her shared appearance with her son has recurrence implications. At a number of moments during this visit, I questioned where I was going, and why I had not even constructed a pedigree. The counselees never gave me permission to engage in a more formal genetic evaluation process. But, without the conversation that did unfold, the couple might never have identified the possibility of risk assessment.

DISRUPTIONS IN THE WORKING RELATIONSHIP

The working relationship, as illustrated in the case report above, may emerge relatively quickly by attending to the emotions, beliefs, and memories that form associations for the patient. These associations occur in varying degrees and can be positive or negative. The ability to observe, hear, and understand them can be critical to maintaining the integrity of the working relationship. The most common disruptions to the working relationship can be categorized into three areas: transference, countertransference, and empathic breaks.

Transference and Countertransference

Transference and countertransference refer to ways of relating in which an *unconscious* template becomes superimposed on a person. For example, consider the basis of our first impressions about another person: "I like her," or "There is something odd about her." We have made an assumption or we may have applied a historical imprint onto the person; this is a transference. It is often difficult to sort out what has stimulated our assumptions or beliefs. It is through self-awareness and analysis of case material that the counselor comes to understand the issues specific to her. In my own career, I have learned that one of my countertransferences is an overestimation of someone else's capacity to take action. This unconscious belief comes from historical experience. The countertransference is most likely to be invoked when I believe I have detected passivity on the part of a female patient. I am convinced in the moment that the counselee shares my vision of her. This can be burdensome to the patient, and she may feel that I do not really understand her. On the other hand, this particular countertransference can have a useful role. It may surround the counselee with the hope to which she herself may have no emotional access.

In the genetic counseling session, transference or countertransference may or may not be evidenced. These reactions are dependent on the dynamic of the personalities involved and, to some degree, the depth of the working connection. These reactions can occur in any single encounter between genetic counselor and counselee.

The counselor's goal in the genetic counseling session is self-awareness in the face of such dynamics. The value of this knowledge is not to provide a therapeutic intervention to eliminate the response but to see its possible impact on the dynamics of the session. Depending on the time frame of a genetic counseling session, it may be possible to invite a discussion that holds the potential for the unconscious to become conscious.

Transference   The term "transference," with its roots in psychoanalytic theory, generally describes patterns of expectations that interfere with relationship building or the client's ability to take actions. These expectations are unconscious and are not easily observable by the patient or counselor. Transference is a ubiquitous phenomenon in which one brings old patterns of expectation to new situations in attempts to create a familiar structure for the event (Basch, 1980). This psychoanalytic concept can be generalized to all counseling experiences. It is further argued that "Transference phenomena emerge in all relationships . . . we bring our experience with important relationships in the past into our current interactions with people—particularly through the relational images we carry with us" (Miller and Stiver, 1997).

Most genetic counseling sessions do not readily evidence a transferential response. When one does occur, however, it usually leaves the counselor feeling some confusion or discomfort: since transference is experienced by the counselor to contain some degree of misperception or distortion, there can be a wish to correct it. However, the decision to correct can be disruptive to the working relationship if the genetic counselor has not appreciated the psychological dynamic. If the transference is confronted and the counselee is unaware of the distortion, the resulting confusion may elicit a more intense response from the counselee.

Countertransference Countertransference is commonly defined as the counselor's reaction to the patient's transference. Countertransference can also encompass the counselor's response to a patient's story. We attempt to understand patient stories with a genuineness that requires us to have our affective states accessible to us. It is possible to be "hooked" by a perception or transference of the patient because of our own unresolved issues.

The countertransference reactions in genetic counseling practice are most often of two types, associative and projective (Kessler, 1992b). Associative countertransference arises in that empathically attuned moment in a session when a patient shares an experience, a loss, a wish, or a story that carries the counselor (usually for a short time) into his or her inner self. This introspection is often accompanied by a review of mental images and a recall of conversations. These remembered emotions may rise to the surface. "Such associations may lead to others, and before the counselor knows it, she may no longer be attending to what the counselee is saying and feeling but to her own internal voice and suffering. Even if the counselor is partially attending, these associations may impede or interfere with the total understanding of the counselee" (Kessler, 1992b). There is a critical difference between an association that allows us to more deeply appreciate a patient's experience and one that takes up time and emotional space within the session. The patient often detects the counselor's distractedness or recognizes that the counselor's comments are personal, rather than about the patient. It is here that the empathic connection is lost, causing the discussion to shift.

Projection is a type of countertransference in which a counselor has made assumptions about the experience of the patient. These assumptions are based on a parallel past experience for the counselor such as loss, suffering, fear, or panic. Usually, the counselor does not recognize these as her own and instead assumes that the feelings are originating from the counselee (Peters, 1997). This type of countertransference carries the potential risk that the counselor will make assumptions about a patient's experience and truncate the empathic attunement, thus disrupting the working relationship.

It is important to realize that the issues of transference and countertransference can originate in the counselor as well as the patient. The counselor's efforts to appreciate this psychodynamic stratum of transference/countertransference responses will offer useful clues that will usually reduce the risk of disrupting the working relationship. The following case offers a glimpse into a possible transference reaction, one that could easily evoke a countertransference.

> The genetic counselor is meeting with a couple who have just learned through an ultrasound exam that their 23-week fetus has a cystic hygroma. The genetic counselor provides information known about cystic hygromas and the areas of uncertainty that surround this finding. The father insists, "You are not telling us everything you know." The counselor is taken aback by this comment and attempts to frame the areas of uncertainty again. The father presses the counselor for detailed information about all possible scenarios, and whenever the counselor attempts to remind the couple that the only possible scenarios have been presented, the father becomes more suspicious that the counselor is withholding information. He continues to repeat, "You know more than you're telling!"
>
> The father appears to be recalling a historical experience or expectation that is interfering with his ability to participate in the counseling session. The experience could have had its origins in a previously frustrating medical visit, or it is possible that the father had early life experiences that resulted in poor information disclosure to him, with resulting trauma upon receipt of unexpected news. We can imagine that the counselor could feel an increasing level of frustration or even anger at the father's response. As a committed genetic counselor, there would be a natural vulnerability to a suggestion that our counseling agenda was to withhold information. It is possible that a counselor who has had personal experience with a combative parent may detach and end the case prematurely, the countertransference to the patient's transference having disrupted the goals of the session.

The tensions that were created limited the counselor's ability to maintain a relationship with the father. The counselor's instinctual response, to reassure the father of her commitment to full disclosure, did not remove the barrier of distrust that the worried parent was enforcing.

In one possible intervention, the counselor might consider transference as playing a part in the father's reaction and ask the father to explain what he believes has not been disclosed. Attempts to appreciate and acknowledge the father's reactions are more likely to be helpful in changing the dynamics and shifting the transference.

An Empathic Break

The experience of a shift in the relationship dynamics or a sudden loss of an empathic connection has a potentially serious impact on the process and goals set at the beginning of the session. This disruption is called an empathic break. Breaks can result from many factors: conscious and unconscious perceptions, including issues of boundaries, autonomy, confidentiality, lack of cross-cultural sensitivity, or simple interruptions in the counseling session (phone calls, a knock on the door, hallway noises, etc.). The recognition of an empathic break can sometimes be difficult, as it is often a momentary experience. Empathic breaks can result from the actions of the counselor as well as the counselee. The counselor may suddenly remember an issue that has not been explored and conversationally thrust it into the session. In her anxiety to include this new topic, the counselor may overlook the unresolved status of the topic already under discussion. The counselee may then feel that the counselor has minimized the issue (or the counselee's concern). This may result in feelings of confusion, stimulating the need to create distance. At this point, the counselee is not psychologically available to continue with the session.

Sometimes the counselee, in the course of storytelling, will share information that may be challenging to the values or perspective of the counselor. The counselor's need to reassess her perspective (even if it takes only a few moments) could signal disapproval or confusion to the counselee. This perception could lead to an empathic break, precluding a more psychologically driven discussion. It is not uncommon for counselees to make attempts to shift from a point of discussion. This is sometimes a test on the part of the counselee to assess what is possible or not possible to share.

The counselor's ability to return to the empathically attuned position following an empathic break may assist in a reconnection. Sometimes the counselor finds it difficult to reconnect at the point at which the disconnection occurred. It is here that the counselor should attempt to "raise her empathy" and once again psychologically embrace the patient and the story. As discussed previously, "raising one's empathy" comes with experience. It is, in my opinion, critical to recognize and address an empathic break in a session. Sometimes acknowledging that a shift has occurred can assure the counselee that his vulnerabilities are being attended to and that the counselor is making a sincere effort to understand.

DISCUSSING DIFFICULT ISSUES AND GIVING BAD NEWS

One of our most important roles as genetic counselors is to sit with a person or family and announce that a genetic evaluation has identified a risk, and that this risk carries implications that can change their understanding of themselves or a loved one, as well as their future. These conversations are frequently experienced as bad news. The psychological response to difficult news varies from person to person. It is, in part, influenced by the gap between what the counselee believes and what is now being presented (Buckman, 1992). The following brief case illustrates such a gap.

The genetic counselor met with a 38-year-old woman of Ashkenazi Jewish descent who was referred by her physician after a discussion of her paternal family history of breast cancer. She was the mother of two boys, 8 and 10. After a review of medical records and risk assessment counseling, the woman elected to undertake testing for the known *BRCA1* and *BRCA2* mutations. Testing revealed that she had one of the known mutations. Upon receiving the information, the woman expressed shock. When this reaction was explored, she explained that she had assumed that since she resembled her mother's side of the family, the chances for being at increased risk were less. Confrontation with the gap between what she had believed was her risk and what was subsequently identified as her risk had left her very shaken.

One of the most critical steps for genetic counselors to manage is the assessment of what counselees believe and understand about the risks they face. This insight allows for structuring a discussion that honors the "gap" and provides adequate time to assist each counselee in her recognition of the issues surrounding the assumed and the real risks.

At the moment of our announcement to the patient or family that we intend to disclose the clinical findings or test results, two reactions are set in motion: the patient awaits the news and the counselor anticipates a response. The counselor needs to appreciate the ways that someone may need to defend against overwhelming feelings and to recognize that various styles of coping will allow a counselee to emerge from the discussion and experience. Achieving the most empathically attuned position with the counselee will, in most cases, minimize the likelihood that the counselee will miss critical information or support. We achieve this partly by intuition, partly through experience, and mostly by taking direction from the counselee.

The approach we use in presenting difficult information must be sensitive to the patient as well as to our own experience of the case. Central to giving bad news is providing information directly and empathically. It is difficult to achieve a balance here: "It is an art to do both at the same time. It is common for medical providers to choose one over the other, the idea is to be straightforward without sacrificing empathic attunement" (Peters, 1997). The context in which we give this information is equally critical. The material that follows, which attempts to guide the counselor in this challenging task, draws from the work of several authors (Ives et al., 1979; Goodman and Abrams, 1988; Buckman 1992; Guest, 1997).

1. Delineate goals of the session. Prior to meeting with the family, delineate the information that needs to be disclosed. Consider what will assist the patient in understanding and formulate a plan of action.
2. Familiarize yourself with the patient's available support structure.
3. Determine who should be present at the time of disclosure.
4. Identify the room or setting in which the meeting will occur. It is essential that the setting provide privacy. There should also be adequate seating for all present. Consider the time frame available, and inform the counselee of any time constraints.
5. In giving difficult information it is important to introduce the subject quickly: "I have news about ————." Attempt to give the most critical information in two to three sentences, using limited genetic/medical terms. Then **STOP** and allow some quiet time for the patient to absorb what you have said.
6. Ask if the patient would like you to provide more details. Be aware of word choice, so that a balance between honesty and comprehension is achieved. It

is important not to respond to the anxieties or sadness in the counseling room by giving facts. Information doesn't comfort everyone.

7. Look to the counselee to determine the pace of the discussion. Avoid excessively reassuring statements. The counselee must be able to cognitively and emotionally integrate the experience before reassuring statements (which are more often an attempt to reassure the counselor than the counselee) can be accepted.

8. Ask, "Do you have questions?" Listen for "buried" questions.

9. Ask, "How can I be helpful? . . . Do you want someone here with you?"

10. Ask about who will be available to offer support after the patient leaves the session.

11. In bringing the session to a close, ask the patient to tell you what she has understood about the information just received. Ask what she will do after leaving the session. Make a plan to follow up on this conversation. Agree on a next step.

12. Finally, consider who is available to you to review the experience of giving difficult news.

Giving difficult information can be particularly challenging when the patient is cognitively delayed. The challenges are most often in our own concern that the information we see as important may not be understood or appreciated for its significance. It is important for all who work with patients with cognitive limitations to remember that the counselees' ability to discuss feelings and their experience of a situation is not impacted by their particular difficulty in understanding medical or genetic information. This lack of convergence can sometimes make the counselor feel less confident in the session, since what aids our feeling of mastery is often our role of information giver. Here our role may be limited to supporting the counselees in their feelings. Further, it may be helpful to remind ourselves that such counselees, being categorized into a stigmatized group, may be "highly motivated to dissociate themselves from outward signs of incompetence" (Finucane, 1997). In other words, they may not acknowledge their lack of understanding. Some of the suggestions provided below for giving difficult news to children may be applied when informing counselees with cognitive delays.

Several additional parameters need to be considered when one is giving difficult news to children or adolescents. Ideally the parents should be present at such discussions, as well as an interdisciplinary team that includes psychologically trained professionals. This is not always possible, however, and the counselor needs to bear in mind the following pointers (Buckman, 1992):

1. Consider the child's age and developmental ability to appreciate information. Structure the information content to the appropriate level. This may require follow-up appointments over several years.

2. Discuss with parents privately the objectives of the meeting with the young counselee. Also identify who will be included in the session.

3. Information must be presented in clear, simple statements. Repeat the information more than once in the session. Be prepared to have the same question raised by the counselee throughout the session.

4. Listen for "magical thinking" on the part of the counselee.

5. Appreciate the psychological challenges the information may have on self and/or sexual identity. Be available to discuss issues of sexuality (Money, 1994).

6. Make plans to have the conversation continue with you and/or with an appropriately trained child mental health professional.

Another area entailing special considerations is work with counselees for whom English is a second language. This is challenging especially when interpreters are inadequate or are family members who are also affected by the information. Additionally, some patients with a psychiatric diagnosis can pose difficulty in forming a working relationship, or it may be difficult to assess their ability to integrate the information provided in the session.

It is worth noting that all the information we present to a family will not necessarily be experienced with strong feelings or result in psychological conflicts. This is exemplified by the case of a 19-year-old man who was referred to the genetics clinic. This college student's complaint of malaise, and decreased energy and libido, led to a medical workup that included a chromosome analysis. The analysis revealed 46,XX. The genetics team met and outlined a suitable protocol to discuss the karyotype and prepare for the anticipated psychological challenges. A psychiatric consultation was arranged prior to the genetics appointment. During the session, the young man reported that he had previously been told about the 46,XX chromosome constitution and had received the possible diagnosis of Klinefelter syndrome. He had researched the condition and decided that the diagnosis did not fit. He self-determined that he might have "a piece of Y floating around somewhere." He "reported that he felt secure as a male and noted that he has never been questioned about his gender based on his appearance" (Blumenthal, 1990/1991).

Reactions to Bad News and Psychologically Challenging Experiences

The information that patients learn in the genetic counseling process has the potential to be psychologically overwhelming and to result in an emotional response. Among the many responses to unexpected news, the most common are denial, anger, fear, despair, guilt, shame, sadness, and grief. These responses elicit in the patient an emotional rather than a cognitive understanding of the implications of the information. The role of the genetic counselor is to provide what can be called "a holding environment," where there is appropriate support, empathy, and time for a response to be expressed and understood.

Denial   Denial, the inability to acknowledge to oneself certain information or news, is a common response when the information elicits shock and fear. This defensive psychological position emerges when other defenses no longer work. The reaction of denial is usually short term and is part of a coping strategy applied to impending or actual loss (Kubler-Ross, 1969). The patient's failure to acknowledge some aspect of external reality that is apparent to others has been further delineated into the constellation of disbelief, deferral, and dismissal (Lubinsky, 1994). The careful delineation of these denial experiences can assist the counselor in choosing more appropriate interventions. To allow for a process that is respectful of any such defense, the interventions should be considered only after the counselee has had adequate time to assimilate the news.

Anger   Anger is a complex emotion that seeks to blame; in its most extreme form, there can be a wish to achieve revenge. Anger can be directed at others or at oneself (Lazarus, 1991). It can be subtle or overt. It may be that there are gender differences in the construction of anger (Miller, 1991b). The patient with an angry response to information is probably experiencing fear, powerlessness, guilt, shame, or extreme anxiety. Anger is also a critical step in the resolution of loss, as the counselee seeks meaning in the need to abandon the hope of a wanted or expected outcome (Kubler-Ross, 1969; Bowlby, 1980).

Patients need the opportunity to understand the meaning of their anger and to transform it. The lifting of anger invariably reveals other feelings, usually sadness. A useful intervention is to invite the patient to discuss the anger. "You sound angry." "Can you tell me more?" It is very important that the counselor remain nondefensive, and especially crucial not to trivialize the anger (Peters, 1997). Anger can appear in the form of blame and may be focused on objects (e.g., a lab test), on persons in the past or future, or on the counselor, who represents the medical community. Anger can sometimes take other forms; a common one is crying. I have found tears to be more common in women, probably reflecting the social unacceptability of anger for women. The inability to shift anger in the genetic counseling session can impede counseling goals or decisions that are being made (Smith and Antley, 1979). Interventions should include witnessing the feeling, honoring its role, and providing a working relationship that through empathic attunement lifts anger (usually for a short time), making it possible to address immediate issues.

Working with angry patients frequently evokes our own reactions that need to be temporarily quelled. If the counselor's feelings continue to be present following the session, it is important to speak with a colleague to understand the meaning of the interaction for the counselor.

Guilt and Shame   Parents of children born with genetic conditions frequently report feelings of guilt and shame (Antley et al., 1984; Weil, 1991; Chapple et al., 1995). The emotions of guilt and shame have different origins and can provide the counselor with information about the patient and possible approaches for discussing the issue. Patients who hold themselves responsible for what they perceive as a negative outcome will frequently attempt to correct their guilt through self-blame, rationalizations, or other intellectualizations. Patients who express shame are offering you an opportunity to appreciate that events have placed "a burden on the self" (Kessler, 1994). Patients experiencing shame will often attempt to reduce the psychological challenges to the self by means of denial and withdrawal. The counselor's ability to witness the guilt and shame response of the patient provides the first step in offering an intervention that shifts the patient from a position of guilt and shame to one of more self-empathic understanding (Kessler, 1984).

Despair, Grief, Depression These are common responses to loss or anticipated loss. Families who learn that their child has a genetic disorder or that a pregnancy has elicited abnormal prenatal test findings will experience a grief process like that of families in which a neonatal death has occurred (Antley et al., 1984). The process includes feelings of shock, anger, and sadness (Kubler-Ross, 1969). Over time, there is gradual acceptance and a resolution. The sadness or sorrow over a loss can still be experienced over the life cycle (Hobdell and Deatrick, 1996). If the grief process becomes interrupted or blocked, as when the individual is coping by denying the loss, depression frequently results. Psychological interventions can be beneficial for some (Leon, 1990). For the majority, resolution of grief takes the form of being "able to make the loss seem meaningful and to gain a renewed sense of purpose and personal identity" (Shapiro, 1993).

Questions Questions in the genetic counseling session may be vehicles for expressions of feelings. By asking "Why me?" or just "Why?", the counselee is attempting to share an emotional reaction and discover the meaning of the event. If emotion is what drives the question, the counselor's efforts to provide information rather than addressing the emotion behind the question will disrupt the "holding environment," and a facilitative moment for appreciating the patient's experience will be lost.

On occasion, this questioning is an attempt to temporally distance oneself from the psychologically painful moment. A psychologically painful moment is not always visible to the counselor. It generally occurs when the "self" feels challenged, and feelings of shame or guilt, and, not too infrequently anger, are stirred. The question may be a request to repeat some information ("Tell me again what will happen"), or it may be directed at the counselor ("Do you have the time to explain this to me?"). It is necessary to respect such expressions of the counselee's need to reduce the psychological tension.

In rare cases, questions may become barriers between the counselor and counselee. It is useful to attempt to understand what stimulated the barrier: Was it an empathic break, or is the counselee simply letting you know that she is tired and needs to stop? It is natural for the counselor, recognizing the emerging barrier and resulting disruption in the working relationship, to try to limit a flow of disruptive questions. The counselor needs to rely on her evolving understanding of the counselee and the emotional content of the session to determine what to do next. There must be a balance: open acknowledgment of the barrier needs to be accompanied by an offer of assistance to explore the emotional tensions that led to the questions. Ability to achieve this balance comes primarily from experience. One frequently employed intervention is to directly ask the counselee to elaborate on the present experience. One can generally assume that an attempt to describe the emotion is permission for the counselor's assistance. When a patient appears to be inarticulate, or says, "I can't put it into words," the counselor should refrain from information giving and start looking at support options. For example, the counselor may perceive a need to allow adequate time for the patient to choose the next step. If this turns out to mean taking a break or scheduling a return visit, it is important to assess the patient for risk factors (level of support, suicide) and to contract for continuation of the discussion.

Projection and Projective Identification On occasion, we meet patients in genetic counseling who demonstrate an unconscious defensive dynamic called projection, here defined as the use of disavowal to defend against challenging feelings or meaning. In other words, projection is an attempt on the part of the patient to displace on to the counselor a feeling that has the potential to burden the "self." Commonly, for example, counselees will repeatedly inquire whether their decision has angered the counselor. In such cases of projection, the patients are angry with their decision or with the necessity of having to make a decision.

Another form of projection is not commonly experienced in the practice of genetic counseling unless there is long-term follow-up and a strong working relationship has formed. **Projective identification** describes a complex defense mechanism that is triggered when the meaning of an event is extremely psychologically challenging to the patient. The experience of the event leaves him feeling very vulnerable. If this patient also has an inherently fragile self-structure, he may attempt to preserve psychological integrity by projecting the feeling to the counselor. The unique aspect of this defense mechanism is that the counselor, who is the "object" of the projected feeling, finds some congruence in the feeling. Therefore, the counselor becomes a temporary repository of the feeling state that is unacceptable to the patient. More specifically, the counselor has an experience of the feeling that she attributes to herself and not the patient. This phenomenon may have occurred in the case presented at the end of this chapter (Mona).

COUNSELEES' COPING STYLES

Many psychologists have studied coping patterns and styles from different perspectives: psychodynamic, cognitive–behavioral, and neurophysiological. They have observed that most people have one or two primary coping styles, which they employ throughout their lives. Coping strategies are employed for problem solving or to change the meaning of what has been experienced (Pargament, 1997). A set of coping styles identified by Lazarus (1991) is summarized as follows:

Confrontive: try to change the opinion of the person in charge.

Distancing: go on as if nothing happened.

Self-Controlling: keep feelings to oneself.

Seek Social Support: engage in conversations in the hope of learning more.

Accept Responsibility: criticize oneself.

Escape–Avoidance: hope for a miracle.

Plan: identify and follow action plan.

Positive Reappraisal: identify existing or potential positive outcomes.

The coping strategies are not defined by the specific emotion; rather, the emotion triggers one of the coping strategies.

Interventions are called for when a counselee's coping strategy does not work. A simple and often effective intervention is to engage patients in describing how they have managed other difficult situations and to encourage them to redeploy some of the same strategies.

COUNSELOR'S SELF-AWARENESS

The psychological dynamics considered so far have, in part, relied on the counselor's ability to identify them and to offer responses that will permit the psychological process to be fully appreciated. This requires a self-awareness of one's vulnerabilities, countertransference triggers, and abilities to take in the experience of others. This self-awareness becomes more critical in our role as purveyors of difficult news, where being witness to a patient's intimate disclosure or emotional response to information can be uncomfortable and may even raise our concern. We therefore have a professional responsibility to participate in periodic evaluations of our self-awareness. This is generally achieved through discussion with colleagues, case discussions in which psychological issues are considered, and sometimes therapy. In our commitment to achieve this level of psychological self-awareness we, as professionals, can then experience the cognitive challenges presented by the psychological material of genetic counseling sessions. Only then are our patients likely to have a genetic counseling experience that is set within a psychological milieu.

THEORIES THAT SURROUND OUR WORK

The theories that surround my work as a genetic counselor come from a coordination of learning experiences. These include postgraduate course work on family therapy and several summer training institutes on topics ranging from self-psychology to group therapy. These courses then led me to read more extensively in psychodynamic theory. The most useful experience has been the discussion of specific genetic counseling cases with practitioners of specific schools of theory or approaches. This has been done by consultation.

I will present a set of theories that address the psychological dynamics of the clinical session that are most applicable to genetic counseling. This theoretical presentation is supplemented with cases from my practice and that of others, to illustrate the extent to which one can explore psychological dimensions within the genetic counseling session. Each counselor develops his or her own set of clinical perspectives through training, experience, and continuing education. I offer these theories only to demonstrate the rich psychological dynamics that exist in a session. I have found that looking through a psychological "lens" has enhanced my appreciation of the complexity of the issues that a genetic concern raises for the counselee. This deeper understanding of the patient only strengthens the essential working relationship. Each genetic counselor must decide the degree to which she chooses to employ a theoretical framework or to engage the counselee in discussions that may shift a psychological barrier or tension. Our choice of theory will be influenced by our knowledge about the case, our own skill level, and time available.

The metaphor of taking a photograph can be applied to counseling work from a theoretical perspective. For photographers, the process is dynamic, as is our meeting with a counselee. A case that is presented to us is like a composition that draws our attention and engages us to take a photo. The combination of lens and aperture that we choose determines what will be in focus at that moment and what will remain in the background. This parallels what we choose to facilitate as discussion points in a session, and how we change the focus throughout the session. The lens we choose can also be fitted with filters that impose cultural tones on the composition. The click of the shutter captures the image, just as in a counseling session we attempt to fully and empathically understand the counselee on several different levels.

The psychological dynamics of a genetic counseling session can be considered from many theoretical perspectives. However, it is important to remember that most formal theories of psychology were constructed from a western cultural perspective. Furthermore, the common paradigm is white, middle-class, heterosexual, and patriarchal in experience. It is for these reasons that the psychological constructions we use will be limited. Not every template will fit every situation or work for the duration of the counseling relationship. This is why one can appreciate the narrative model as an essential psychosocial tool.

Central to all genetic counseling encounters are the principles defined by Carl Rogers, which provide the framework of practice for the genetic counselor. Rogers pioneered a form of psychotherapy that he described as a "non-directive," "client-centered," or "person-centered" approach. The central hypothesis is that "a climate of facilitative psychological attitudes" promotes self-understanding and changes in attitude and behavior (Kirschenbaum and Henderson, 1989). Carl Rogers argued that for change to occur, three attitudes need to exist in the therapeutic relationship: genuineness, empathic understanding, and unconditional regard of the counselee.

Rogers published his description of this counseling approach in 1942. In addition to stressing the critical role of the facilitative experience, he argued for the centrality of a "non-directive" approach:

> The non-directive viewpoint places a high value on the right of every individual to be psychologically independent and to maintain his psychological integrity. . . . In the field of applied science, value judgments have a part, and often an important part, in determining the choice of technique. (Rogers, 1942)

The historical influences of the eugenics movement and other sociopolitical events led to a quick acceptance of this theoretical framework. The willingness of the genetic counseling community to embrace this approach assisted in the greater public acceptance of genetic counseling. More recently, there have been increasing numbers of papers attempting to remind us of the origin and true meaning of nondirectiveness within the genetic counseling context (Kessler, 1992a; Fine, 1993; Wolff and Jung, 1995; Bernhardt, 1997; Michie et al., 1997; Whipperman-Bendor, 1997).

Other theoretical perspectives (the various "settings" on the psychological lens) offer the counselor a way to appreciate who the counselee is, as an individual, within his or her family, and as a member of the larger community. I prefer to develop my understanding of the counselee from the smallest "aperture" to a panoramic perspective. The smallest aperture is the psychological makeup of the individual. In the process of the counseling session, a broader view is required that includes relational issues, family dynamics, and sociocultural context. In using this approach with many families over the years, I have experienced an understanding that might not have emerged without a theory-based perspective. Looking through these lenses does not imply that we are initiating a psychotherapeutic relationships with a counselee. A decision to engage in a psychotherapeutic dialogue requires mutual consent and process, as described later in this chapter.

I have selected the three most useful clinical perspectives that I employ in my daily practice. The self-psychological perspective provides a way to consider someone's psychological integrity. The relational model offers an understanding about how connections foster coping and relational skills. The family systems approach describes families in action. Each theory offers a way to "see" the genetic counseling patient.

Self Psychology

This psychoanalytic approach (derived from Freud's theory and methods) was formulated by Heinz Kohut in the 1960s. His writings, and the contributions of many others, have offered a clinical developmental perspective on the emergence of a "self" and what is required for the "self" to thrive. Kohut departed from classic analytic theories that identified the origin of behaviors in biological drives. He was convinced that the early childhood relational context with a parent or caregiver ("object") forms the structure of the self. He referred to these experiences as "self objects." Unlike other analytic theorists of the time, he argued that the lifelong vigor of the self is dependent on maintaining a relational context and having experiences that fortify the self. He believed that the psychological challenges that face us can be traced back to disruptions in self object experiences.

Kohut's theory of self psychology has provided a way to appreciate the psychological cohesion of the counselee. This can be useful in anticipating responses to difficult news and in the evaluation of suicidal ideation, as demonstrated in the following case report.

> The patient (V.B.) was a 36-year-old married Caucasian woman who had extreme distress about her abnormal maternal serum α-fetoprotein result. V.B. expressed ambivalence about the pregnancy, identified marital conflicts, and was preoccupied with bodily sensation. She reported difficulty concentrating at work and lack of sleep. In an attempt to appreciate the psychological tensions of V.B., the counselor explored past and present relationships and learned that the patient saw herself as isolated, unattractive, and a failure. She also reported that she was withdrawn in childhood and that her present social support was minimal. On direct inquiry about suicide ideation, V.B. revealed that she had considered killing herself and the fetus if amniocentesis detected anything seriously wrong.

In the foregoing case (Green-Simonsen and Peters, 1992; Peters, 1994), depression and extreme anxiety are not solely attributable to a genetic diagnosis. Rather, these responses are more descriptive of the counselee and her psychological makeup. V.B.'s feelings of unattractiveness, sense of failure, and focus on bodily sensation could suggest to us a "self" that may have had limited self-object experiences. The genetic counselor appropriately appreciated that the "self" cohesion might not be strong. News of a fetal aneuploidy could shatter a fragile cohesion. The "self" attempts to regain cohesion through stages of rage. It is difficult to know how the rage will be expressed (i.e., verbally or through an act that places the person at risk). Direct questioning of this distressed patient revealed a fantasy of suicide.

To promote shifts in self-understanding and to enhance psychological cohesion, the self psychologist uses the classic psychoanalytic tool of interpretation. Kohut's work is distinguished by interpretations that are, in part, derived from empathic attunement with the counselee. The treatment process employs the psychoanalytic concepts of "working through the transference." This step in the analytic process requires long-term counseling during which the counselee begins to see the therapist through a transference, and in an organized way, the therapist and counselee come to identify the transference. A series of interpretations bring the therapist back into focus. This process aims to bring psychological relief and to increase cohesion of the self (Kohut, 1971, 1977, 1984).

The classic psychoanalytic process does not fit with genetic counseling, since the sessions are limited. However, Kohut's contributions offer evidence that the counselor's commitment to empathic listening and understanding can lead to a more effective experience. The theory also can assist the counselor in gaining deeper appreciation for the psychological construction of personalities.

Self in Relation Theory

The theoretical work that has emerged since the 1970s from Jean Baker Miller and her colleagues at the Stone Center, at Wellesley College, has strongly influenced my ability to appreciate what psychological events promote change within the counselee. The theory presented is derived from the experiences of women. The 1970s and 1980s saw a criticism of psychological theories that attempted to generalize across gender lines and cultural perspectives. The writings of Jean Baker Miller, Carol Gilligan, Belenky, Clincy, Goldberger, and Tarule set the stage for collaborative discussions on learning from women about their psychological development and dilemmas. What emerged was an evolving theory of "self in relation" that allows for greater appreciation of relational context for women and, not surprisingly, for men as well (Bergman, 1991).

The Stone Center writers argue that the development of "self" emerges from relational interactions with early caregivers and later with others in adult life. Central to their thinking is a de-emphasis on individuation and separation as psychological requirements for ego formation. Further, the ability to respond to critical life events can be influenced by the level of connectedness with self and others. The genetic counselor's ability to "connect" and understand the counselee's connections and disconnections enhances the counseling session. It is through the counselor's empathic understanding of the events that led to the disconnections that the counselee begins to experience self-empathy. Self-empathy is a requirement for change or action. In other words, psychological vulnerabilities can be reduced through a process that explores connectedness. This exploration is done from a position of empathic attunement and mutuality. The shifts in psychological tensions are often evidenced in what Jean Baker Miller has termed "zest." The actions and attitudes that once could not be considered or accomplished become more possible.

In the genetic counseling setting, relational theory assists in appreciating the psychological barriers of men and women who seem to have difficulty determining a course of action. I have also found this work to be very useful in assisting counselees with protracted grief reactions or feelings that they could not participate in a decision process.

The following case illustrates how disconnections in several relationships led to progressive destabilization of the client's ability to continue to manage her medical concerns.

Susan, a 33-year-old woman with Treacher Collins syndrome, has been followed by the local craniofacial evaluation team since she was 18 years old. Her family arranged for her to be seen by the team to discuss reconstructive surgery. Susan asked the surgeons to change how she looked. At that time, she convinced the medical group to proceed with surgery by telling them disturbing stories of stares and comments from the public as she rode the bus to work. I was a member of this evaluation team and had met with the family and Susan to discuss the diagnosis of Treacher Collins. At that time of our discussion, Susan's mother described a number of obstetrical events that resulted in pregnancy complications and may have contributed to Susan's diagnosis of mild mental retardation. Later that year, after a lengthy surgical procedure, Susan refused invitations to be seen by the craniofacial team for follow-up.

Several years later, Susan and her sister made an appointment to see me. The appointment began with Susan recalling the discussions we had almost 15 years ago. I asked Susan and her sister what had prompted this meeting. I was a bit surprised to learn that they wanted me to provide psychological support to Susan. They had come to me after unsuccessful efforts to identify in the community a therapist or counseling program that worked with persons with a diagnosis of mental retardation.

I made an attempt to better understand the basis of Susan's distress. Susan was able to describe to me, and I observed, extreme anxiety and obsessive-compulsive behaviors. She also added that she had been having difficulty swallowing so she had been limiting her diet to pudding and mashed potatoes. I explained to Susan that I would try to help her, but I would need to first hear more about her situation and to establish whether my schedule would allow me to work with her. Susan began to cry and asked repeatedly why she was having these difficulties.

I then asked the obvious question, "Have you discussed this with your doctor?" Susan's sister quickly volunteered that Susan would not go. "She refuses!" added the sister. I asked Susan to tell me why she did not want to go to the doctor. Susan slowly explained her experiences at the time of her craniofacial surgery and a hysterectomy for dysmenorrhea. I came to understand that she was angry with her present physician because her concerns were not being addressed ("He won't listen!"). Susan's sister interrupted with a description of family frustrations, indeed, anger, over Susan's behavior. The sister, who is closest in age to Susan, felt that "maybe if Susan talked to someone, it might help." As the session came to a close, I came to believe that Susan and I could bridge her medical distrust and find some relief for her severe anxiety and difficulty in swallowing. Susan agreed to come again.

The second session, which Susan's sister did not attend, was primarily devoted to hearing and appreciating Susan's experience. This was not always easy, as she would find herself in the middle of a story and be unable to complete it. I repeatedly acknowledged how helpful her stories and memories were in understanding why it was so hard to get medical guidance now. Whenever possible, I introduced the need to make a referral for medical evaluation and the option of exploring medication that could assist in alleviating her extreme anxiety and possibly eliminating the aphagia. Susan responded to all such suggestions by crying and asking me to promise that I would not make her see a physician. I continued to remind her that she was the person who would make the appointment. The session led to two more. The third session focused on her feeling of "being different" and her sense that "no one listens." I explored with her what would be required for her to make an appointment with her doctor, specifically, more social support. The session ended with her saying "Maybe I will go."

During the fourth session, Susan called the doctor's office and requested that the doctor call her at home. She thought that she could explain her medical concerns better if she spoke to him ahead of time. This intention having been identified, Susan scheduled and kept an appointment with the physician, who prescribed medication that subsequently reduced Susan's anxiety and obsessive behaviors.

This case demonstrates that in the face of family disconnection around her increasing anxiety about medical appointments, a patient's sense of difference and abandonment was heightened. I accepted Susan's challenge of "no one listens." I believed that by providing empathic attunement, it would be possible to stimulate Susan's self-empathy. This in turn allowed her to appreciate that it was hard to go to her medical appointment and ask for help. Finally, the case also hints at the complexities of working with counselees who are intellectually impaired. Until recently, little has been published about the developmental, cognitive, and cultural aspects of persons with mental retardation (Finucane, 1997).

Family System Theories

The counselee and his or her family are central to all genetic counseling sessions. Our consideration of the family begins with those present in the room and extends to other more distant members of the pedigree. Our efforts to collect medical information in the pedigree construction process illuminates relationships, disconnections, and intergenerational interactions. The bringing forth of a family context into the genetic counseling session influences many discussions.

Our ability to appreciate families and their perspectives is critical in most cases. The birth or prenatal identification of a child with a recognized genetic condition brings many life cycle and developmental challenges to the family as a whole. Families in which an adult-onset genetic condition exists are deeply influenced by the meaning and implications thus imparted to the family members.

"Family" is conceptualized "as an open system that functions in relation to its broader sociocultural context and that evolves over the life cycle" (Walsh, 1982). The diversity of family structures and norms requires the counselor to learn about each particular family. It is through each one's story and concerns that an assessment of whether a family is experiencing a loss of function can be made. This assessment must include an appreciation of family norms that are influenced by society or an ethnic subculture (Walsh, 1982). Family systems theorists have studied families to understand the processes that influence cohesion and function, focusing on family structure, rules, boundaries, and intergenerational beliefs and attitudes. Models of intervention have then been developed to assist families who experience loss of function. A table of the major models of family therapy, their goals and interventions is provided in the appendix to this chapter.

Two family systems models, the family systems-illness model and a systemically based psychotherapeutic technique, have been applied to families facing genetic concerns. These models offer an integrated approach, which permits consideration of all the elements that are stimulating the family dynamic or response.

Family Systems-Illness Model   In the mid-1980s the writings of John Rolland began to describe the family unit as it addressed the issue of illness over time. Today, this work is known as the family systems-illness model. The model provides a perspective on the interactive processes of the psychosocial demands of the illness, family beliefs, and family functioning. The inclusion of belief systems in a family response is essential to the facilitation of effective coping and adaptation to illness (Kleinman, 1988; Rolland, 1994). This model also considers anticipation of future loss, an experience common to families faced with genetic disease. The following case presentation highlights this.

Lori and Richard were referred to the genetics clinic by Lori's obstetrician after her fourth first-trimester miscarriage. This young couple worked together in her family-owned business. While taking the family history, the genetic counselor learned that Lori's mother had had seven successive miscarriages before giving birth to Lori, and that the family had adopted a son after Lori's birth. Lori's mother was an only child. Lori's grandmother had had several miscarriages, and there was a belief that the women in the family were destined to lose many pregnancies before a live birth would occur.

Information was provided to Lori about miscarriages and the possible role of a chromosome translocation. The couple refused chromosome testing, since they had confidence that there would eventually be a child. Lori confidently stated, "My mother had seven [miscarriages] and then I was born."

Several months passed and Lori had another miscarriage. Her physician arranged chromosome testing on the products of conception, and the analysis revealed an unbalanced translocation. The results were given to the couple, who then agreed to testing for themselves. Chromosome analysis identified Lori as carrying a balanced translocation. Lori and Richard returned to the genetics clinic to learn more about the results. The session was filled with the couple's continued optimism and reminders of the "family pattern."

Lori and Richard experienced several more miscarriages after our appointment. Eventually, the total number of pregnancy losses came to seven. Lori entered her eighth pregnancy with excitement and optimism. Richard reported becoming less confident after the last pregnancy loss. Lori attempted to reassure her husband with statements like "It makes sense that all 'unbalanceds' would end in a miscarriage," "It already happened to us," and "That's what miscarriages are for." To our surprise, as the pregnancy approached the 16th week of gestation, Lori and Richard requested prenatal testing. Results of fetal testing identified an unbalanced translocation. Lori began the session, "It wasn't supposed to happen like this: the eighth was going to be okay" (Shapiro and Djurdjinovic, 1990).

> *In the face of possible loss, creating meaning for an illness that preserves a sense of competency is a primary task for families. In this regard, a family belief about what and who can influence the course of events is fundamental.*
> —*Rolland, 1990*

Lori's family had normalized the multiple, successive pregnancy losses as the pattern that occurs prior to the birth of a child. There is also a suggestion that after the birth of one child, no other pregnancies are sought and families would expand through adoption. As Lori experienced each of her seven losses, she had turned to the family expectation about reproduction to find some meaning to her experience of loss. With the prenatal determination that the eighth pregnancy represented an unbalanced translocation, Lori's disbelief went beyond her grief into feelings of failure.

Postscript: The ninth pregnancy resulted in a healthy girl with a balanced translocation.

Systemically Based Psychotherapeutic Technique   Eunpu (1997b) presented a model of case formulation that links individual, interactional, and intergenerational issues as they may apply to a family with a genetic concern. She advocates the combined use of the genogram with an intersystem case formulation to provide a thorough psychosocial assessment of the individual and family. In utilizing this approach, the counselor is able to identify intergenerational beliefs and attitudes about a genetic condition. The counselor's knowledge about patterns of communication and relational dynamics offers a genetic counseling experience that is relevant to the family structure and honors their belief system.

The Health/Illness Beliefs and Attitudes Genogram is an appropriate tool for use with an individual from a family affected by a genetic condition (Eunpu, 1997b). The following are questions from the genogram to be asked of the counselees to assess health, illness beliefs, and attitudes:

1. Who was reported as affected (sick/ill) in your family?
2. Who believed himself or herself to be affected?
3. What messages did you receive from family members about the diagnosis?
4. What were the attitudes of family members to those who were/are affected?
5. What obligations or loyalties were created by the presence of an affected family member?
6. What is the meaning of doing something different regarding the genetic risk in your family?

The following case example uses an intersystem genogram guideline. Amy T. is seen for familial cancer risk assessment. Amy is 24 years old and engaged. She has been followed by the local breast center because of fibrocystic breast disease and a family history of breast cancer. At her last mammogram, she requested prophylactic mastectomies. A genetic consult was arranged.

Amy attended the first session accompanied by an unaffected maternal aunt. The genetic counseling session explored the reasoning that had prompted her request for surgery and reviewed her family history. Three of Amy's five maternal aunts were reported to have had breast cancer in their mid-thirties. One maternal cousin also had been diagnosed in her thirties. The maternal grandmother was initially reported to have had "stomach cancer." However, review of her medical records confirmed ovarian cancer. Amy's mother was in her fifties and had no history of a cancer diagnosis.

Discussion about the family revealed that Amy and her aunt believed that the only way to "escape" cancer was to have their breasts removed. Amy went on to report that her mother had had prophylactic mastectomies at 34 years of age (this was not disclosed when I asked "Has your mother had any history of cancer?") and that the aunt attending the session had had prophylactic mastectomies last year, when she turned 30 years of age. The aunt told stories of caring for her dying sisters who were 12 to 16 year older than she. She said that "It was stupid not to do this." Amy's mother was the first to proceed with prophylactic surgery, and the women believed that was why she was "cancer free" at 50 years of age. This led the remaining aunts to seek surgery, and now Amy was being encouraged to take this step.

This partial glimpse into the family tells of an emerging belief about how to protect against breast cancer. The decision to have surgery also assures survivors and preserves family ties. The trauma of the surviving sisters (Amy's mother and aunts) has also had a strong influence. All these factors will be part of Amy's decision. This case argues for the value of bringing the family together in family cancer discussions. The genetic counselor's ability to engage the survivors in the discussions may offer perspectives and options not yet considered in this family. This case supports the effectiveness of blending the pedigree with principles of the genogram through an intersystem case formulation.

PSYCHOSOCIAL CASE DISCUSSION

The ability of counselors to master the lens, aperture, and filters of the psychological issues present in each genetic counseling session requires an organized analysis of content and experience. Genetic counselors who are drawn to the psychological aspects of the field and the psychosocial implications for counselees will seek additional reading and training. One learning method that is underutilized is the case discussion format, sometimes referred to in the mental health community as "supervision." The supervision process in clinical practice assures a continuing education experience as well as an ongoing dialogue about the counselor's self-awareness regarding vulnerabilities, tensions, and frustrations that may impact counselees. Mental health providers have a professional commitment to participate in regular discussions of case material. These discussions can occur on an individual basis with a clinician recognized to have more experience than the counselor. Group case discussion is another common format.

The aim of case discussion is to guide the counselor to a deeper and more complex understanding of the issues that were presented by the counselee. Moreover, it allows counselors to examine themselves as part of the psychological milieu that stimulates a response from the counselee. Supervision or case discussion is best achieved in a structured professional relationship with a clinician or group, where trust and confidentiality are preserved. This may require that the clinician selected be outside the institution where the counselor practices.

Psychosocial case discussion is not often a part of the practice of genetic counselors. These discussions are different from clinical case reviews. Recent years have seen the emergence of psychological case discussion groups, which meet as an ancillary program of the annual professional meeting of genetic counselors. Recently, a group of genetic counselors formed in New England and plans to continue meeting over an extended period (Kennedy, 1997). Participation in such groups as part of training and continuing education should be integrated into the professional life of a genetic counselor.

How to Prepare a Case for Discussion

The decision of which case to bring to a case discussion rests most often on the reactions of the counselor to the patient or to discussions in the session. Sometimes it is useful to discuss the case that stays with you over several days. It is not always necessary to know prior to case presentation what specifically about the case captured your attention. The case discussion experience usually brings that understanding into focus. When preparing the case for discussion, the following questions may be useful guides.

1. Why did you decide to discuss this case instead of another?
2. What were your beliefs about the case prior to the meeting?
3. What was your immediate reaction to the counselee?
4. Was there a comment or behavior that drew particular attention? Can you recall the specific statement made by the counselee? What was your understanding of that comment or behavior?

5. Describe the counselee.
   a. First impression
   b. Social history
   c. Relational history
   d. Family of origin history
   e. Ethnocultural affiliation
   f. Describe your understanding of the counselee's psychological, relational, and family structures.
   g. Describe the reactions or emotions that were presented in the session.
   h. Explain how you understood the origin of these emotions or reactions.
   i. Did you have any preconceived ideas about the counselee and the issue that drew your attention?
6. Have you had a similar experience?
7. What did the case remind you of?
8. Does the counselee remind you of anyone?
9. What did you feel in the session, when, and in response to what cues?
10. What theoretical perspective most closely fits the dynamics of this case?

STAGES OF ONGOING GENETIC COUNSELING PRACTICE

Genetic counseling cases can be carried over several sessions because of a client's numerous questions, wish to review material presented during the initial session, or need to return for follow-up. Alternatively, the client may find the genetic counselor a "safe" person with whom to share the evolving experience of new genetic knowledge or results. Such cases should be viewed as an unfolding experience and generally can be followed through the subsequent stages and processes.

Stage 1: This state assesses why a particular question or psychological concern needs to be addressed as part of the genetic counseling process. It helps to determine the counselee's interest and capacity to undertake an exploration and to identify which particular counseling approach should be considered in the process. Also during this stage, parameters of the session are set, such as time frame, duration, and fee. The counselor and the client mutually determine the criteria by which the process will be evaluated—the expected outcome and the likely time frame.

Stage 2: This stage can be called the "talking cure," a term coined by one of Freud's patients, Anna O., to describe her experience with psychoanalysis. This expression can be generalized to all interactive experiences in which the telling of a story and feelings offers relief. The client experiences a transition from a present state of crisis to a more stable and resolved perspective. A foundation forms that develops the working relationship and provides the platform for continued empathic understanding. It is within this experience that the counselee will gain understanding and find increased self-empathy.

Stage 3: This is the phase in which the evaluation criteria are reviewed and a plan for ending is negotiated. The plan may include a referral to a mental health professional for ongoing counseling. Pay close attention to feelings about ending, as issues of loss often surface.

THE UNFOLDING OF A GENETIC COUNSELING CASE

The following case reflects the potential of in-depth psychological analysis of a genetic counseling patient. This case, which is presented in two parts, involves many sessions, an option that may not be available for some, but it is important to consider that the first part of the case occurred over only three sessions. Part I of this case could have occurred in most prenatal clinic settings. Most clinical settings that address issues of prenatal testing make some provision for follow-up. I also believe that some of the face-to-face conversations I had with my patient could have as easily occurred over the phone. Regardless of the number of sessions and the availability of physical meetings, our connections have an impact on patients. I selected this case because it illustrates the ongoing complexities that are revealed beyond the time-limited genetic counseling session. In addition, I hope it argues for the option of extended counseling in some circumstances.

I brought this case to a discussion group after I experienced feelings that I can only describe as "losing interest." My feeling state did not necessarily match my cognitive activities around this case, and I recognized that something was happening. As you read this case, consider employing the lenses and perspectives presented in this chapter and make conjectures about the patient and the counselor.

The names and other identifiers have been changed to protect the confidentiality of the couple. The case notes offered are descriptive, with quotations when appropriate. Part I of the case is presented through the phases of a typical genetic counseling session (Kessler, 1992c). Part II exemplifies the stages of extended genetic counseling.

The Case of Mona

Part I: "It's my fault"

Intake Phase

Mona was a 42-year-old Caucasian woman who was referred following an abnormal sonogram at 21 weeks of gestation. The patient had declined counseling and amniocentesis at 16 weeks of pregnancy. Following the sonogram, a number of findings suggested a possible chromosome abnormality. Mona's physician performed an amniocentesis, and a referral was made to our program. The couple felt they did not want to be seen before the cytogenetic results were back. This was Mona's second pregnancy; she and her husband, Steve, had a 3-year-old son. Steve, a successful dairy farmer, was 45 years old. Mona had owned a bakery for 8 years, until 5 months ago, when she had had an opportunity to sell the business. The couple lived in a rural community approximately an hour from the genetics clinic.

Encounter Phase: The Phone Call

A phone call was made to the couple to review what had transpired over the previous 48 hours. The phone discussion reviewed what they understood about the sonogram findings and what the amniocentesis could diagnose. Steve reported concern about possible test results. The majority of Mona's questions focused on the time frame for results.

The Session

I met with the couple 6 days after my initial call. This session was arranged after I had phoned a second time to advise Mona and Steve that the amniocentesis results were back and that a problem had been identified (the results revealed trisomy 21). I asked them to come in.

The session progressed through the process of giving difficult news, exploring the counselees' understanding of Down syndrome, and eliciting their experiences with people with disabilities. The session also considered the clinical concerns revealed by the ultrasound and possible associated anomalies. The options available to the couple were outlined. I provided them with a booklet written for parents making choices following the prenatal diagnosis of Down syndrome.

Relationship Phase

I explored the counselees' support system and learned that their families of origin were not nearby and that Mona's relationship with her mother was difficult. Their "family" of choice was an elderly couple who lived next door. I noted that in the session there was little physical contact between the couple. Also Mona and Steve were restrained in their descriptions of their thoughts and feelings.

Feeling Phase

Toward the end of this session Mona openly showed her sadness and shared that she felt responsible: "It's my fault." Steve made no attempt to alleviate her acceptance of responsibility. Steve did not comment on his own feelings but looked sad.

Decision-Making Phase

During our discussion of available options, the issue of pregnancy termination at 22 weeks of gestation was raised. When Mona showed some interest by asking where they would have to travel, Steve interrupted her by stating "We can handle it [the prospect of a child with Down syndrome]; it's God's will." This preempted any attempt to continue this discussion. The option of meeting with a parent support group felt overwhelming to Mona. Steve said "It's too soon." The decision to pursue sonogram studies was easily reached, and travel requirements for a high resolution sonogram were explained.

Summary Phase

The medical/genetic issues were repeated. The available options were again reviewed, and a plan was mutually constructed to offer follow-up of the pregnancy and to meet and review our discussion. A second appointment was arranged for later in the week.

Follow-Up Phase

Mona came alone to the follow-up appointment. When I inquired if there was a reason for her husband's absence, Mona responded by saying, "Steve did not feel it was important. . . . He doesn't think I should waste your time. . . . You must be very busy." In this session, she told me about reading the booklet I had provided and revealed that she had decided to name the unborn child. Throughout the session she looked at the clock and asked if it was all right for her to be there, then quickly inquiring "Can I go?" I would encourage her to stay, since it seemed she had some things she wanted to talk with me about. The session ended with a plan for her to return in 3 weeks and possibly for Steve to join her.

I received a phone call from Mona 2 weeks later. She told me that she had miscarried over the weekend. On Friday she had begun to bleed and by Sunday morning she was in the hospital. She said, "I just wanted you to know." When I asked how she and Steve were doing, she explained that he was out with their son. She went on to say that she was alone when she delivered. "Steve doesn't like hospitals. . . . I'm OK." The conversation ended abruptly with her saying that she was tired and needed to go. A follow-up appointment was scheduled for 2 weeks.

The Follow-Up Session

Mona again came alone. She said Steve was busy. The session was filled with questions like "Why did my child die?" The questions would be followed by "I was just thinking it was going to be OK. . . . I think it must be my fault." I attempted to explore these statements, and Mona cried. She did acknowledge feelings of grief and anger, ambivalence for the child, and fears about the diagnosis. I listened carefully and made very few comments. I wanted to provide Mona with as much opportunity as possible to share her experience. Several times in the session she looked at the clock and asked if it was OK to be there, following up with a question about whether I was busy.

The session ended when Mona began to openly grieve. She stood up from her chair and said, "I need to go now! . . . You have other people who need you. . . . My problem is gone now . . . it's no longer genetic. . . . Can I go now?" As Mona began to leave the room I asked her to wait a moment and discussed what I could offer her next. She thanked me for helping me and said, "I need to go now. . . . Good bye."

I decided to wait a couple of weeks to call to see how she was doing. I also attached a note to the genetic counseling letter encouraging the couple to seek support during this time. I provided the names and phone numbers of support groups and the perinatal loss group.

Part II: "Life is OK"

Stage 1

Two weeks went by and Mona called and asked if she could come in and see me. At the session she told me that she had been plagued by migraine headaches. She went on to say how sad she had been. In our discussion about her sadness, she described intense longings to see her fetus again. I attempted to determine what level of support she had had since we last met. She had been home with her son.

Her husband had been working late, and there was very little time for them to talk. She came to the office with a large shopping bag in which she had several books on pregnancy loss and a book describing children with Down syndrome. She told me that she had been reading them and felt that they had been helpful. The remainder of the session was devoted to considering the grief process and what would be useful to her during the coming months. She asked if she could come in and see me once in a while. I told her that would be possible. We discussed an interval of 2 to 3 weeks. We determined that our sessions would strive to provide support as she struggled with her feelings about the diagnosis and the loss. I also explained that we would look at ways for her and Steve to have some conversations about their loss. We would expect an improved sense of well-being to emerge within the next two and half months. A fee was discussed, and the next appointment scheduled.

It was my assessment that Mona had made a connection with me in spite of her protestations. Mona frequently brought me bakery items and, of course, did keep coming back. She had been socially isolated since she closed the bakery. Her social support consisted of Steve and the couple next door. She did report having conversations with the neighbors about her sadness. She was frustrated by their statements of relief that she and Steve would not have the burdens of a child with a disability. She did not feel that attending a group would be possible for her. I hoped to offer her a safe environment where her full range of feelings and fantasies could be heard over the next couple of months.

In addition to providing opportunities to explore Mona's grief and associated meanings, I imagined that our work would include consideration of some relationship issues. I also hypothesized that her routine queries ("Are you busy?" "Can I go?") were somehow related to issues of connection to others. I planned to explore her history around connections and her access to self-empathy (self in relation theory).

Stage 2

The sessions that followed became a blend of discussions about Mona's feelings of loss and statements of guilt and an inability to share her feelings with her husband. I soon came to observe a repetitive ritual. Each session had at least one moment when Mona looked at the clock, asked if it was OK to be there, then assured me that she was going to be OK, asked if I was busy, and concluded with "I should go."

Over the weeks Mona told me of other experiences of loss, which included the death of her grandmother. I learned more about her family of origin. Both Mona's parents had worked more than one job and so she had spent most of her childhood with her grandmother, who had cared for Mona and her brother. Mona remained in telephone contact with her brother, who lived in another part of the country. Mona had almost no contact with her mother and was quite contemptuous of her. She would answer questions about her but would always limit the discussion.

I also learned that Mona had a teaching degree in math. She had taught at the high school level and at the local community college. She had left teaching because migraine headaches forced her to miss too many classes. During this session I came to appreciate Mona's long history of migraines. She went on to tell me that she had been evaluated by a number of specialists and had been given various medications to control the frequency and pain of the attacks. At the conclusion of this session Mona told me that her child's death was linked to the migraines and left the room without more discussion.

The next session began with the familiar clock-watching ritual and then Mona painfully told me that she had a history of abusing one of her pain medications. She believed that no one knew. This "secret" is what she had meant by "It was my fault." Mona went on to tell me that she felt good only when she was taking this medication. "Without it, life is [just] OK . . . and I miss being happy." Her physician had become suspicious, however, and had refused to write additional prescriptions for this drug. Mona was presently taking a newer medication that did not offer the mood benefits of her favored medication. Future sessions considered teratogen data, addiction concerns, and her present use of the new medication. Mona assured me that she had not used this medication prior to conception of the Trisomy 21 pregnancy. She added that she was continuing to avoid use, since this prescription drug was so central to her guilt. The effects of the "secret" were explored in terms of some of the relationship dynamics. She remained adamant about not telling her husband.

Later sessions included active discussions of Mona's feelings about pursuing another pregnancy. Steve had left that decision to her. He reportedly said it would be fine either way. Issues of recurrence, concerns about having only one child, and what Mona saw as goals for herself became more central to our work.

As this phase of our work progressed, I found myself losing interest. This was troublesome to me, since I could see Mona's grief resolving and her increasing ability to achieve moments of self-empathy about her history of medication abuse. This personal response was counterintuitive, since I was continually interested by what emerged.

Stage 3

As the sessions reached a 3-month period, Mona reported that she was able to feel more confident about her decision not to pursue another pregnancy. She continued to battle with migraines and wished that she could return to the medication she had once abused. We discussed our original contract and hoped-for outcomes. I facilitated a discussion of what Mona might achieve if she continued a counseling process. Mona did not feel she needed to continue and did not hold out much hope that life could be anything but "just OK." We continued to meet over several more weeks. Our last planned session considered how her connection to me could be maintained outside a formal counseling experience.

Postscript

Mona has sent notes on occasion to let me know she was [just] "OK." She decided to pursue another graduate degree and has been attending her graduate classes.

CONCLUSIONS

Let me invite you again to make conjecture about the patient and the counselor. By doing so you will not only appreciate what was going on in the session, but begin to understand how you might react. I hope that this exposure to basic ideas and theories of psychological practice opens the door to further knowledge. Once we acknowledge the emotional realm in our counseling practice, it will become more accessible, and we will become more competent. This can only make the genetic counseling experience more positive for our patients and for ourselves.

The typical genetic counseling session interweaves many of the issues discussed in this chapter. The most effective and direct approach is to have clients tell their stories. The degree to which a counselor and a patient choose to explore the psychological content in the session will depend on the goals that were mutually set, the skills of the counselor, and the time available. It is a disservice to the patient to fail to be fully cognizant of the psychological framework that surrounds genetic counseling discussions. In a medical climate where other professionals are seeking roles with families facing genetic testing and diagnosis, it becomes more imperative that genetic counselors demonstrate and remain committed to the full scope of genetic counseling. Time constraints and the role expectations set in the definition of genetic counseling force some counselors to attend to psychological aspects last, if time is remaining. However, to see genetic counseling as linear steps in a process inaccurately deconstructs an interactive dynamic where the psychological is parallel to the genetic and medical discussions.

In settings where counselors have the opportunity to meet families along with mental health professionals, such as social workers and psychologists, genetic counselors must still continue attending to the psychological dynamic. Our mental health colleagues can bring a special richness to the genetic counseling session, but we do not want to distance ourselves from our psychological skills—this would remove the core of our professional role.

I am committed to the belief that the vitality and longevity of the genetic counseling profession is very much tied to psychological practice. I am not proposing that we see ourselves as mental health professionals, only that we remain committed to the earliest visions of our profession. This will require that professional competencies include more psychological skill sets and that forums of continuing education encourage participation in psychological case discussions. I believe a more process- and relationally based definition will honor all the experiences in the genetic counseling session. It is the genetic counselor who provides the unique knowledge base that allows for the unfolding of genetic and medical information in a psychologically attentive way.

ACKNOWLEDGMENTS

I thank Peg Johnston for her encouragement and emotional support during the months this chapter was written. I also have greatly benefited from her editorial talents. In addition, I thank Debbie Eunpu for her comments, June Peters for reading this manuscript and offering thoughtful suggestions, and Diane Baker for her direction.

REFERENCES

Antley RM (1979) The genetic counselor as facilitator of the counselee's decision process. *Birth Defects Orig Artic Ser* 15(2).

Antley RM, Bringle RG, Kinney KL (1984) Downs syndrome. In: Emery AH, Pullen IM (eds), *Psychological Aspects of Genetic Counseling*. London: Academic Press, pp. 75–94.

Basch MF (1980) *Doing Psychotherapy*. New York: Basic Books.

Basch MF (1988) *Understanding Psychotherapy*. New York: Basic Books.

Bellet PS, Maloney MJ (1991) The importance of empathy as an interviewing skill in medicine. *JAMA* 266(13):1831–1832.

Belenky MF, Clincy BM, Goldberger NR, Tarule JM (1986) *Women's Ways of Knowing: The Development of Self, Voice, and Mind*. New York: Basic Books.

Bergman SJ (1991) Men's psychological development: A relational perspective. *Work in Progress, Stone Center Working Paper Series,* Wellesley, MA.

Bernhardt BA (1997) Empirical evidence that genetic counseling is directive: Where do we go from here? *Am J Hum Genet* 60:17–20.

Blumenthal D (1990/91) When reality differs from expectations. *Perspect Genet Couns* 12(4):3.

Bowlby John (1980) *Loss, Sadness and Depression*. New York: Basic Books.

Bordin ES (1979) The generalizability of the psychoanalytic concept of the working alliance. *Psychother Theory Res Prac* 16(3):252–260.

Brock SC (1995) Narrative and medical genetics: On ethics and therapeutics. *Qual Health Res* 5(2):150–168.

Buckman R (1992) *How to Break Bad News*. Baltimore: Johns Hopkins University Press.

Chapple A, May C, Campion P (1995) Parental guilt: The part played by the clinical geneticist. *J Genet Couns* 4(3):179–192.

Corgan RL (1979) Genetic counseling and parental self-concept change. *Birth Defects Orig Artic Ser* 15(5C):281–285.

Djurdjinovic, L (1997) Generations lost: A psychological discussion of a cancer genetics case report. *J Genet Couns* 6(2):177–180.

Eunpu DL (1997a) Generations lost: A cancer genetics case report commentary. *J Genet Couns* 6(2):173–176.

Eunpu DL (1997b) Systemically-based psychotherapeutic technique in genetic counseling. *J Genet Couns* 6(1):1–20.

Fine B (1993) The evolution of nondirectiveness in genetic counseling and implications of the Human Genome Project. In: Bartels DM, LeRoy BS, Caplan AL (eds), *Prescribing Our Future. Ethical Challenges in Genetic Counseling*. Chicago: Aldine De Gruyer, pp. 101–117.

Fine BA, Baker DL, Fiddler MB and ABGC Consensus Development Consortium (1996) Practice-based competencies for accreditation of and training in graduate programs in genetic counseling. *J Genet Couns* 5(3):113–121.

Finucane B (1997) Acculturation in women with mental retardation and it's impact on genetic counseling. *J Genet Couns* 7(1):31–47.

Frazer FC (1976) Current concepts in genetics: Genetics as a health care service. *NEJM* 295:486–488.

Gilligan C (1982) *In a Different Voice: Psychological Theory and Women's Development.* Cambridge, MA: Harvard University Press.

Goodman JF, Abrams EZ (1988) Giving bad news to parents. In: Ball S (ed), *Strategies in Genetic Counseling,* Vol 1. New York: Human Sciences Press, pp. 137–154.

Goffman E (1963) *Stigma.* Englewood Cliffs, NJ: Prentice-Hall.

Gratz H (1997) Personal communication.

Greene Simonsen DM, Peters J (1992) A successful blending of genetic counseling and psychotherapy. *Perspect Genet Couns* 14(2):5.

Guest F (1997) Giving bad news. Presented at Trends in Health Care, Scottsdale, AZ.

Hand-Mauser ME (1989) Techniques in systemic family therapy: Application for genetic counseling. In: Ball S (ed), *Strategies in Genetic Counseling,* Vol 2. New York: Human Sciences Press, pp. 93–120.

Hecht I, Holmes LB (1972) What we don't know about genetic counseling. *N Engl J Med* 287:464.

Heimler A (1990) Group counseling for couples who have terminated a pregnancy following prenatal diagnosis. *Birth Defects Orig Artic Ser* 26(3):161–167.

Hobdell E, Deatrick JA (1996) Chronic sorrow: A content analysis of parental differences. *J Genet Couns* 5(2):57–68.

Ives EJ, Henick P, Levers MI (1979) The malformed newborn: Telling the parents. *Birth Defects Orig Artic Ser* 15(5C):223–231.

Jordan JV (1991a) Empathy and self boundaries. In: Jordan JV, Kaplan AG, Miller JB, Stiver IP, Surrey JL (eds), *Women's Growth in Connection.* New York: Guilford Press, pp. 67–80.

Jordan JV (1991b) Empathy, mutuality, and therapeutic change: Clinical implications of a relational model. In: Jordan JV, Kaplan AG, Miller JB, Stiver IP, Surrey JL (eds), *Women's Growth in Connection.* New York: Guilford Press, pp. 283–289.

Jordan JV, Surrey JL, Kaplan AG (1991) Women and empathy: Implications for psychological development and psychotherapy. In: Jordan JV, Kaplan AG, Miller JB, Stiver IP, Surrey JL (eds), *Women's Growth in Connection.* New York: Guilford Press, pp. 27–28.

Kenen RH, Smith ACM (1995) Genetic counseling for the next 25 years: Models for the future. *J Genet Couns* 4(2):115–124.

Kennedy A (1997) Personal communication.

Kessler S (1979) *Genetic Counseling.* New York: Academic Press.

Kessler S (1992a) Psychological aspects of genetic counseling. VII. Thoughts of directiveness. *J Genet Couns* 1:9–17.

Kessler S (1992b) Psychological aspects of genetic counseling. VIII. Suffering and countertransference. *J Genet Couns* 1(4):303–308.

Kessler S (1992c) Process issues in genetic counseling. *Birth Defects Orig Artic Ser* 28(1):1–10.

Kessler S, Kessler H, Ward P (1994) Psychological aspects of genetic counseling. III. Management of guilt and shame. *Am J Med Gent* 17:673–697.

Kessler S (1996) Presentation to the annual education meeting of the National Society of Genetic Counselors, San Francisco, CA.

Kirschenbaum H, Henderson VL (eds) (1989) *The Carl Rogers Reader.* Boston: Houghton Mifflin.

Kleinman A (1988) *The Illness Narratives.* New York: Basic Books.

Kohut H (1971) *The Analysis of the Self.* New York: International Universities Press.

Kohut H (1977) *The Restoration of the Self.* New York: International Universities Press.

Kohut H (1984) *How Does Psychoanalysis Cure?* Chicago: University of Chicago Press.

Kubler-Ross E (1969) *On Death and Dying.* New York: Macmillan.

Lazarus RS (1991) *Emotion and Adaptation.* New York: Oxford University Press.

Leon IG (1990) *When a Baby Dies.* New Haven, CT: Yale University Press.

Lerman C (1997) Psychological aspects of genetic testing: Introduction to the special issue. *Health Psychol* 16(1):3–7.

Levine C (1979) Genetic counseling: The client's viewpoint. *Birth Defects Orig Artic Ser* 15(2):123–135.

Lubinsky MS (1994) Bearing bad news: Dealing with mimics of denial. *J Genet Couns* 3(1):5–12.

Marteau T, Richards M (eds) (1996) *The Troubled Helix: Social and Psychological Implications of the New Human Genetics.* Cambridge: Cambridge University Press.

Matloff ET (1997) Generations lost: A cancer genetics case report. *J Genet Couns* 6(2):169–180.

McGoldrick M, Gerson R (1985) *Genograms in Family Assessments.* New York: Norton.

McGoldrick M, Anderson CM, Walsh F (eds) (1989) *Women in Families.* New York: Norton.

Michie S, Mareau TM, Bobrow M (1996) Genetic counseling: The psychological impact of meeting patients' expectations. *J Med Genet* 34(3):237–241.

Michie S, Bron F, Bobrow M, Marteau TM (1997) Nondirectiveness in genetic counseling: An empirical study. *Am J Hum Genet* 60:40–47.

Miller JB (1976) *Towards a New Psychology of Women.* Boston: Beacon Press.

Miller JB (1991a) The development of women's sense of self. In: Jordan JV, Kaplan AG, Miller JB, Stiver IP, Surrey JL (eds), *Women's Growth in Connection.* New York: Guilford Press, pp. 11–26.

Miller JB (1991b) The construction of anger in women and men. In: Jordan JV, Kaplan AG, Miller JB, Stiver IP, Surrey JL (eds), *Women's Growth in Connection.* New York: Guilford Press, pp. 181–196.

Miller JB, Stiver IP (1997) *The Healing Connection: How Women Form Relationships in Therapy and in Life.* Boston: Beacon Press.

Money J (1994) *Sex Errors of the Body and Related Syndromes: A Guide to Counseling Children, Adolescents and Their Families,* 2nd ed. Baltimore: Paul H. Brookes.

Pargament KI (1997) *The Psychology of Religion and Coping: Theory, Research, Practice.* New York: Guilford Press.

Parkes CM, Laungani P, Young B (1997) *Death and Bereavement Across Cultures.* London: Routledge.

Peters J (1993) Interface of genetic counseling and psychotherapy: Take back the hour! Unpublished.

Peters J (1994) Suicide prevention in the genetic counseling context. *J Genet Couns* 3(3):199–213.

Peters J (1997) Personal communication.

Rolland JS (1990) Anticipatory loss: A family systems developmental framework. *Fam Process* 29(3):229–244.

Rolland JS (1994) *Families, Illness, and Disability.* New York: Basic Books.

Rogers CR (1942) *Counseling and Psychotherapy: New Concepts in Practice.* Boston: Houghton Mifflin.

Rosenwald GC, Ochberg RL (eds) (1992) *Storied Lives.* New Haven, CT: Yale University Press.

Shapiro HC (1993) *When Part of the Self Is Lost: Helping Clients Heal After Sexual and Reproductive Losses.* San Francisco: Jossey-Bass.

Shapiro HC, Djurdjinovic L (1990) Understanding our infertile genetic counseling patient. *Birth Defects Orig Artic Ser* 26(3):127–132.

Smith RW, Antley RM (1979) Anger: A significant obstacle to informed decision making in genetic counseling. *Birth Defects Orig Artic Ser* 15(5C):257–260.

Sorenson JR, Swazey JP, Scotch NA (1981) Effective genetic counseling: Discussing client questions and concerns. *Birth Defects Orig Artic Ser* 17(4):51–77.

Swinford A, Phelps L, Mather J (1988) Countertransference in the counseling setting. *Perspect Genet Couns* 10(3):1–4.

Surrey JL (1991) The self-in-relation: A theory of women's development. In: Jordan JV, Kaplan AG, Miller JB, Stiver IP, Surrey JL (eds), *Women's Growth in Connection.* New York: Guilford Press, pp. 51–66.

Targum SD (1981) Psychotherapeutic considerations in genetic counseling. *Am J Med Genet* 8:281–289.

Thayer B, Braddock B, Spitzer K, Irons M, Miller W, Bailey I, Rosenbaum B, Blatt RJR (1990) Development of a peer support system for those who have chosen pregnancy termination after prenatal diagnosis of a fetal abnormality. *Birth Defects Orig Artic Ser* 26(3):149–156.

Tichenor V, Hill CE (1989) A comparison of six measures of working alliance. *Psychotherapy* 26(2):195–199.

Wachbroit R, Wasserman D (1995) Clarifying the goals of nondirective genetic counseling. *Rep Inst Philos Public Policy* 15(2, 3):1–6.

Walsh F (1982) *Normal Family Processes.* New York: Guilford Press.

Weil J (1991) Mother's postcounseling belief about causes of their children's genetic disorders. *Am J Hum Genet* 48:145–153.

Whipperman-Bendor L (1996/1997) Nondirectiveness: Redefining our goals and methods. *Perspect Genet Couns* 9(4):1, 9.

Wolf ES (1979) Transferences and countertransferences in the analysis of disorders of the self. *Contemp Psychoanal* 15(3):577–594.

Wolf ES (1988) *Treating the Self.* New York: Guilford Press.

Wolff G, Jung C (1995) Genetic counseling in practice: Nondirectiveness and genetic counseling. *J Genet Couns* 4(1):3–26.

Major Models of Family Therapy

TABLE 6.1 Normality, Dysfunction, and Therapeutic Goals

Model of Family Therapy	View of Normal Family Functioning	View of Dysfunction/Symptoms	Goals of Therapy
STRUCTURAL			

Minuchin
Montalvo
Aponte
	1. Boundaries clear and firm.
	2. Hierarchy with strong parental subsystem.
	3. Flexibility of system for
	  a. Autonomy and interdependence
	  b. Individual growth and system maintenance
	  c. Continuity, and adaptive re-structuring in response to changing internal (develop-mental) and external (environ-mental) demands.
		Symptoms result from current family structural imbalance:
		  a. Malfunctioning hierarchical arrangement, boundaries
		  b. Maladaptive reaction to changing requirements (developmental, environmental)
			Reorganize family structure:
			  a. Shift members' relative positions to disrupt malfunctioning pattern and strengthen parental hierarchy
			  b. Create clear, flexible boundaries
			  c. Mobilize more adaptive alternative patterns

STRATEGIC			

Haley
Milan team
Palo Alto group
	1. Flexibility.
	2. Large behavioral repertoire for
	  a. Problem resolution
	  b. Life cycle passage
	3. Clear rules governing hierarchy (Haley).
		Multiple origins of problems; symptoms maintained by family's
		  a. Unsuccessful problem-solving attempts
		  b. Inability to adjust to life cycle transitions (Haley)
		  c. Malfunctioning hierarchy: triangle or coalition across hierarchy (Haley)
		Symptom is a communicative act embedded in interaction pattern.
			Resolve presenting problem only: specific behaviorally defined objectives. Interrupt rigid feedback cycle: change symptom-maintaining sequence to new outcome.
			Define clearer hierarchy (Haley).

BEHAVIORAL–SOCIAL EXCHANGE

Liberman
Patterson
Alexander

1. Maladaptive behavior is not reinforced.
2. Adaptive behavior is rewarded.
3. Exchange of benefits outweighs costs.
4. Long-term reciprocity.

Maladaptive, symptomatic behavior reinforced by
  a. Family attention and reward
  b. Deficient reward exchanges (e.g., coercive)
  c. Communication deficit

Concrete, observable behavioral goals: change contingencies of social reinforcement (interpersonal consequences of behavior):
  a. Rewards for adaptive behavior
  b. No rewards for maladaptive behavior

PSYCHODYNAMIC

Ackerman
Boszormenyi-Nagy
Framo
Lidz
Meissner
Paul
Stierlin

1. Parental personalities and relationships well differentiated.
2. Relationship perceptions based on current realities, not projections from past.
Boszormenyi-Nagy: Relational equitability.

Lidz: Family task requisites:
  a. Parental coalition
  b. Generation boundaries
  c. Sex-linked parental roles

Symptoms due to family projection process stemming from unresolved conflicts and losses in family of origin.

1. Insight and resolution of family of origin conflict and losses.
2. ↓Family projection processes.
3. Relationship reconstruction and reunion.
4. Individual and family growth.

FAMILY SYSTEMS THERAPY

Bowen

1. Differentiation of self.
2. Intellectual/emotional balance.

Functioning impaired by relationships with family of origin:
  a. Poor differentiation
  b. Anxiety (reactivity)
  c. Family projection process
  d. Triangulation

1. Differentiation.
2. ↑Cognitive functioning.
3. ↓Emotional reactivity.
4. Modification of relationships in family system:
  a. Detriangulation
  b. Repair cutoffs

(continued)

TABLE 6.1 (Continued)

Model of Family Therapy	View of Normal Family Functioning	View of Dysfunction/Symptoms	Goals of Therapy
EXPERIENTIAL			
Satir Whitaker	Satir 1. Self-worth: high. 2. Communication: clear, specific, honest. 3. Family rules: flexible, human, appropriate. 4. Linkage to society: open, hopeful. Whitaker: Multiple aspects of family structure and shared experience.	Symptoms are nonverbal messages in reaction to current communication dysfunction in system.	1. Direct, clear communication. 2. Individual and family growth through immediate shared experience.

7

Multiculturalism and the Practice of Genetic Counseling

Anne Greb

One of the distinguishing features of genetic counseling is the importance placed by those who provide such services on considering the whole family. This emphasis is necessary in part because genes are transmitted through families and because certain genetic conditions are revealed through the family history. We also appreciate that within a family the members may be affected by, at risk for, or caring for someone with a genetic condition. We provide patients and their families with information and support so that they may make informed decisions about childbearing, prenatal testing, pregnancy termination, medical management, presymptomatic testing, and lifestyle choices. In part, how patients make these decisions, adapt to genetic conditions, and communicate within the family about genetic issues will depend on their cultural background.

Every family's background is multicultural, and no two families share exactly the same cultural roots. Different families faced with the same genetic diagnosis can make different decisions and use different coping mechanisms to respond to the situation. This is in part because of the spectrum of beliefs and value systems that exists in our society. The genetic counseling process elicits and respects these distinctions.

Culture is the most significant factor shaping our values, beliefs, and customs (McGoldrick et al., 1996). It is derived from our family and social community and is

A Guide to Genetic Counseling, Edited by Diane L. Baker, Jane L. Schuette, and Wendy R. Uhlmann.
ISBN 0-471-18541-8 (cloth), 0-471-18867-0 (paper). Copyright © 1998, Wiley-Liss, Inc.

passed from one generation to the next. Just as we strive to understand the genes in a family when making a genetic diagnosis, we also strive to understand the role of culture as a factor in the decision-making process and in shaping the coping skills of our patients and their families.

This chapter examines our understanding of culture, ethnicity, and minority group status. It discusses ways of supporting multiculturalism through self-awareness of our own cultural milieu, including the culture of western medicine within which most genetic counselors practice, and through analysis of a case example. Illustrations of some culturally influenced behavior related to life cycle events and roles are presented. This chapter is not intended to be comprehensive in its coverage of the many issues relevant to multiculturalism, but it does touch on some aspects of culture and individuality that should be considered in the practice of genetic counseling. Even with the guidance provided here, culture must always be appreciated from the perspective of the individual patient and family before you.

The following case example illustrates how conflicts can arise if there is failure to recognize the differences between the cultural backgrounds of patients and those providing genetic counseling. This case illustrates the need to appreciate the cultural values of the patient, the genetic counselor, other care providers, and the health care system of which they are all part. The patient and her family are Arabic (although Wahby is not their real name), and the genetic counselor is white and of European ancestry. This case is presented from the perspective of the genetic counselor and is referred to throughout the chapter.

CASE EXAMPLE

The genetic counselor received a consult from the neonatal intensive care unit (NICU) regarding a baby born the week before with multiple anomalies; the infant's karyotype revealed trisomy 13. The neonatology staff was having difficulties communicating to the family the medical reasons for an option that needed to be considered, namely, removing the baby from a respirator. It was hoped that by meeting with the genetics staff, the parents could arrive at a better understanding of the diagnosis and its prognosis. The medical geneticist asked the genetic counselor to begin the consultation and stated his availability to participate if he was needed.

At the NICU the counselor found the neonatologist familiar with the case, who quickly filled in some important details. This was the first pregnancy for the parents, an Arabic couple, Mr. and Mrs. Wahby. Mrs. Wahby, age 21, and Mr. Wahby, age 34, are first cousins. Mrs. Wahby had had one early prenatal visit but only recently started seeing her obstetrician when complications with the pregnancy began. At the early prenatal visit Mrs. Wahby had been referred for genetic counseling because of the consanguinity. According to the neonatologist, she seemed willing to be counseled, but failed to keep two appointments in the genetics clinic.

Mrs. Wahby went into preterm labor, and attempts to stop her labor failed. At 31 gestational weeks an emergency cesarean section was performed because of irregularities in the fetal heart rate. Shortly after birth it was noted that the male infant had a bilateral cleft lip and palate, hypotelorism, and polydactyly. The baby required oxygen, and he was subsequently placed on a ventilator. Blood sent for a chromosome analysis revealed a 46,XY,–14,+t(13;14) karyotype, consistent with the diagnosis of trisomy 13, as the result of an unbalanced translocation (Gardner and Sutherland, 1996).

The neonatologist told the counselor that working with the family had been extremely difficult, adding that she felt that they doubted her professional expertise. An entourage of the extended family and friends were constantly at the hospital, insisting that everything be done to help the baby. At times the presence of so many people interfered with the care of the newborn. The neonatologist felt that the family members were in denial about the severity of the diagnosis. In fact, relatives were refusing to tell Mrs. Wahby that the baby was in serious trouble. Matters really started to deteriorate when the family was approached about removing the baby from the respirator. The Wahbys even seemed to be offended when someone from the neonatology staff asked them if they wanted to consult with a Moslem cleric. The well-meaning neonatologist had run out of ideas about how to interact with this family. She was hoping that if the family were fully informed about the grave prognosis associated with trisomy 13, relations would improve. Some of the family members spoke English, but some did not.

The genetic counselor thanked the neonatologist for involving genetics in the case and left to go find the family. She ran through in her mind how she might explain, in terms the group would understand, the cytogenetic basis for the problems in the baby and the undeniably poor prognosis associated with the specific diagnosis.

In a waiting room the genetic counselor found eight members of the Wahby family. She asked for one of the parents, but was greeted by a grandparent, Mr. Wahby's father. She first asked how the family was doing and expressed her condolences about the severity of the infant's condition. Next she explained who she was and said that she was there to talk with them and to answer their questions about the cause of the baby's condition. The grandfather was cordial and seemed interested in speaking with her, but insisted that his son and daughter-in-law be excluded from the meeting. The genetic counselor was surprised by this request and emphasized the importance of having the parents present. The senior Mr. Wahby was adamant that his daughter-in-law be protected from any upsetting information, but finally it was agreed that the baby's father would be summoned. Then, with both sets of grandparents and the father of the newborn, the meeting began. The genetic counselor assumed that in spite of the family's reluctance today, she would talk with Mrs. Wahby at another time.

After a discussion about the nature of chromosomes and the relevance of the extra chromosome to problems in a baby, the genetic counselor explained the serious consequences, including shortened life span, associated with trisomy 13. Throughout the conversation the family spoke to one another in Arabic and did not, even when invited, ask the genetic counselor any questions about the information presented. At one point they did have a question—about prenatal testing in future pregnancies—which they directed to a male medical student, rotating on the wards, who had joined them partway through the session. By this time the counselor was aware that she had not connected with the younger Mr. Wahby, nor with either set of grandparents. She also recognized that the techniques she customarily used to create a sense of shared purpose with a family were ineffective in this situation.

She felt it was important to mention that the baby's chromosome abnormality may have been inherited and that it would be important at some point to talk about obtaining a blood sample from both parents to analyze their chromosomes for the presence of a balanced translocation. The younger Mr. Wahby said that he and his wife would do whatever the doctor told them. This response also surprised the counselor, since she thought she had made it clear that the testing would be up to the couple, not the hospital or the staff. In an attempt to investigate the presence of a Robertsonian translocation, the genetic counselor then sought to obtain some family history information. However, the family appeared offended and, rather than answer specific questions, vehemently stated that both sides of the family were in excellent health.

At this point the family members appeared either sad or angry, and one of the grandmothers was crying. The genetic counselor was aware that this interaction with the family had deteriorated to the point that their limited trust had been lost. She had covered the necessary clinical information, so decided to end the session. She gave them her card and encouraged them to call her if they had any questions. They insisted they did not have any questions and that they would relay the gist of the discussion to Mrs. Wahby.

After charting on the case, the genetic counselor returned to her office to ponder why she had not managed to help the Wahby family understand and cope with their situation. She reflected that right from the beginning, she had failed to establish rapport with the family members. She realized that because of the translocation, follow-up would be necessary, and she wondered what she might be able to do to facilitate better interaction with this family.

The next week the genetic counselor ran into the neonatologist, who reported that interactions with the Wahby family had gone from bad to worse. Indeed, as the baby's condition deteriorated the family members became quite hostile toward the medical staff. The genetic counselor regretted anew her inability to establish a trusting relationship with this family and wondered what she would do differently in the future if she should become involved in a similar case.

The anxiety, confusion, and frustration felt by the family and the health care professionals in this case can be partially attributed to a lack of appreciation for, and knowledge about, different cultural beliefs and value systems. It is important to understand what is meant here by the term "culture" and why culture is intrinsic to behavior, our own as well as the patient's. Thus we must become aware of our cultural heritage, how it affects us personally and professionally, and how we can begin to develop a multicultural outlook.

WHAT ARE CULTURE AND ETHNICITY?

Defining culture and ethnicity is difficult because neither term has a single meaning and often the two are used interchangeably. The following definitions give some general structure to their meaning. **Ethnicity** pertains to a group's sense of identification surrounding common characteristics, such as physical traits, religion, history, and/or common ancestry (Schaefer, 1996; Spector, 1996). From a genetic perspective, we think of ethnicity in biological terms to differentiate various groups based on their characteristic set of allele frequencies.

Culture is a collection of nonphysical traits, such as values, beliefs, attitudes, and customs shared by a group of people (Schaefer, 1996; Spector, 1996). Thought of broadly, culture includes demographic factors such as age, gender, occupation, and place of residence. It can also encompasses social, educational, and economic dimensions (Pedersen, 1994). Culture determines a good deal of what we believe, think, and do in both obvious and subtle ways. It is a medium for understanding personhood and social relationships, and it changes as our position in the life cycle changes.

A sense of belonging to an ethnic or cultural group provides an understanding of our roots and is a deep psychological need (McGoldrick et al., 1996). There is evidence that cultural identification and values are retained for many generations after immigration (Greeley, 1969, 1978, 1981). Even if unconsciously, the culture of our ancestors contributes to our beliefs and value systems.

Multiculturalism refers to an ability to appreciate the values, beliefs, and behavior of cultures other than our own (Pedersen, 1994; Falicov, 1995). To be multicultural, we must look at how we are the same and how we are different at the same time. Multiculturalism is fundamentally complex and requires a high tolerance for ambiguity, since many perspectives within and between individuals must be recognized and appreciated.

Race is a somewhat artificial category in which individuals are classified according to observable physical characteristics, such as skin color (Schaefer, 1996). Since many such characteristics are noticeable, race is often mistakenly used to define an individual's ethnicity or culture. As it turns out, attempting to classify individuals by physical characteristics is subjective (Barker, 1992), and for all practical purposes race is more a social, rather than a biological, classification (Schaefer, 1996). In the United States racial groups are classified into five minimum categories: American Indian or Alaskan native, Asian, black or African American, native Hawaiian or other Pacific Islander, and white. Two categories are then used to classify ethnicity: Hispanic or Latino, and not Hispanic or Latino. This classification reflects a recent change made by the Office of Management and Budget in an attempt to reflect the increasing racial and ethnic diversity in the United States. This new classification will be used by the Bureau of the Census in the 2000 decennial census (Bureau of Census, 1997).

WHAT IS A MINORITY GROUP?

A **minority** group is generally a group whose members have substantially less control over their own lives than members of a **majority** group (Schaefer, 1996). As a result, the minority group is subordinate to the majority group in terms of power and privilege. A population numerically smaller than another does not necessarily constitute a minority group.

Minority groups often exist because of characteristics of race, ethnicity, religion, or gender. Such groups typically have distinguishing physical or cultural traits, which may be the basis for their unequal treatment. However, people who are older, have physical disabilities or chronic illnesses, or are homosexual are examples of other minority groups. Minority groups are aware of their subordinate status and may have a strong sense of solidarity and feeling of "us versus them." For a variety of reasons, both majority and minority groups tend to socialize and marry within their respective populations (Schaefer, 1996).

Migration is the general term used to describe the movement of a set of people and is often what brings about the existence of minority groups within a population. Emigration and immigration are two sides of the migration coin. **Emigration** describes leaving one country to settle in another; while **immigration** means coming to live in a country in which one is not a native. The process of migration can be extremely disruptive, with consequences that are felt for several generations (McGoldrick et al., 1996). Whether an individual came alone or with others, and the reasons for leaving, all play important roles in how people adjust following migration. Immigrants may abandon much of their ethnic and cultural heritage to try and accommodate to their new circumstances. As a result, they may surrender part of their identity.

Assimilation or **acculturation** describes what happens when a minority individual or group takes on the characteristics of the dominant group (Schaefer, 1996; Spector, 1996). Individuals may abandon their cultural traditions to conform to a different, sometimes even antagonistic, culture. **Pluralism** is the term applied when numerous distinct ethnic and cultural groups coexist within one community. It implies an attitude of mutual respect among individuals from diverse groups (Schaefer, 1996; Spector, 1996). Pluralism encourages minorities to express their own cultural values and beliefs without the expectation of encountering prejudice or hostility. Assimilation can eliminate cultural diversity, whereas pluralism helps to preserve it.

In the United States, Caucasians are the majority population. When the English colonists arrived here in the seventeenth century, they shaped basic American social and political institutions. White Anglo-Saxon Protestant (WASP) lifestyles and values, and the English language, became the dominant culture. Even today, many of those initial values are intrinsic to the majority white culture. The British, however, were not the only ethnic group to immigrate to this country. Across the nation there is evidence of the variety of places from which immigrants have come. An ethnically and culturally diverse nation is the living legacy of migration.

The United States has always been characterized by great cultural diversity, and the face of the country continues to change as we approach the twenty-first century. The Bureau of the Census projects the number individuals who are white and non-Hispanic will significantly decrease by the year 2050 and the number of individuals who are Hispanic and Asian will increase (Bureau of the Census, 1991). The United States is currently experiencing the greatest rise in immigration in over 100 years. Table 7.1 shows the distribution by regions of the 10 groups that have experienced the largest percentage change in population growth in the United States from 1980 to 1990 (Bureau of Census, 1991).

As we strive to practice multiculturalism, it is important to step back and remember when and why people migrate. Although people from various cultural back-

TABLE 7.1   Distribution of Specific Groups in the United States Based on the 1990 Census and the Percent Change in Those Population from 1980 to 1990

Group	Northeast		Midwest		South		West	
	Number	Change (%)	Number	Change (%)	Number	Change (T)	Number	Change (%)
Afghan	4,684	475	1,092	425	4,982	627	8,875	722
Cambodian	30,176	1,222	12,921	472	19,279	650	85,035	853
East (Asian) Indian	285,103	136	146,211	72	195,525	116	188,608	162
Ethiopian	2,679	759	2,770	694	7,961	741	6,833	632
Hmong	1,731	390	37,166	1,237	1,621	1,218	49,564	2,446
Iranian	5,917	3,360	1,482	2,180	4,275	1,996	2,275	4,680
Korean	182,061	167	109,087	76	153,163	116	354,538	130
Laotian	15,928	241	27,775	108	29,262	273	76,049	249
Russian	75,984	10,366	18,354	13,395	12,608	11,908	47,847	11,319
Vietnamese	60,509	144	51,932	42	168,501	119	333,605	178

Source: Adapted from Bureau of the Census, Economics and Statistics Administration (1991).

grounds do not always welcome one another, our country is still the greatest refuge for people all over the world from political, religious, and cultural persecution. To practice multicultural genetic counseling, we need to understand and appreciate our own individual heritages and cultural backgrounds, those of the families we serve, as well as those of our colleagues.

DEVELOPING MULTICULTURALISM

Multiculturalism begins with the process of becoming aware of our own cultural heritage and how it affects our values and assumptions about human behavior. Next we must actively try to understand the perceptions of those with whom we interact who are culturally different from us. Lastly, as genetic counselors, we need to develop appropriate intervention skills for use in working with patients who are culturally different from us. Because multiculturalism is complex and the patients we interact with will forever be diverse, developing multiculturalism is an ongoing process.

BEGINNING TO UNDERSTAND WHAT HAPPENED IN THE CASE EXAMPLE

The Culture of Western Medicine

We have as part of our identity not only the culture we inherited from our ancestors, but also the health care provider culture we learned during our professional education. The theories and practice of western medicine, including medical genetics, are based on scientific discoveries, human and animal studies, empirical data, practi-

tioner experiences, and trial and error. Western medical culture values patient autonomy, practitioner honesty, and open communication, and it accommodates the assessing of the efficacy of medical care (Spector, 1996). In this culture it is believed
that the person seeking medical care should make the health care decisions and has a
right to all the information that is available about the diagnosis and treatment of the
presenting condition. The methodical process of informed consent dictates that the
patient be informed of the exact nature of the illness and the risks and benefits of
treatment, as well as the existence of alternative treatments (Barker, 1992). Western
medicine highly values patient autonomy and truthful communication.

Those involved in the provision of clinical genetic services are deeply committed
to this sense of patient independence and autonomy, as well as to the principle of full
disclosure of information (including ambiguous and uncertain information). Genetic
counseling is based on a patient-centered approach and depends on patients who are
active participants and capable of decision making. We must realize that this premise
may directly conflict with the beliefs and values of other cultures. Although genetic
counselors are committed to providing vital information essential to informed decision making, those receiving it may have no context or practical use for the information being provided (Andrews, 1994). Additionally, the information may directly
challenge their belief and value systems, resulting in confusion and anxiety.

Step 1: Self Awareness

The genetic counselor completely misunderstood the Wahby family's way of coping
with the infant's diagnosis. From her perspective, the "reality" of the diagnosis and its
poor prognosis were paramount issues that needed to be addressed with the family.
She did not appreciate that the family was not seeking to comprehend the scientific or
medical explanations of what had caused the newborn's condition. Thus the counselor
failed to recognize the need to shift from her focus on the explanations for health and
disease. She was perplexed by the new parents' apparent need to deal with many of the
issues by involving a large extended family, whose presence was perceived as making
the care of the infant difficult for the NICU staff. Also perplexing was the family's in-
sistence on sheltering the mother of the child from much of the information the genetic counselor was attempting to provide. Additionally, the counselor privately
recognized that she had little patience with people who voluntarily came to this country and then either did not learn English at all or chose not to speak it in the presence
of those who were ignorant of their primary language. It initially struck her as rude
when the family spoke Arabic during the counseling session, leaving her out of the
discussion. Additionally, she felt that family members were demanding and seemingly
unappreciative of the efforts of the staff and herself. And finally, the one question the
family had asked was directed to a medical student, thus implying that he knew more
about clinical genetics than the board-certified counselor.

Becoming aware of our cultural heritage requires identifying and understanding
attitudes, beliefs, and feelings toward cultural differences, not only on an intellectual,
but also on an emotional level. In particular, we must confront our biases (some authors call this racism), keeping in mind that they can be unintentional, and may even

be present in well-meaning and caring individuals (Pedersen, 1994). This self-examination process can be extremely threatening. If we presume we are free of assumptions and biases, we have already underestimated the impact of our own culture and socialization. If we are to value and respect the differences of others, however, we must learn to recognize our biases, feelings, fears, and any associated guilt. Once we are able to do this, the effects of our values and biases on how we interact with our culturally different patients will start to become apparent, and we can learn to minimize them in our interactions with patients.

The most important part of developing multiculturalism is understanding our own cultural and ethnic heritage (Sue and Sue, 1990; Weil and Mittman, 1993; Pedersen, 1994). Although the genetic counselor in our case example showed an unusual degree of cultural naiveté, taking a look at her European roots may facilitate an objective view of her opinions and value system. Looking at heritage in a historical context may help us understand how and why certain behaviors and values evolved. We might begin this process by taking a social/cultural history of our extended family.

- When did my ancestors immigrate to the United States and for what reason?
- How old were they when they immigrated?
- Did they come alone or with other family members?
- Was the family name changed?
- How long was the native language maintained?
- What was their religious affiliation in their country of origin, and has this religion been maintained?
- Did they marry individuals from within or outside their cultural and religious group?
- What were some of their beliefs about the causes of illness and death?
- Are there family members with disabilities or chronic illness?
- How many of the cultural practices and customs of my ancestors does my family observe today?
- How much contact does my family have with extended family members?
- Have I read reference books, articles, and even novels about my ancestors' country of origin, and the reasons individuals from that country immigrated to the United States?

Step 2: Expanding Our Knowledge

Multiculturalism encompasses an awareness of our own values and attitudes, as well as an understanding of the values and attitudes of those we encounter who are culturally different (Pedersen, 1994). This does not mean that we must share their values, only that it is necessary to appreciate them in a nonjudgmental manner. Such an appreciation begins by knowing something about the cultural groups with whom we interact. Since culture is so significant a factor in determining who an individual is and his or her belief and value systems, it will inevitably affect the way people view reproduction, pregnancy termination, birth defects, presymptomatic status, and chronic disease (Scheper-Hughes, 1990). Culture will influence how a family copes with a genetic condition.

The family life cycle and its cultural rituals play a significant role as a family copes with a genetic condition. It is important to look at how different cultural groups vary in the rituals surrounding, and the emphasis placed on, various family life cycle transitions. The family life cycle includes birth, transition into adulthood, marriage, pregnancy, raising children, divorce, job loss, illness, retirement, and death. As individuals move within the family life cycle, they are inherently put in touch with their cultural roots (McGoldrick et al., 1996). This may be a source of support, but it may, instead, trigger conflict. For example, in Irish culture death is the most important life cycle transition, and as a result the wake is highly celebrated and invites drinking, joking, and storytelling. The traditional Irish view is that the world is full of suffering and death brings a better afterlife. Traditional Italian, Asian, and Polish cultures all emphasize marriage, whereas the Jewish culture generally emphasizes the bar or bat mitzvah, a transition from childhood to adulthood that many western cultures do not recognize.

Let us consider more closely examples of how culture can influence behaviors and attitudes about the family structure, gender roles, marriage, and the role of the children. It is important to remember that the behaviors and attitudes described here generally apply to "traditional" normative behaviors and do not necessarily apply to all individuals in a particular cultural group. In the light of these varied perspectives on health and disease, it is important for the counselor who is collecting medical and family history information to spend some time asking patients how they understand the cause and best treatment of their conditions. Some questions might include:

What do you think caused the problem?

When do you think it started?

What do you feel would be the best treatment?

How has this illness or condition affected you and your life?

How has it affected your family and social community?

What do you fear most about your condition?

Family Structure   The definition of "family" differs from one cultural group to another. Although the dominant Euro-American concept refers to the nuclear family, many other cultures include a larger network of relatives and perhaps even community members. In many non-Euro-American families, the individual is often expected to consider the needs of the family first and those of himself or herself second. In return, the family provides protection, emotional and economic support, and identity. Under these circumstances major decision making is under the control of the family, not the individual.

The African-American family is representative of a defined family that includes the extended kinship (Spector, 1996). Family members often live nearby and rely on one another in times of need. Frequently child care responsibilities are shared among extended family members. It is not unusual for a child to be informally adopted by a family member who is "better off" than the child's parents (McGoldrick et al., 1996).

Many American Indian tribes made no distinctions between blood and nonblood family members (Sue and Sue, 1990). In these tribes there is no word for "in-law." A sister-in-law is thought of as a sister. In contrast, the Greeks recognize definite boundaries between blood relatives and community members not related by blood. Because of these deep feelings about "bloodline," many couples of Greek ancestry are disinclined to choose adoption as an option for having children.

Gender Roles   It is important to recognize the different roles and values different cultures bestow on men and women. In traditional Asian Indian culture, women are generally devalued relative to men (McGoldrick et al., 1996). Females represent a burden for their father and brothers, who must provide a dowry for a future wedding. If a family allows a daughter to pursue an education, it may be with the goal of making her more marketable as a bride. Hindu women were expected to marry early and produce children, especially sons. The birth of a son was highly celebrated and gave a woman a way to gain power economically and spiritually (Fisher, 1996; McGoldrick et al., 1996). The tendency to view women as devalued in traditional Asian Indian culture is also an example of the danger of generalizing any cultural attitude, since in Indira Gandhi India has already had a woman chief of state.

In many cultures, traditional lines of authority are patriarchal: males have authority over females and parents over children. Historically Korean women were expected to obey their fathers before marriage, their husbands during marriage, and their sons in old age. This patriarchal line of authority is changing as Korean women living in the United States become "Americanized" and work outside the household (McGoldrick et al., 1996).

In the African-American culture, families are generally matriarchal, with women regarded as the source of strength and head of the family (Sue and Sue, 1990; Schaefer, 1996). This system was traditional in much of Africa, where women not only raised children, but provided food, organized community events, and were leaders and rulers. In larger African-American families, older female children are expected to fulfill some of the child care responsibilities.

Marriage   In cultures where family well-being has priority over that of the individual, marriages often are arranged to benefit the family. In these cultures, marriage is more an alliance between two families than a romantic union between two people. In traditional Vietnamese families, great effort is made to ensure that both families are compatible based on social status, cultural background, and religious beliefs (McGoldrick et al., 1996). Divorce is very rare in Vietnamese society and is acceptable only if adultery is committed by the woman.

The practice of consanguinity is well accepted in many Middle Eastern cultures (Hoodfar and Teebi, 1996). Historically this stems from geographical isolation, economic benefits (keeping land within the family), and psychosocial benefits (knowing the spouse and his or her family) (Panter-Brick, 1991). The increased risk for unfavorable genetic conditions is seldom a deterrent, since inheritance is not always accepted as a cause of medical problems unless the family history makes it obvious. Also, it is recognized that other family members have married relatives and had normal children.

Over half the people living in the United States today marry outside their ethnic group (McGoldrick et al., 1996). Although this may indicate that Americans are quite tolerant of cultural differences, intermarriage does complicate issues that arise when families cope with dual cultural identities during times of stress. In fact, there is some evidence that couples of different cultural backgrounds have more difficulty adjusting to marriage (McGoldrick et al., 1996).

Role of Children   All cultures highly value children, but their role and socialization differs among cultures. Almost always, children are expected to respect and obey their parents. However, the Euro-American idea of the "democratic" family, where parents tend to involve their children in shared decision making, is a strange concept in much of the world, where adults do not negotiate with, or explain decisions to, children. In many Asian groups, elders are greatly respected, not openly disagreed with, and children are expected to remain silent in their presence (Fisher, 1996).

Sometimes the value and expectations of male and female children are seen differently. Male children are extremely important to Korean families, to carry on the family name (Fisher, 1996). Males are also highly valued in traditional Asian Indian and Middle Eastern families (DeGenova, 1996; Fisher, 1996). The higher value placed on boys is routinely verbalized in these cultures.

Parent–child relationships can become strained if family roles are reversed, as happens when children are asked to function as interpreters during a medical encounter. The "generation gap" often becomes wide for traditional families who have immigrated to the United States as the U.S.-born children become "Americanized."

The phases of the family cycle will vary among different cultural groups. What may be considered age-appropriate behavior in the Euro-American culture may not be acceptable elsewhere. For example, in Mexican-American families early and middle adulthood extends as a longer stage of interdependence between parents and children. This cultural pattern can mistakenly be interpreted by outsiders as overdependence (Falicov, 1995).

Bereavement   As we understand how culture influences the role of the family and its members in their passage through the stages of the family life cycle, we can begin to appreciate how families cope with issues related to bereavement. Every culture has rituals surrounding birth and death. Moreover, the impact of religion and superstitions on a family's coping mechanisms is culturally determined, as are people's views of health and illness.

Meaningful rituals supported by a culture give both family and community a venue for the expression of the deeply felt emotions associated with the loss of a family member (Shapiro, 1996). Each culture approaches the experience of loss and bereavement in ways that are consistent with its beliefs about life, death, and the afterlife. The proper relationship between the dead and the living, managing the intense emotions evoked by loss, resuming family functions, and rebuilding social roles are some of the areas cultural rituals help to address.

Rituals Surrounding Birth and Death   Many rituals practiced at times of birth and death have ancient origins. In earlier times people believed that they had to appease evil spirits and prevent them from interfering with their lives (Spector, 1996). For example, circumcision traditionally is performed to release the child from dangerous supernatural powers in his early life. Once the sacrifice has been made, the child can enter a worldly existence. Orthodox Jews practice circumcision on the eighth day of life. The Moslems of Palestine circumcised their sons on the seventh day in the tradition that Mohammed established.

Traditional rituals are performed to protect the dead, the family of the deceased, and the community from evil spirits. In many cultures bodies are cared for in specific ways (ritual washing) and graves are specifically prepared to facilitate an after-death journey. Other rituals involve appeasing ghosts so that they will not harm surviving family members.

An extreme example from a western perspective is the practice of suttee among members of the Hindu culture who have strong beliefs about karma and destiny: a *satī* (from which the word "suttee" is derived) is a Hindu widow who is willingly cremated along with her dead husband as an indication of her devotion. The underlying belief is that the widow brings good luck to herself, her husband, and the husband's entire family. In Hindu culture, death is seen as an event for both mourning and celebration. Death is the end of one life cycle and the beginning of another (McGoldrick et al., 1996).

In the Jewish faith, death is viewed as the end of life rather than the beginning of another phase. The body is buried usually within 24 hours. "Sitting *shiva*" is a seven-day period of mourning that includes visits to the home by family and friends. A memorial prayer, the *Kaddish,* is said daily until the one-year anniversary of the death. Eleven months following the death marks the end of the mourning period, and the family reconvenes at the cemetery for the dedication, or unveiling, of the tombstone (Fisher, 1996).

In certain cultures, such as the Lebanese, individuals are expected to be very expressive with their emotions in response to death. Emotional outbursts, usually in the form of wailing and crying, are expected at the funeral and are seen as a sign of respect for the deceased. The Lebanese will often stay with a body until the funeral director has arrived—even if this does not happen until the next day (McGoldrick et al., 1996).

Religion and Superstitions   Culture and religion are clearly related, since often one's religion determines one's cultural group, and vice versa. Religion, therefore, plays a vital role in helping individuals and their families cope with a genetic condition. Religious teachings present a meaningful philosophy, and following a religious code will hopefully result in spiritual harmony and physical and spiritual health.

Religious beliefs will affect a person's responses to events relating to health and illness. It is important not to assume that all individuals from the same culture believe in the same religion. Also, individuals within the same religious group may vary in their interpretation of and participation in religious ceremonies and beliefs. Table 7.2 summarizes the beliefs of 12 religious traditions with respect to health-related events (Spector, 1996).

Superstitions are beliefs that are upheld by magic and religion. Believers are convinced that performing an action, wearing a protective charm, or eating a given food will have an influence over life events. The belief in the "evil eye" is a superstition found in many parts of the world, such as southern Europe, the Middle East, and

TABLE 7.2 Selected Religions' Responses to Health Events

Baha'i
"All healing comes from God."

Abortion	Forbidden
Artificial insemination	No specific rule
Autopsy	Acceptable with medical or legal need
Birth control	Can choose family planning method
Blood and blood products	No restrictions for use
Diet	Alcohol and drugs forbidden
Euthanasia	No destruction of life
Healing beliefs	Harmony between religion and science
Healing practices	Pray
Medications	Narcotics with prescription
	No restriction for vaccines
Organ donations	Permitted
Right-to-die issues	Life is unique and precious—do not destroy
Surgical procedures	No restrictions
Visitors	Community members assist and support

Buddhist Churches of America
"To keep the body in good health is a duty—
otherwise we shall not be able to keep our mind strong and clear."

Abortion	Patient's condition determines
Artificial insemination	Acceptable
Autopsy	Matter of individual practice
Birth control	Acceptable
Blood and blood products	No restrictions
Diet	Restricted food combinations
	Extremes must be avoided
Euthanasia	May permit
Healing beliefs	Do not believe in healing through faith
Healing practices	No restrictions
Medications	No restrictions
Organ donations	Considered act of mercy; if hope for recovery, all means may be taken
Right-to-die issues	With hope, all means encouraged
Surgical procedures	Permitted, with extremes avoided
Visitors	Family, community

Roman Catholics
"The prayer of faith shall heal the sick, and the Lord shall raise him up."

Abortion	Prohibited
Artificial insemination	Illicit, even between husband and wife
Autopsy	Permissible
Birth control	Natural means only
Blood and blood products	Permissible
Diet	Use foods in moderation
Euthanasia	Direct life-ending procedures forbidden
Healing beliefs	Many within religious belief system
Healing practices	Sacrament of sick, candles, laying-on of hands
Medications	May be taken if benefits outweigh risks

TABLE 7.2 *(Continued)*

Roman Catholics—*(Continued)*

Organ donations	Justifiable
Right-to-die issues	Obligated to take ordinary, not extraordinary, means to prolong life
Surgical procedures	Most are permissible except abortion and sterilization
Visitors	Family, friends, priest
	Many outreach programs through Church to reach sick

Christian Science

Abortion	Incompatible with faith
Artificial insemination	Unusual
Autopsy	Not usual; individual or family decide
Birth control	Individual judgment
Blood and blood products	Ordinarily not used by members
Diet	No restrictions
	Abstain from alcohol and tobacco, some from tea and coffee
Euthanasia	Contrary to teachings
Healing beliefs	Accepts physical and moral healing
Healing practices	Full-time healing ministers
	Spiritual healing practiced
Medications	None
	Immunizations/vaccines to comply with law
Organ donations	Individual decides
Right-to-die issues	Unlikely to seek medical help to prolong life
Surgical procedures	No medical ones practiced
Visitors	Family, friends, and members of the Christian Science community and healers, Christian Science nurses

Church of Jesus Christ of Latter Day Saints

Abortion	Forbidden
Artificial insemination	Acceptable between husband and wife
Autopsy	Permitted with consent of next of kin
Birth control	Contrary to Mormon belief
Blood and blood products	No restrictions
Diet	Alcohol, tea (except herbal teas), coffee, and tobacco are forbidden
	Fasting (24 hours without food and drink) is required once a month
Euthanasia	Humans must not interfere in God's plan
Healing beliefs	Power of God can bring healing
Healing practices	Anointing with oil, sealing, prayer, laying-on of hands
Medications	No restrictions; may use herbal folk remedies
Organ donations	Permitted
Right-to-die issues	If death inevitable, promote a peaceful and dignified death

(continued)

TABLE 7.2 *(Continued)*

Church of Jesus Christ of Latter Day Saints—(Continued)	
Surgical procedures	Matter of individual choice
Visitors	Church members (elder and sister) family and friends
	The Relief Society helps members

Hinduism	
"Enricher, Healer of disease, be a good friend to us."	
Abortion	No policy exists
Artificial insemination	No restrictions exist but not often practiced
Autopsy	Acceptable
Birth control	All types acceptable
Blood and blood products	Acceptable
Diet	Eating of meat is forbidden
Euthanasia	Not practiced
Healing beliefs	Some believe in faith healing
Healing practices	Traditional faith healing system
Medications	Acceptable
Organ donations	Acceptable
Right-to-die issues	No restrictions
	Death seen as "one more step to nirvana"
Surgical procedures	With an amputation, the loss of limb seen as due to "sins in a previous life"
Visitors	Members of family, community, and priest support

Islam	
"The Lord of the world created me—and when I am sick, He healeth me."	
Abortion	Accepted
Artificial insemination	Permitted between husband and wife
Autopsy	Permitted for medical and legal purposes
Birth control	Acceptable
Blood and blood products	No restrictions
Diet	Pork and alcohol prohibited
Euthanasia	Not acceptable
Healing beliefs	Faith healing generally not acceptable
Healing practices	Some use of herbal remedies and faith healing
Medications	No restrictions
Organ donations	Acceptable
Right-to-die issues	Attempts to shorten life prohibited
Surgical procedures	Most permitted
Visitors	Family and friends provide support

Jehovah's Witnesses	
Abortion	Forbidden
Artificial insemination	Forbidden
Autopsy	Acceptable if required by law
Birth control	Sterilization forbidden
	Other methods individual choice

TABLE 7.2 *(Continued)*

Jehovah's Witnesses—(Continued)

Blood and blood products	Forbidden
Diet	Abstain from tobacco; moderate use of alcohol
Euthanasia	Forbidden
Healing beliefs	Faith healing forbidden
Healing practices	Reading scriptures can comfort the individual and lead to mental and spiritual healing
Medications	Accepted except if derived from blood products
Organ donations	Forbidden
Right-to-die issues	Use of extraordinary means an individual's choice
Surgical procedures	Not opposed, but administration of blood during surgery is strictly prohibited
Visitors	Members of congregation and elders pray for the sick person

Judaism
"O Lord, my God, I cried to Thee for help and Thou has healed me."

Abortion — *Any time /Anyplace*	Therapeutic permitted; some groups accept abortion on demand
Artificial insemination	Permitted
Autopsy	Permitted under certain circumstances; all body parts must be buried together
Birth control	Permissible, except with orthodox Jews
Blood and blood products	Acceptable
Diet	Strict dietary laws followed by many Jews— milk and meat not mixed; predatory fowl, shellfish, and pork products forbidden; kosher products only may be requested
Euthanasia	Prohibited
Healing beliefs	Medical care expected
Healing practices	Prayers for the sick
Medications	No restrictions
Organ donations	Complex issue; some practiced
Right-to-die issues	Right to die with dignity
	If death is inevitable, no new procedures need to be undertaken, but those ongoing must continue
Surgical procedures	Most allowed
Visitors	Family, friends, rabbi, many community services

Mennonite

Abortion	Therapeutic acceptable
Artificial insemination	Individual conscience; husband to wife
Autopsy	Acceptable
Birth control	Acceptable
Blood and blood products	Acceptable
Diet	No specific restrictions

(continued)

TABLE 7.2 *(Continued)*

Mennonite—*(Continued)*

Euthanasia	Not condoned
Healing beliefs	Part of God's work
Healing practices	Prayer and anointing with oil
Medications	No restrictions
Organ donations	Acceptable
Right-to-die issues	Do not believe life must be continued at all cost
Surgical procedures	No restrictions
Visitors	Family, community

Seventh-Day Adventists

Abortion	Therapeutic acceptable
Artificial insemination	Acceptable between husband and wife
Autopsy	Acceptable
Birth control	Individual choice
Blood and blood products	No restrictions
Diet	Encourage vegetarian diet
Euthanasia	Not practiced
Healing beliefs	Divine healing
Healing practices	Anointing with oil and prayer
Medications	No restrictions
	Vaccines acceptable
Organ donations	Acceptable
Right-to-die issues	Follow the ethic of prolonging life
Surgical procedures	No restrictions
	Oppose use of hypnotism
Visitors	Pastor and elders pray and anoint sick person
	Worldwide health system includes hospitals and clinics

Unitarian/Universalist Church

Abortion	Acceptable, therapeutic and on demand
Artificial insemination	Acceptable
Autopsy	Recommended
Birth control	All types acceptable
Blood and blood products	No restrictions
Diet	No restrictions
Euthanasia	Favor nonaction
	May withdraw therapies if death imminent
Healing beliefs	Faith healing; seen as "superstitious"
Healing practices	Use of science to facilitate healing
Medications	No restrictions
Organ donations	Acceptable
Right-to-die issues	Favor the right to die with dignity
Surgical procedures	No restrictions
Visitors	Family, friends, church members

North Africa (Lipson and Meleis, 1983; Spector, 1996). This is a belief that it is possible to project harm, either illness or other misfortune, by gazing or staring at another person. The nature of the evil eye is defined differently by different cultures. However, commonly it is believed that power comes forth from the eye or mouth, suddenly striking the victim. The person casting the evil eye may not be aware of this power, and the victim may not know the source. The harm caused by the evil eye is believed to be preventable or curable by means of rituals or symbolic items.

Health and Illness   Generally it is important for people to understand what illness is, how it comes about, what can prevent it and cure it, and why it affects some people and not others. Culture influences the understanding of health and illness (Strauss, 1990). Illness can be interpreted as punishment for having violated religious codes and morals (Spector, 1996).

Traditional explanations for health and illness are complex and often incorporate a view of the interrelationship of body, mind, and spirit. The *body* includes all physical properties such as sex, age, nutrition, metabolism, and genetics; the *mind* includes all cognitive processes, such as thoughts, memories, and feelings; the *spirit* includes learned spiritual practices, dreams, intuition, and metaphysical forces. This traditional model defines health when body, mind, and spirit are in a state of balance. Illness, on the other hand, is the imbalance of body, mind, and spirit (Spector, 1996).

Methods of treatment and prevention of illness are also influenced by an individual's culture. "Alternative therapies" is a term used in western culture to refer to medical practices not generally accepted by western medicine or practiced in hospital settings. These therapies can include folk medicine (the use of herbs, plants, minerals, and animal substances) and magicoreligious medicine (the use of charms, holy words, and holy actions). These forms of medicine have existed for a long time and are still practiced throughout the world. Alternative therapies may be used exclusively or together with western therapies. Other examples of what by western standards are considered alternative therapies can include acupuncture, massage therapy, chiropractic medicine, imagery, exercise, biofeedback, hypnosis, megavitamin therapy, and spiritual healing (Spector, 1996).

In some traditional Hispanic-American communities good health is thought to be related to good luck or else a gift from God (Spector, 1996). In either case, health is not to be taken for granted, and people are expected to maintain a balance by eating the proper foods and working the proper amount of time. Illness is believed to result from either punishment or an imbalance in the individual's body. Prevention is accomplished through prayer and the wearing of religious medals and protective charms. Treatments often include using certain herbs and spices.

In the Hispanic community individuals may use the services of a folk healer, or curandero, either exclusively or in combination with western health care (Spector, 1996). The curandero is a holistic healer who provides treatment and support for social, physical, and psychological problems. He (female healers are called curanderas) is considered a religious figure and maintains a close personal relationship with patient and family alike. The curandero is a part of his patients' community, speaks their language, and understands their culture.

Traditional health beliefs among African Americans include reliance on a moderate lifestyle as the basis for good health. Protecting the body from cold, keeping it clean, and eating a proper diet are important for health maintenance (Spector, 1996). Elaborate beliefs about the nature of blood and its functions have been described (Snow, 1983). Blood may be spoken of as good or bad, clean or dirty, thick or thin, high or low, and sweet or bitter. The old and the young are thought to have thinner blood, which makes them more vulnerable to illness. Blood is also seen as a vehicle through which germs and other impurities can circulate in the body. The terms "high" and "low" blood designate changes in blood volume and shifts in blood location (e.g., too much blood in the head), which are thought to result from an improper diet. These culturally defined high and low blood conditions are often confused with hypertension.

Illnesses among traditional African Americans are attributed to "natural" or "unnatural" causes (Snow, 1983): natural illnesses result from the effects of cold, dirt, and improper diet; unnatural illnesses result from witchcraft and are thought to reflect social conflict. Behavioral and gastrointestinal illnesses are often considered unnatural. However, any illness can be "unnatural" if it does not respond to ordinary treatment, and thus it may be believed that a "fix" or "hex" was administered in the victim's food or drink. Physicians are felt to be especially unable to deal with unnatural illnesses (Snow, 1983).

Another belief traditionally found among African Americans is that God punishes sin with illness (Fisher, 1996). Such punishment may take the form of an illness that is sudden and dramatic or one that is visible, incurable, and long-lasting. Sick infants and children are sometimes considered to be victims of their parents' misdeeds. Birth defects and mental retardation may be felt to be the result of the mother's bad actions or thoughts during pregnancy. It is also believed that God heals, and religion is very important in the recovery or coping mechanisms.

Cultural Barriers to Medical Care   Many barriers restrict the access of certain minority populations to the health care system; these include poverty, language and nonverbal communication patterns, unrealistic expectations of the health care providers, inaccurate perceptions of disease etiology, lack of transportation, and culturally based perceptions of stigma. If people believe the cause of their illness is different from or more complicated than the explanation provided by western medicine, they probably will not seek the help of western health care providers. If a cultural group stigmatizes a particular disease or condition, individuals are less likely to admit that they, or a family member, have the condition and, therefore, may not seek medical care. This is often the case with mental illness. Also, expectations about the role of the health care provider can differ among individuals from different cultural groups. In the absence of an understanding between the patient and a western health care provider about the patient's expectations from seeking medical care and what constitutes a clinically responsible medical relationship, the patient may not comply with the medical recommendations or return for care.

The cycle of poverty is a substantial barrier preventing individuals from obtaining health care services. Individuals with a lower income often live in deteriorating and overpopulated housing and do not receive adequate nutrition. This results in a lack of preventive care and a higher incidence of illness. Increased illnesses results in less vigorous intellectual and physical development, decreased production in the workforce, and continued lower wages. Individuals who are poor often cannot afford health care or the cost of transportation to medical appointments (Spector, 1996). Individuals whose medical expenses are covered through some form of transfer payment and are able to find transportation to their medical appointments often still have difficulty dealing with a health care system that is not user friendly.

Language and communication, though essential tools for genetic counselors, potentially represent the greatest obstacle in the provision of clinical genetic services to individuals who are culturally different (Punales-Morejon and Penchaszadeh, 1992; Andrews et al., 1994). It is important that we not only *accurately* and *appropriately* communicate our messages to our patients, but also that we *accurately* and *appropriately* receive their messages (Sue and Sue, 1990). Communication includes language, verbal and nonverbal behaviors, and appropriate distance. Even if the same language is spoken by the genetic counselor and the patient, different speech patterns and gestures can hinder communication. Nonverbal forms of communication such as body language, eye contact, tone of voice, and silence are interpreted differently by different cultural groups. The use of language interpreters has limitations, since their communication style also affects the interaction. An interpreter can censor information or render an inaccurate translation. Also, translation adds time to the genetic counseling session. Communication differences, however, can be bridged by the use of carefully selected and competent interpreters (Faust and Drickey, 1986; Fisher, 1992; Haffner, 1992). Table 7.3 gives some suggestions for selecting an interpreter (Fisher, 1996).

TABLE 7.3 Suggestions for Selecting an Interpreter

Choose an individual trained in medical terminology.

If such an individual is not available, select someone who is not a member of the patient's family to avoid family biases or influences.

Choose an interpreter of the same sex as the patient, of comparable age, or older than the patient. Information about sensitive or intimate topics is more readily given to such people. Older individuals signify authority and trust.

Talk with the interpreter beforehand to establish an approach.

Have a quiet, unhurried approach. Take time to seat everyone. Learn to use a few key introductory words in the patient's language.

When speaking or listening, watch the patient and not the interpreter. Use short, simple statements.

Address the patient in the second person (i.e., "you"). Ask how the patient wishes to be addressed.

Do not expect word-for-word translation.

Expect long conversations between the patient and interpreter, but do not accept "no" and "yes" for responses at the conclusion of them. Ask for substantial interpretation.

Ask the interpreter what the patient's expectations, fears, and concerns are.

If the interpreter speaks several languages, inquire of the interpreter whether the patient may hold any animosity for the ethnic group the interpreter may be viewed as representing.

CONSIDERING THE PERSPECTIVE OF THE FAMILY IN THE CASE EXAMPLE

Let us return to the case example and consider it from a cultural perspective, thinking about how the genetic counselor might have more productively directed her interactions with the family members. Although tailored to the specific case outlined earlier, this approach can provide a framework for other cases. So that we can better understand the Wahby family's frame of reference, let us begin by looking at some of the belief and value systems common to Middle Eastern cultures.

Middle Eastern Culture

Middle Easterners include individuals from Arab and non-Arab countries. Arabic-speaking countries include Algeria, Bahrain, Egypt, Iraq, Jordan, Kuwait, Lebanon, Libya, Morocco, Oman, People's Republic of Yemen, Qatar, Saudi Arabia, Syria, Tunisia, United Arab Emirates, and Yemen Arab Republic. Non-Arab countries in the region include Iran and Israel. Iranians are of Indo-European origin and speak Farsi. A great many Middle Easterners practice one of several forms of Islam, but residents also include Christians, Jews, and (mainly in Iran) Zoroastrians and Baha'is. The political turmoil in the Middle East during recent decades has resulted in a large exodus of Arabs and Iranians to the United States.

The Middle Easterners who will be encountered by U.S. genetic counselors differ not only with respect to their country of origin but in other demographics as well: whether they are from an urban or rural area, how long they have been in the United States, and their religion. Individuals from Middle Eastern countries come from a wide range of social and economic backgrounds which usually are related to their educational level (Lipson and Meleis, 1983). With these variations in mind, shared values and behaviors that may affect a Middle Easterners' relationship with a genetic counselor or other health care professions can be considered.

Individuals from the Middle East care deeply about, and are intensely involved with, their family and close friends (DeGenova, 1996). People outside this intimate circle may be viewed skeptically or mistrusted. The family is the paramount resource for coping during times of crisis, such as with illness. The family structure is patriarchal (Lipson and Meleis, 1983; DeGenova, 1996). The family looks up to men before women and to older members before younger for advice and guidance. Immigrants from the Middle East who are in the United States without this critical social network undoubtedly will feel lonely and isolated.

Already we are beginning to understand the context of some of the Wahby family members' behavior. The role of and prolonged presence of the extended family at this time becomes more apparent. Also, it may have been a culturally defined gender assumption that caused the grandfather to ask a question of the male medical student rather than of the female genetic counselor.

Middle Easterners pay attention to much more than verbal messages when trying to make sense out of certain events or circumstances (Lipson and Meleis, 1983). They need to know a lot about another individual before a relationship can develop. They are often offended by the American preference to "get down to business" before a relationship has been established (Lipson and Meleis, 1983). Punctuality is valued differently in the Middle East than in the United States (Lipson and Meleis, 1983). Schedules are quickly readjusted if another, more important matter arises. Thus a person may be late for a medical appointment, or miss it altogether, without notice or apology.

This cultural perspective may partially explain why the family reacted so strongly to discussions about removing the baby from a ventilator. The abruptness with which the topic was introduced might have been offensive, the busy NICU environment notwithstanding. This mind-set may also partially explain why the genetic counselor was unable to establish a satisfactory relationship with the family during one session.

Middle Easterners tend to have great respect for western medicine, although they often retain folk beliefs and practices such as the use of "hot" and "cold" foods (Pliskin, 1992). Mental illness is highly stigmatized in the Middle East, resulting in the avoidance of psychiatrists and other counseling professionals (Lipson and Meleis, 1983). It is commonly believed God controls illness, health, life, and death. Traditionally, its people have found security and peace of mind in the belief that God has a reason for all the events in their lives (Swinford and El-Fouly, 1987; Panter-Brick, 1991).

Preventive and anticipatory care is often neglected in the Middle East, since each person's fate is seen as part of God's master plan. For the same reasons, preparations for birth and death do not occur: a clergyman is seldom summoned to the bedside of a dying person (Lipson and Meleis, 1983). Traditional Middle Easterners feel that an attempt to outmaneuver God is likely to bring on misfortune.

Consideration of this perspective suggests alternative ways for approaching the Wahby family when it became necessary to discuss removing the baby from a respirator. Perhaps acknowledging the cultural perspective just elaborated, and sharing it with the staff, would have eased tensions in the NICU.

The role of the family during illness and death is to act as a clearinghouse of information (Lipson and Meleis, 1983). This is especially critical for information related to a grim diagnosis. For those who believe that only God knows how bad a prognosis is, to give up hope would mean to doubt God's power to help (Lipson and Meleis, 1983; Meleis and Jonsen, 1983). To inform a family member about a bad prognosis is considered at best tactless, and potentially unforgivable, since it could bring disaster to that person. Hope allows the family to cope. Thus every effort is made to shield vulnerable members from unwelcome news.

This information helps us to understand that the extended family members' reluctance to involve either parent in the genetic counseling and their sheltering of the mother from the reality of her son's diagnosis represented an attempt to care for both parents. Additionally, the NICU staff's suggestion of a visit from a Moslem clergyman may have been offensive insofar as it implied that the staff had given up on the child.

The family exerts itself to make sure that a patient receives the best care possible (Lipson and Meleis, 1983). Moreover, family members are expected to show concern by not leaving the bedside and by constantly showering the patient with care and affection. Such assertiveness can make it difficult for western health care professionals to maintain institutional routines and can sometimes interfere with patient care.

Perhaps the neonatologist's suspicion that the family did not respect her judgment was due to the fact that she was a staff physician, not the head or chair of the department, and the Wahbys may have downgraded her accordingly.

Traditional Middle Eastern cultural norms and values may make it difficult for health care professionals to obtain adequate information from patients (Lipson and Meleis, 1983). Such resistance stems partly from a characteristic mistrust of strangers and consequent resistance to disclosing detailed personal information. The practice of obtaining complete family and medical history information is seen as an intrusion unless the relevance of such material to the current medical problem is clearly understood (Meleis and Jonsen, 1983). Often questions are answered in a manner calculated to please the interviewer while absolving the family from responsibility for the illness. Although Middle Easterners greatly respect the advances and technology of western medicine and may attempt to appease an interviewer, they often cannot understand why so many "irrelevant" questions or tests need to be performed to arrive at a diagnosis.

The tendency to regard personal information as irrelevant to a medical diagnosis, coupled with the desire to clear the family of any responsibility for the baby's condition, may explain why family members brushed off the counselor's attempts to obtain a family history and why they quickly stated that both sides of the family were healthy.

Traditional Middle Easterners revere authority figures, neither questioning them nor volunteering information contradictory to what a respected person—such as a physician—has said (Lipson and Meleis, 1983). On the basis of their experience and education, physicians are accorded status as experts. They are expected to make decisions about health care for their patients, and they are held accountable for those decisions. Thus the possibility of options and the concept of informed consent are foreign to many Middle Easterners.

This culturally based belief that it is inappropriate for patients and families to assume responsibilities for medical management could be the source of the grandfather's statement that the family would do "whatever the doctor recommended," with regard to further cytogenetic testing.

The process of obtaining informed consent brings up other troublesome issues for many Middle Easterners (Meleis and Jonsen, 1983). Practitioners of western medicine usually go about acquiring such consent in a professional and succinct manner. Thus the whole process contradicts the traditional Middle Easterner custom of establishing a relationship with someone before getting down to business. Health care professionals who ask for written consent during crises are perceived as cold and calculating, and therefore are to be mistrusted. Additionally, in a culture that considers verbal agreements to be binding, a request for written confirmation of a verbal accord suggests mistrust and puts personal honor at stake (Meleis and Jonsen, 1983).

In Islam, death is considered a normal continuation of the life cycle which should not be feared. On the day of death, the body of the deceased is buried in the ground without embalming. The body is treated with great respect and is not displayed. Cremation is not permitted, and usually autopsies are discouraged, although such procedures may be performed to determine cause of death or to increase scientific knowledge (Swinford and El-Fouly, 1987). Since death is seen as a continuation of life, Islamic teachings encourage family members to resume a normal routine as soon as possible (Bates and Rassam, 1994). In the cases of fetal or neonatal death, the family may see and hold the dead baby. It is important that issues of respect for the remains of infants and others be handled sensitively by health care professionals.

Middle Eastern culture is highly family oriented and children are very important, especially boys. Childlessness is greatly feared, and when it occurs is blamed on the woman. Infertility can ultimately leads to divorce (Young and Lieber, 1987). The fear of having a child with a birth defect is also great and seen as punishment for a previous wrongdoing. The use of prenatal diagnosis is supported by the teachings of Islam, which encourage the pursuit of knowledge and scientific research (Swinford and El-Fouly, 1987). Pregnancy termination is permissible up until 120 days (19 weeks of gestation) because it is believed that until this time the soul has not emerged (Swinford and El-Fouly, 1987; DeGenova, 1996). Birth control is allowed when necessary such as to prevent the birth of children with genetic conditions. Adoption is usually not accepted, since Islamic law states that every child must be called by the biological father's name (Bates and Rassam, 1994). For the same reason, artificial insemination and ovum donation are not allowed (Swinford and El-Fouly, 1987).

Perhaps much of the anxiety, confusion, and frustration felt by the family and health care providers in the case example could have been alleviated if the genetic counselor, the medical student, and the neonatologist had had minimal understanding of some of the practices and beliefs of Middle Eastern culture. Health care professionals often come to know families at the time of a tragic situation. This can result in powerful bonds as each tries to help the other cope with and grow from the situation. When the relationship between the health care provider and the family is strained, further sadness and despair can develop. If culturally different people understand that they have the same expectations even though their behaviors may seem to clash, recognition that everyone is approaching the same goal from different directions is possible. The family and health care providers in the case example had the same goal—to do what was best for the baby and his family. However, because they did not understand each other's values and behaviors, they were not able to work together to reach this goal. Unfortunately, the family in this case refused to return for additional genetic counseling sessions and further follow-up information is not known.

DEVELOPING SKILLS

Skill development ultimately entails developing a respect for each patient's belief system. This allows us to accurately interpret our patients' behavior and to understand their culturally learned coping practices and support networks. When working with families who are culturally different, it is important to keep in mind the values we hold as a genetic counseling profession and how these might clash with the values of others. We must also recognize institutional barriers that prevent families from utilizing clinical genetic services. Work with culturally different families calls for an understanding of alternative family structures, hierarchies, values, and beliefs. We should be knowledgeable about other cultures' communities and their resources. Seeking out traditional healers and religious and spiritual leaders in a given community is sometimes helpful. The use of a translator with cultural knowledge about the family's background as well as an appropriate professional background is another valuable aid. The referral of the family to a bilingual genetic counselor may be appropriate, too, although often this is not possible. We should also be aware of any effects discrimination may have or have had on a patient's community. Lastly, a fundamental skill of multiculturalism is recognizing that there are great *individual* differences within all cultural groups in beliefs, attitudes, and behaviors. Great care must be taken not to generalize and stereotype.

We must integrate into the foundation of our professional practice an appreciation of culture that blends an awareness of history, geography, religion, and politics. This knowledge will allow us to further broaden the scope of our work with families. But first, we must be secure in our own identity, for only then can we act with openness toward people of different ethnic and cultural backgrounds.

REFERENCES

Andrews LB, Fullarton JE, Holtzman NA, Motulsky AG (eds) (1994) *Assessing Genetic Risks: Implications for Health and Social Policy.* Washington, DC: National Academy Press.

Barker JC (1992) Cultural diversity—Changing the context of medical practice. *West J Med* 157:248–254.

Bates D, Rassam A (1994) *Peoples and Culture of the Middle East.* Englewood Cliffs, NJ: Prentice-Hall.

Bureau of the Census, Economics and Statistics Administration (1991) Press releases on 1990 Census data CB91-100 (April), CB91-215 (June). Washington, DC: U.S. Department of Commerce.

Bureau of the Census, Economics and Statistics Administration (1997) Press release on results of research on questions on race and Hispanic origin, CB97-83 (May). Washington, DC: U.S. Department of Commerce.

DeGenova MK (1996) *Families in Cultural Context, Strengths and Challenges in Diversity.* Mountain View, CA: Mayfield Publishing.

Falicov CJ (1995) Training to think culturally: A multidimensional comparative framework. *Fam Process* 34:373–388.

Faust S, Drickey R (1986) Working with interpreters. *J Fam Pract* 22:131–138.

Fisher NL (1992) Ethnocultural approaches to genetics. *Pediatr Clin North Am* 39:55–64.

Fisher NL (1996) *Cultural and Ethnic Diversity, A Guide for Genetic Professionals.* Baltimore: Johns Hopkins University Press.

Gardner RJM, Sutherland GR (1996) *Chromosome Abnormalities and Genetic Counseling.* New York: Oxford University Press.

Greeley, AM (1969) *Why Can't They Be Like Us?* New York: Institute of Human Relations Press.

Greeley AM (1978) *The American Catholic.* New York: Basic Books.

Greeley AM (1981) *The Irish Americans.* New York: Harper & Row.

Haffner L (1992) Translation is not enough—Interpreting in a medical setting. *West J Med* 157:255–259.

Hoodfar E, Teebi AS (1996) Genetic referrals of Middle Eastern origin in a western city: Inbreeding and disease profile. *J Med Genet* 33:212–215.

Lipson JG, Meleis AI (1983) Issues in health care of Middle Eastern patients. *West J Med* 139:854–861.

McGoldrick M, Giordano J, Pearce JK (1996) *Ethnicity and Family Therapy,* 2nd ed. New York: Guilford Press.

Meleis AI, Jonsen AR (1983) Ethical crises and cultural differences. *West J Med* 138:889–893.

Panter-Brick C (1991) Parental responses to consanguinity and genetic disease in Saudi Arabia. *Soc Sci Med* 33:1295–1302.

Pedersen P (1994) *A Handbook for Developing Multicultural Awareness,* 2nd ed. Alexandria, VA: American Counseling Association.

Pliskin KL (1992) Dysphoria and somatization in Iranian culture. *West J Med* 157:295–300.

Punales-Morejon D, Penchaszadeh VB (1992) Psychosocial aspects of genetic counseling: Cross-cultural issues. *Birth Defects Orig Artic Ser* 28:11–15.

Schaefer RT (1996) *Racial and Ethnic Groups,* 6th ed. New York: HarperCollins College Publishers.

Scheper-Hughes N (1990) Difference and danger: The cultural dynamics of childhood stigma, rejection, and rescue. *Cleft Palate J* 27:301–310.

Shapiro ER (1996) Family bereavement and cultural diversity: A social developmental perspective. *Fam Process* 35:313–332.

Snow LF (1983) Traditional health beliefs and practices among lower class black Americans. *West J Med* 139:820–828.

Spector RE (1996) *Cultural Diversity in Health and Illness,* 4th ed. Stamford, CT: Appleton & Lange.

Strauss RP (1990) Culture, health care, and birth defects in the United States: An introduction. *Cleft Palate J* 27:275–278.

Sue SW, Sue S (1990) *Counseling the Culturally Different,* 2nd ed. New York: Wiley.

Swinford AE, El-Fouly MH (1987) Islamic religion and culture: Principles and implications for genetic counseling. *Birth Defects Orig Artic Ser* 23:253–257.

Weil J, Mittman I (1993) A teaching framework for cross-cultural genetic counseling. *J Genet Couns* 2:159–169.

Young M, Lieber C (1987) Vitiligo in an Arab family. *Birth Defects Orig Artic Ser* 23:177–180.

8

A Guide to Case Management

Wendy R. Uhlmann

The phone rings. The person calling is:

- a 22-year-old man whose mother and several relatives have Huntington disease
- an oncologist whose 42-year-old female patient has an extensive family history of cancer
- a 24-year-old pregnant woman who took several medications in a suicide attempt
- a 36-year-old.pregnant woman whose 29-year-old brother has Down syndrome
- a pediatrician whose newborn patient has multiple congenital anomalies
- a 45-year-old man whose 17-year-old son is suspected of having Marfan syndrome.

Case management begins the minute you pick up the phone and continues long after the patient has left your clinic. It encompasses the preparatory work done prior to seeing a patient, the management of the counseling session, and the follow-up initiated subsequent to the clinic visit. The abilities to think critically, assess, anticipate, and communicate are key to successful case management. As the preceding examples illustrate, genetics clinics see patients for a wide variety of indications, requiring that the genetic counselor be able to modify case management accordingly. This chapter focuses on how the genetic counselor prepares to counsel patients, perform a risk assessment, arrange genetic testing, and link patients with support groups. These and

A Guide to Genetic Counseling, Edited by Diane L. Baker, Jane L. Schuette, and Wendy R. Uhlmann.
ISBN 0-471-18541-8 (cloth), 0-471-18867-0 (paper). Copyright © 1998, Wiley-Liss, Inc.

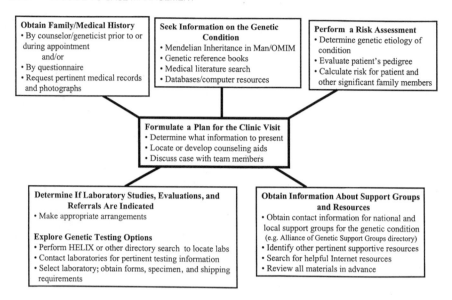

Figure 8.1 *Overview of case preparation.*

other major components of case preparation are depicted in Figure 8.1. The extent of this case preparation is somewhat unique to genetics and is critical for providing accurate information, assessing risks, and appropriately evaluating patients. It is because of the vast number of genetic conditions, combined with the rapid advances being made in molecular genetics and testing, that case preparation is a challenging and important part of a genetics clinic visit.

Managing the clinic visit and coordinating appropriate follow-up are important components of case management for genetics patients, as they are for any specialty. The different aspects of case management (Figure 8.2), including dealing with time constraints, delivering test results, communicating effectively with other team members and specialists, and coordinating appropriate follow-up, are described. Other important aspects of case management, including obtaining a family history, using databases, writing patient and physician letters, and chart documentation are discussed in greater detail in other chapters in this book. The topics in this chapter are presented in a manner that walks you through the steps, much as you would proceed with an actual case. Many details are included because successful case management is all about keeping track of details and balancing several tasks simultaneously.

THE INITIAL INTAKE

Case preparation begins with the initial intake. Your first phone contact with a patient can provide a wealth of information, even if you are just obtaining standard information for scheduling a clinic visit. For example, important insights can be gained from listening to the patient's words and tone of voice, which can provide the beginnings of your psy-

---

Manage the Clinic Visit
- Describe to patient what visit will entail and who will be involved
  - Ascertain patient issues
- Update team members about pertinent information you have obtained
- Keep the session on track
- Be aware of how patient is processing information and adapt counseling and education as indicated
- Provide care within allotted time frame
  - For patients needing more time, determine if session can be extended or whether return visit is needed
- Close session by summarizing key points and recommended follow-up

---

Coordinate Follow-up
- Coordinate referrals, diagnostic evaluations, and testing
- Provide consulting specialists with pertinent information about the genetic condition and patient's medical/family history
- Establish with patient how results of tests/evaluations will be communicated
- Document information and interactions pertaining to case in patient chart
- Obtain and report results
- Send letters to referring physician and patient as indicated

---

Figure 8.2 *Components of case management.*

chosocial assessment and assist in your case preparation. Was the patient anxious? . . . angry? . . . confused? Is the diagnosis a new one, or has the patient been dealing with this genetic condition for a long time? How the patient responds to questions during an intake can also provide information about the patient's level of education and whether there is a language barrier. A patient who seems to have a cognitive impairment may require a longer appointment, and it may be necessary to present information at a more basic level. For a patient who speaks little English, you may need to make arrangements for an interpreter unless he or she will be accompanied by someone who can translate.

Sometimes it is not the patient who calls to schedule the appointment, but a parent, partner, caregiver, or referring physician's office. Since often in genetics the actual patient may in fact be a family unit, it is important to identify the patient and to determine whether additional family members will be coming to the appointment. If other family members are coming, they may also need physical examinations; thus more time will have to be scheduled for the appointment, and a larger room may be necessary. However, depending on the issues to be discussed, it is not always appropriate for all family members to be present in the clinic room. Family members may themselves be seeking information and may have agendas different from that of the patient. Therefore, it may be preferable to see family members at a follow-up visit, and this matter is best determined and communicated prior to the initial appointment.

The most important information to ascertain during the intake is the indication for the clinic visit, which will be used to establish who sees the patient and to help you determine what information to obtain in advance from the patient, the medical literature, genetic testing laboratories, and support groups. Sometimes patients are uncertain about the reason for their referral to genetics or unclear about what the visit will entail. It is important to provide enough information to ensure that patients understand the reason for the appointment without doing counseling over the phone. Usually it is helpful to briefly describe in general terms the typical components of a genetics clinic visit (e.g., taking a family history, performing a risk assessment). The patient also should be reassured that any questions will be addressed during the clinic visit.

Depending on where the genetic counselor works and the indication for the clinic visit, the counselor may not have contact with the patient prior to the appointment. Therefore, the genetic counselor may need to rely on the staff member who was responsible for the intake to communicate pertinent information if the intake was not routine.

OBTAINING FAMILY HISTORY INFORMATION

The patient's family history is essential to genetic counseling. Family histories are usually obtained verbally by the genetic counselor/geneticist prior to or during the clinic visit and/or through a questionnaire that a patient completes. Depending on the indication for the appointment, the genetic counselor may contact a patient by phone to obtain limited family history information or even the entire pedigree. Telephoning a patient in advance can be helpful, but also time-consuming and costly. Before calling a patient, you should determine the minimum information that will be needed in your preparation to counsel the patient. Asking a few "targeted" questions—to identify the specific genetic condition(s), who in the family has the condition(s), and how these individuals are related to the patient—can provide important information for your case preparation. Obtaining a complete family history and constructing a pedigree are discussed in detail in Chapter 2.

Requesting Photographs

Viewing photographs can be a very important component of a genetics clinic visit and can assist in establishing a diagnosis, particularly when dysmorphic features are present. Photographs of the patient at different ages are useful, as the features of a genetic condition can change over time. Requesting photographs of other family members can help in establishing whether a feature that is consistent with a genetic condition (e.g., hypertelorism) is actually a familial trait. Letting patients know that they can simply bring in photo albums may make it easier to fulfill your request for photographs.

OBTAINING MEDICAL RECORDS

Medical records often need to be obtained from the patient and frequently from family members as well. Reviewing medical records is important for confirming a diagnosis, establishing differential diagnoses, and determining what diagnostic studies or evaluations need to be performed. Such a review may be critical for providing an accurate risk assessment, as well. In general, you need to request records that are pertinent to the clinical indication or diagnosis. If medical records cannot be obtained, it may limit the information that can be provided.

If medical records are needed, patients and/or referring health professionals should be informed at the time the appointment is scheduled, to provide sufficient time for sending the records. It is helpful to provide the patient a medical records release form to facilitate obtaining records; such forms are generally available in hospitals and clinics. If you do not have a standard medical records release form, the patient must supply certain specific written information (e.g., name, birthdate, types of records requested, signature consenting to release of records, date) for records to be released. While it is often possible to obtain within minutes medical records requested by fax, up to several weeks may be needed to process such requests, especially if the records are located in off-site storage or on microfilm. Patients can sometimes expedite obtaining medical records by going to the hospital or doctor's office to personally pick up the records.

Obtaining medical records from pertinent family members is not always possible. Authorization for release of medical records needs to be obtained from the individual or next of kin (if deceased), or from a legal guardian. Patients may be reluctant to request medical records from relatives because they do not want it known that they are undergoing evaluation or treatment. Relatives in turn may be reluctant to release their medical histories, preferring to keep such information confidential. Since patients may have gone to great effort to obtain their own or relatives' medical records, it is important to suggest that they retain copies for themselves.

It is helpful to patients to be told specifically what medical records they need to obtain, especially since medical charts can be quite extensive. For example, the 22-year-old man whose mother and other relatives have Huntington disease ideally needs to obtain a neurology report, genetic test results, or other diagnostic medical documentation, since several neurodegenerative conditions have symptoms similar to those of Huntington disease. Obtaining documentation of the diagnosis of Huntington disease in one relative is generally sufficient. However, for the 42-year-old woman with an extensive family history of cancer, it will be important to obtain medical records from several affected relatives to confirm their diagnoses of cancer. The types of cancer, ages of onset, and other factors can make a significant difference in the risk assessment and determination of which gene to potentially screen. When there is a family history of a chromosome abnormality, such as the case of the pregnant woman whose 29-year-old brother has Down syndrome, a copy of the cytogenetics report should be requested. When there is a history of a structural chromosome abnormality or rearrangement (e.g., deletion, translocation) for which a patient or a pregnancy is being evaluated, it is optimal to have a copy of the karyotype in addition to the cytogenetics report for counseling and assessing risks. It is also important to provide these records to the cytogenetics laboratory to focus their analysis, since some chromosome abnormalities and rearrangements are subtle.

Patients being seen in a prenatal genetics clinic because of an abnormal screening test result, ultrasound evaluation, or infectious workup need to bring in reports documenting these findings. Reviewing these reports can significantly impact your counseling; screening test results may be abnormal simply because incorrect information was entered. For patients being seen because of concerns about medications taken during pregnancy, you need to ascertain the names of the medications, the dosages, and when during pregnancy the medications were taken. Determination of teratogenic risk depends on the type of agent and stage of fetal development at the time of exposure. For the 24-year-old woman who took multiple medications in a suicide attempt, the types of medications may not be known. If a patient is not sure of the names of the medications but has some of the actual pills, a pharmacist may be of assistance in establishing their identity.

Information similar to that obtained about medications is also needed from patients concerned about occupational exposures (e.g., to chemicals) during pregnancy. For patients who have had radiologic or other diagnostic studies during pregnancy, it is important to ascertain the types of study, the dates on which studies were performed, the number of films taken, and the radioisotopes used. If determining the radiation dose to the pregnancy requires you to contact a radiologist or nuclear physicist at the hospital that performed the tests, be sure to allow enough time before the patient's appointment to enable such calculations to be done. If the patient has had a stillbirth or neonatal death, an autopsy report can be vital for assessing recurrence risks.

Reviewing medical records can provide important information about what tests have already been performed, reducing the likelihood of unnecessary duplicate studies. This review is also helpful in determining whether additional evaluations need to be scheduled. For example, a patient being seen because of a diagnosis of Marfan syndrome generally needs an echocardiogram and ophthalmology evaluation. Patients being followed for von Hippel–Lindau syndrome regularly require monitoring with MRI and CT scans.

A review of medical records can also make a critical difference in performing a risk assessment, particularly when an assessment is based on relatives' diagnoses. When medical records are reviewed, it often becomes evident that inaccurate medical information about relatives has been passed down through the family. In addition, a patient may have only vague information about a relative (e.g., "My grandmother had a female cancer."). Records that establish whether the cancer in question was breast, ovarian, or uterine could have significant implications for risk assessment and genetic testing.

Medical records can also indicate how the genetic condition has impacted the patient or family member. Looking at the number of hospitalizations, procedures, and tests, and reading through the notes and summaries, can yield a sense of what the course of the genetic condition has been, how long the patient has been dealing with the condition, what the patient has already experienced, and whether the patient has been compliant with past medical recommendations. This information may form the basis for beginning an exploration of psychosocial issues (e.g. "I had a chance to review your medical records and can see that you have really been through a lot."). A notation in the medical records that a patient has not been compliant in the past may influence how you present medical recommendations. Sometimes medical records will allude to the patient's disposition (pleasant, angry, noncommunicative, etc.). It is important not to attempt to predict from these subjective comments how a patient might respond to a genetics clinic visit. Remember that a patient's disposition is largely influenced by what is occurring in their life and depends on the focus of an appointment (e.g. a patient may be angry because they were just informed about a diagnosis).

Therefore, reviewing medical records is often a key component of case preparation and can provide information that is critical for genetic counseling. Other team members may rely on the genetic counselor to make them aware of important medical information. Marking pertinent records with Post-it notes can alert your team members to records they should specifically review. This system also makes it easier for you to return to these records without having to sift through the entire chart each time. Writing out a timeline that summarizes significant dates of hospitalizations, evaluations, and diagnostic tests can also be of help to you and your team members and will be useful in providing the medical summary for dictations.

PREPARING A CASE

The extent to which you prepare a case will depend on the clinical indication and what your team generally prefers to ascertain prior to evaluating a patient in the clinic. Not all the components depicted in Figure 8.1 will be performed with each case, and the preparation needed will largely depend on whether the genetic condition is familiar to you or unknown. If you have seen a patient for the same genetic condition, you may be able to reach into your files or your memory and pull out the information you need. However, given the rapid advances being made in genetics, it is important to make sure your information is current and not to rely on information from your files that may be outdated. In preparing a case, it is also important to balance what is needed with the possibility that the patient will not keep the appointment.

As you prepare a case, try to think about it from the patient's perspective. What questions would you have? What would you want to learn? More than likely, your patients will want to have information provided on the topics listed in Table 8.1. Other

TABLE 8.1   Basic Information to Ascertain for Cases

Genetic etiology of condition
    Mode(s) of inheritance
    Chromosome location(s)
    Molecular genetics
    Penetrance
Incidence and carrier frequency
Clinical features
    Variable expressivity?
    Anticipation?
Age of onset, natural history, and life span
Testing
    Diagnostic
    Carrier
    Presymptomatic
    Prenatal testing
Surveillance, management, and treatment options
Support groups and other resources

TABLE 8.2   General Questions to Guide Your Case Preparation

Given the clinical indication, what questions do you think the patient has?
Why do you think the patient is seeking genetic counseling at this time?
What family and medical history information would be important to obtain?
Are there specific medical records that should be requested from the patient?
    From relatives?
Given the patient's family history and the inheritance of the condition, what risk
    figures would you provide the patient? Relatives?
Is genetic testing available? Is it clinically indicated?
Are there other tests that should be performed? Should the patient be referred to
    other specialists for evaluation?
Is research being conducted that would be an option for the patient to consider?

TABLE 8.3   Basic Information to Address in Prenatal Cases

Risks for chromosome abnormality/birth defects/genetic condition
    Age-associated risks and baseline risks
    Description of common chromosome abnormalities/birth defects
    Causes (e.g., nondisjunction)
Options for prenatal diagnosis
    How procedures are performed (step by step)
    Timing during pregnancy
    Risk for pregnancy loss/other procedural complications
        Baseline risks for pregnancy loss
Limitations of testing
Follow-up testing/evaluations
Time frame for results
Communication of results
Costs

general questions to guide your case preparation are listed in Table 8.2. If prenatal testing is an option, the topics presented in Table 8.3 will need to be addressed. For prenatal genetics patients who have a personal or family history of a genetic condition, you should be prepared to answer the following basic questions:

Is there a risk to the pregnancy?
Are there options for reducing risk?
Is prenatal testing available?

In other words, anticipating patients' questions and preparing accordingly is key to successful case management. As you prepare a case and address anticipated questions, it is important to keep in mind that you must be able to back up your information with reliable medical sources. Helpful resources for locating information essential to your case preparation are discussed below.

SEEKING INFORMATION ON GENETIC CONDITIONS

You have just been scheduled to counsel a patient for a genetic condition you can barely spell, hardly pronounce, and know absolutely nothing about. Genetic counselors often find themselves in this situation, since there are several thousand inherited genetic conditions. The unknown, however, can easily become known if you know where to seek information. Where do you begin? Victor A. McKusick's *Mendelian Inheritance in Man: Catalogs of Human Genes and Genetic Disorders*, which was originally published in 1966 and has since gone through many editions, is considered to be the bible in our profession. It is also accessible on the Internet, where it is regularly updated and known as OMIM (Online Mendelian Inheritance in Man). Within minutes, you can find out about clinical features, inheritance, molecular genetics, diagnosis, and management for the specific genetic condition. The information is summarized from the major articles that have been published on the particular genetic disorder, and provides a historical perspective as well. OMIM has a clinical synopsis section and a "mini-OMIM" summary, which are useful starting points when you need to access information quickly, respond to a telephone inquiry, or determine what to ask when obtaining a targeted family history. The clinical synopsis section which summarizes the clinical features can also be made into a handout for distribution at case conferences. OMIM also provides links to cited articles and to visual images of different genetic conditions.

Smith's Recognizable Patterns of Human Malformation, originally published in 1970, likewise has been revised several times. This resource has photographs and information about genetic conditions including chromosome abnormalities, skeletal dysplasias, storage disorders, connective tissue disorders, and other conditions involving dysmorphic features; the clinical features, natural history, and etiology of the genetic conditions are presented. Two other key resources found in most genetics clinics are *The Metabolic and Molecular Bases of Inherited Disease* and *Emery and Rimoin's Principles and Practice of Medical Genetics*, which contain comprehensive information about a broad range of genetic conditions. Access to a medical dictionary is a must for defining unfamiliar terms for both yourself and your patients. There are also several clinical genetics computer resources and databases, and these are discussed in Chapter 15.

Conducting a Search of the Medical Literature

Conducting a search of the medical literature is an important step in making sure that you have the most current information about a genetic condition. One way to conduct a thorough and efficient search is to access MEDLINE, a comprehensive biomedical literature database maintained and regularly updated by the National Library of Medicine. Commercial software packages are available for searching the MEDLINE database, or it can be accessed by the Internet through PubMed. If you have never used a medical literature search program, you should check with a reference librarian or Internet site for instructions, as there are tips and shortcuts for conducting searches. A search of the medical literature is an important part of case preparation, and it is helpful to share the results of your search and pertinent articles with team members. Often patients are reassured to learn that you have searched the medical literature to obtain the most up-to-date information about their genetic condition, especially in cases of less common conditions.

PERFORMING A RISK ASSESSMENT

Patients often seek information about risks to themselves and their children, and about implications for other relatives. The first step in conducting a risk assessment is to determine the pattern of inheritance for the genetic condition. If the pattern is known, calculating the risk usually is straightforward. If the mode of inheritance is not known, you may find yourself seeking data from empiric studies. In conducting a risk assessment, it is important to take the variables listed in Figure 8.3 into account and to remember to take into consideration that there may be more than one mode of inheritance for the specific genetic condition.

Knowing a patient's age and the typical age of onset for the genetic condition can be useful in determining whether the patient's risk should be increased or decreased. For example, a patient whose father had Huntington disease is at 50% risk for this condition. If the patient turns out to be 65 years old and asymptomatic, however, the risk is significantly less than 50%, since the mean age of onset of Huntington disease is around 40 years (Hayden, 1981). Carrier risk can sometimes be modified by eval-

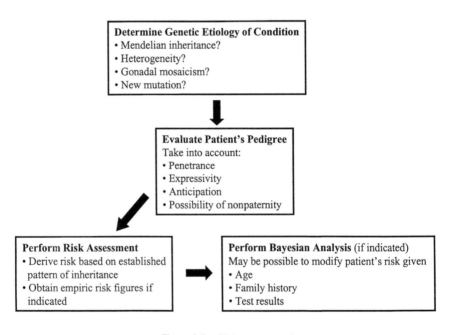

Figure 8.3 *Risk assessment.*

uating the family history and, depending on the condition, taking into account the number and gender of unaffected individuals. Risk can also be modified depending on test results. For example, two-thirds of carriers of Duchenne muscular dystrophy have elevated serum creatine kinase (CK) levels (Thompson et al., 1991). If a patient who is at risk for being a carrier of Duchenne muscular dystrophy has a normal serum creatine kinase level, carrier risk for this condition is reduced. Often a key step in risk assessment is performing a Bayesian analysis, a calculation that takes into account the patient's baseline (a priori) risk and modifies it based on their age, clinical status, test results, or family history. To illustrate how important Bayesian analysis can be in risk assessment, consider the case depicted in Figure 8.4 and described as follows:

> Ann, age 30, comes to see you for preconception genetic counseling. She is concerned about her risk of having a son with Duchenne muscular dystrophy (DMD), since both her uncles, a maternal great-uncle, and a maternal first cousin were affected. DMD is an X-linked recessive condition. Given this family history, Ann's maternal grandmother would be an obligate carrier of DMD, Ann's mother would

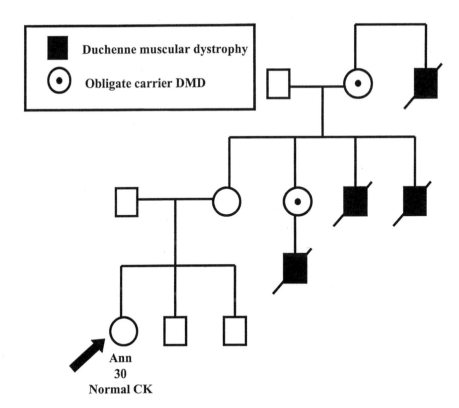

Figure 8.4   *Pedigree of Duchenne muscular dystrophy (DMD) case. By performing a Bayesian analysis, patient's (Ann's) DMD carrier risk can be reduced from 25% to 3.5%. For discussion, see text and Tables 8.4 and 8.5.*

have a 50% risk for being a carrier, and therefore Ann would have a 25% risk for being a carrier of DMD. However, Ann's mother's carrier risk for DMD is actually less than 50% because she has two unaffected sons. By performing a Bayesian analysis (Table 8.4) her risk can be reduced from 50% to 20%; Ann's risk would be half this risk, which is 10%.

It turns out that Ann has had a normal serum creatine kinase test. Performing another Bayesian analysis that includes this modifying information can reduce her carrier risk further, as indicated in Table 8.5.

Therefore Ann's risk of being a DMD carrier has been significantly reduced (from 25% to 3.5%) by performing a Bayesian analysis. Ann's risk of having a son with Duchenne muscular dystrophy in a given pregnancy is less than 1% [3.5% risk that she is a carrier multiplied by 50%, the chance for having a son, multiplied by 50%, the chance a son would receive the X chromosome with the DMD gene mutation = 0.875%].

TABLE 8.4   Bayesian Calculation of Carrier Status for Duchenne Muscular Dystrophy Using Family History Information

Probability	Ann's mother is a carrier	Ann's mother is not a carrier
Prior probability	0.5[a]	0.5[a]
Conditional probability (2 unaffected sons)	0.25[b]	1[b]
Joint probability	0.125	0.5
Posterior probability	0.125/(0.125 + 0.5) = 0.2 = 20%	0.5/(0.5 + 0.125) = 0.8 = 80%

[a]Ann's mother's prior probability of being a carrier of DMD is 50% (50% chance that she inherited the X chromosome with the DMD gene mutation from her mother, who is an obligate carrier = 0.5). Probability that Ann's mother is not a carrier is also 0.5 (50% chance that she inherited the X chromosome without the DMD gene mutation from her mother).

[b]If Ann's mother is a carrier of DMD, each son would have a 50% chance of receiving the X chromosome with the DMD gene mutation and a 50% chance of being unaffected. The probability that both sons would be unaffected if Ann's mother is a carrier is 0.25 [0.5 × 0.5 = 0.25]. If Ann's mother is not a carrier of DMD, she would have a 100% (= 1) chance for having unaffected sons.

TABLE 8.5   Bayesian Calculation of Carrier Status for Duchenne Muscular Dystrophy Using Test Results

Probability	Ann is a carrier	Ann is not a carrier
Prior probability	0.10	0.90
Conditional probability (normal CK test)	0.33[a]	1
Joint probability	0.033	0.90
Posterior probability	0.033/(0.033 + 0.90) = 0.035 = 3.5%	0.90/(0.90 + 0.033) = 0.965 = 96.5%

[a]Two-thirds of DMD carriers have elevated CK levels. Therefore, there is a one-third (or 0.33) chance that Ann could be a carrier of DMD and have a normal CK level.

Performing a Bayesian analysis requires careful thought in terms of the numbers plugged into the calculations, and the reader is referred to genetics reference books to learn the intricacies of this statistical tool.

Providing counseling and risk assessment for patients referred because of a structural chromosome abnormality or rearrangement can be challenging. Any chromosome and any part of a chromosome can be involved. Chromosome rearrangements can occur in a multitude of ways and therefore there may be a limited to nonexistent number of patients with the identical chromosome abnormality or rearrangement reported in the medical literature. While you may be able to locate reports on patients who have similar chromosome abnormalities, through a database or search of the medical literature, it is important to emphasize the limitations of drawing conclusions from a small patient sample. Often you will be able to provide patients only with a range of risk figures that will partially depend on whether the chromosome rearrangement was ascertained because of a history of pregnancy loss, after a child was born with birth defects, during an infertility workup, or after a stillbirth. Risk assessment for chromosome rearrangements is complex and involves taking into account the breakpoints involved, reasoning through the meiotic outcomes, and determining which possibilities might be viable.

The necessity for taking the time to perform an accurate risk assessment cannot be overemphasized. The risk figures you provide could significantly impact the decisions of patients and other family members. You should be able to clearly explain to patients how a risk figure was derived. If you are using empiric studies and find that risk figures differ from study to study, try to determine the reasons for the differences and consider whether the studies have limitations. If more than one risk figure is applicable, this information should be provided to the patient. In summary, make sure that the risk figures you provide are accurate, clearly communicated, and documentable.

COORDINATING GENETIC TESTING

An important part of case management is exploring genetic testing options, evaluating their applicability to your case, presenting the complexities of genetic testing in a manner that facilitates patients' understanding and decision making, and coordinating arrangements for testing. Critical to genetic testing is the counseling that takes place before testing is performed. It is important to keep in mind that the availability of genetic testing does not obligate its use.

The focus of this section will be the coordination of testing for cases requiring DNA or biochemical analysis. Coordination of cytogenetic studies is not included, since most genetics clinics either have a cytogenetics laboratory at their institution or have an established contract with an outside laboratory to perform routine and specialized cytogenetic studies [e.g., fluorescent in situ hybridization (FISH)]. Your institution may also have preferred laboratories to which specimens are sent for more "routine" genetic tests (e.g., cystic fibrosis, fragile X syndrome, sickle cell anemia, Tay–Sachs disease, Huntington disease). However, given the large number of genetic conditions and the limited number of laboratories offering testing for a specific condition, you often will find yourself in search of a laboratory.

Finding a Laboratory

Some laboratories have established reputations for testing for certain genetic conditions that you can learn about through your genetic counselor/geneticist colleagues and by reading the genetics literature. There are printed and computerized directories listing laboratories and the tests that are offered. The most comprehensive and widely used directory at present is HELIX, a computerized directory of national and international laboratories offering testing for genetic conditions. A HELIX search for a specific genetic condition quickly yields a list of diagnostic and research laboratories, as well as contact information, general testing methodology (e.g., linkage analysis, direct mutation analysis, biochemical assay), and indication as to whether prenatal diagnosis is available. The use of HELIX is restricted to health care professionals (including genetic counseling students) and is accessible to registered users by phone, fax, or the Internet.

It is important to stress that directories list laboratories but do not evaluate or guarantee the sensitivity or specificity of the testing performed. Laboratories need to be contacted directly to establish what is offered, as illustrated by the following case:

Jill, age 26, requested genetic testing for adult polycystic kidney disease (APKD) because she was at 50% risk for this condition. Since her affected mother was in need of a kidney transplant, Jill, who was asymptomatic and recently had a normal renal ultrasound, was being considered as a potential donor. Thus she wanted to ascertain whether she had inherited an APKD gene mutation. At the time Jill was seen, there were two known APKD genes: *PKD1* (on chromosome 16) and *PKD2* (on chromosome 4). A directory search yielded a list of seven laboratories offering linkage analysis for APKD, without specifying whether testing was for *PKD1*, *PKD2*, or both genes. One laboratory was eliminated from consideration because it provided testing on a research basis only. Two laboratories were located outside the United States and would be considered only if genetic testing could not be performed in the country, since sending samples internationally is more time-consuming. The four remaining laboratories were contacted and all offered testing for *PKD1*. One laboratory, however, offered testing for *PKD1* and *PKD2* at a fee similar to that charged by the other laboratories for just *PKD1*.

Blood samples from Jill's family members were sent to the laboratory offering linkage analysis for both genes. The linkage analysis results indicated that Jill had a 99% chance of being a carrier of *PKD2* (risk reduced to 94% by factoring in normal renal ultrasound). Clearly, these results had significant implications for Jill, and yet her carrier status for APKD would not have been ascertained if samples had been sent to a laboratory offering testing for *PKD1* only. This case illustrates the importance of keeping up to date about advances in genetic testing, contacting laboratories directly to confirm what testing is actually offered, and selecting a laboratory whose testing will provide the most information for your patient.

What do you do if you have searched all directories and not found a laboratory offering genetic testing for a specific genetic condition? It could be that there is no testing for the genetic condition, or that testing exists on a research basis only. To further explore whether research testing is available, check *Mendelian Inheritance in Man*/OMIM or conduct a search of the medical literature to ascertain the names of researchers currently working on the condition of interest. Presumably, these researchers either will have up-to-date information or will know who might be contacted about testing. Since articles generally include the authors' institutional affiliations, you can locate researchers easily enough.

Time, perseverance, and multiple phone calls to different researchers may be needed to locate a laboratory offering testing on a research basis. Members of the National Society of Genetic Counselors may be able to locate a laboratory by entering queries on the NSGC LISTSERV. If you determine that genetic testing is not available, you will want to document your search efforts and communicate this to the patient and the referring physician.

Arranging Genetic Testing with Research Laboratories

When testing is available for a genetic condition both clinically and on a research basis, it is usually preferable to have the test performed by a clinical laboratory. Research laboratories may not necessarily handle specimens with the same degree of care required by clinical laboratories. If genetic testing is arranged through a research laboratory, it is important to keep in mind that the laboratory's focus is research, and testing performed in this setting may be subject to some limitations. Questions to consider when arranging genetic testing through a research laboratory are presented in Table 8.6.

When genetic testing is performed by a research laboratory, it should be established up front what will be provided to you and the patient. Since the distinctions between testing done on a research and on a clinical basis will not be obvious to all patients, these must be clearly explained. Unlike a clinical laboratory, which strives to complete testing in the shortest time possible, a research laboratory may need sev-

---

TABLE 8.6   Questions to Consider When Arranging Genetic Testing Through a Research Laboratory

What is the anticipated time frame for the analysis?

If a result is obtained, will the laboratory provide you with the result? Will a written report be issued?

Would the research laboratory be willing to share the results with a clinical laboratory?

How long will samples be retained by the laboratory? Will samples be accessible to the patient in the future?

Would the research laboratory be willing to do prenatal testing if requested?

Will the sample be divided and shared with researchers at other laboratories?

Will it be necessary to obtain the approval of an institutional review board for testing to be performed by this laboratory? Are there consent forms for the patient to complete?

Will any costs associated with participating in the research study (e.g., obtaining and shipping samples) be covered by the research laboratory?

eral months or longer to analyze a sample. A result may not be obtained, or may not be fully interpretable given the limitations of the evolving technology. However, some laboratories performing testing on a research basis will make results available in a timely fashion and even provide a written report. Other research laboratories may provide information that a mutation has been identified in the family yet prefer that the finding be shared with a clinical laboratory so that the patient's status can be confirmed and a formal report issued by a clinical testing laboratory. Other laboratories will make it clear that samples are being collected for research purposes only, and that there is no intention of communicating results. This is consistent with most institution review board (IRB) requirements for research studies that test results not be made available to patients until the clinical validity of the test is well established.

The issue of whether samples provided to a research laboratory will be available to the patient in the future for prenatal or other diagnostic testing should be addressed. Particularly if the patient is of reproductive age, it is important to clarify this issue so that the patient does not have unrealistic testing expectations when a pregnancy is under way. A research laboratory may be reluctant to proceed with prenatal testing because of the implications of a patient making a decision about a pregnancy based on testing that has not been approved for clinical use. Therefore, it is important to be sure early on that patients understand the limitations of participation in research studies.

The Logistics of Sending Specimens

In addition to selecting a laboratory, genetic counselors will often be involved in making the appropriate arrangements to send samples for analysis. Unlike other clinical tests, ordering genetic testing is usually more complex than just checking off a box on a form. Table 8.7 lists the general information you will need to obtain from a laboratory prior to sending a sample. A more comprehensive list of questions can be found in the *Molecular Genetics Laboratory Questionnaire* of the National Society of Genetic Counselors (1995). When several laboratories offer testing for the same genetic condition, you are likely to find yourself doing some comparison shopping to select the laboratory that will best meet your patients' needs. Given the rapid evolution of genetic testing, it is important to query even a familiar laboratory about any changes in test methodology, cost, or paperwork since you last sent a specimen.

Many laboratories have a genetic counselor on staff, making it possible to arrange testing with someone who speaks the "same language" you do. If a laboratory is unknown to you, you may want to inquire about its accreditation status and the director's certification (e.g., is the director a certified geneticist?). Some laboratories include lists of their professional staff and scientific advisory board in their information packets.

Most samples for genetic testing will need to be sent by overnight courier at room temperature. Laboratories may provide shipping materials, and some incorporate shipping fees in their genetic testing fees. Laboratories usually need to be notified in advance that a sample is being sent, and a tracking number should be provided in case the shipped sample is not received. Special arrangements may be required for samples sent on Friday, since the laboratory may use a different address for weekend deliveries or may not accept samples on Saturday. Laboratories that run assays on specific days and batch samples for testing may prefer to receive samples only on certain days.

TABLE 8.7   Information to Ascertain When Arranging Genetic Testing

Specific gene(s)/mutations/markers analyzed
Techniques utilized (e.g., PCR, Southern analysis, sequencing)
   Direct analysis (e.g., mutation analysis)
   Indirect analysis (e.g., linkage analysis)
   Functional assays
   Other methods
Accuracy and limitations of results
   Sensitivity, specificity
Specimen requirements
   Specimens accepted
      (e.g., blood, tissue, cultured cells, tissue block, cheek swab)
   Type of tube/flask and amount of specimen required
   Prenatal specimens accepted?
      Control sample, maternal or parental blood samples needed?
   Sample from affected family member required?
Shipping instructions
   Shipping temperature (e.g., room temperature, cold packs, dry ice)
   Address and phone number
   Shipping courier service and account number
   Days of the week specimens accepted
   Laboratory provision of kits to send specimens
Time frame for results
Communication of results
Fees and billing options
   Institution/clinic/physician billed?
   Patient's insurance/patient billed directly by laboratory?
   Full or partial payment required with sample?
   Forms of payment accepted
      (e.g., credit cards, personal or cashier's check)
Paperwork
   Requisition
   Consent form
   Pedigree
   Clinical information/evaluations required prior to testing
Availability of patient/professional literature
Retention of samples upon completion of testing

If an affected family member is known to have had genetic testing, efforts should be made to send samples to the same laboratory and "keep families together." This is important because laboratories generally request a sample from an affected family member to confirm the presence of a detectable mutation prior to testing an at-risk family member. If no affected family members are living, inquire about whether genetic testing on tissue block samples from a deceased relative is an option.

Laboratories may request a copy of the pedigree and a clinical summary for review prior to accepting a sample for genetic testing or at the time the sample is submitted. Certain evaluations may also be required before testing can proceed. For example, patients requesting presymptomatic genetic testing for Huntington disease need to have genetic counseling, a neurologic evaluation, and a psychologic evaluation (International Huntington Association and World Federation of Neurology Research Group on Huntington's Chorea, 1994).

A requisition and a consent form signed by the patient will likely need to accompany a sample submitted for genetic testing. Copies of all paperwork sent with a sample should be kept in the patient's chart. Since genetic testing can be costly and insurance reimbursement limited, laboratories may require full or partial payment with the specimen. Arranging genetic testing for patients with Medicaid or Medicare coverage can pose challenges because some laboratories will accept this form of payment only from patients who reside in the same state.

Arranging Genetic Testing in Cases Involving Linkage Analysis

Lisa, age 25, requests genetic testing for hemophilia A (factor VIII deficiency), an X-linked recessive condition. Her two maternal uncles and maternal great-uncle died of hemophilia A, confirmed by medical records. Lisa's unaffected brother lives in California, their parents live in Florida, and Lisa's maternal grandmother, who has terminal cancer, is in a hospital in Oregon. Lisa's maternal grandmother, an obligate carrier, had direct genetic testing for the common inversion mutation that is seen in about 45% of severe hemophiliacs (Antonarakis et al., 1995) and does not carry this mutation (Figure 8.5). Can linkage analysis be performed? From whom should blood samples be obtained? What if a family member refuses to provide a blood sample? How do you go about obtaining blood samples from family members in different states? Who will be billed for the testing?

Linkage analysis involves tracking markers in close proximity to or within a disease gene and is generally performed when the disease-associated mutation(s) are unknown. Blood samples are needed from specific family members to establish by linkage analysis whether the disease gene has been inherited. In the case described above, there is no living relative with hemophilia, but linkage analysis can still be performed to try to establish Lisa's carrier status. You could first obtain blood samples from Lisa, her mother and her maternal grandmother to determine whether the X-chromosome Lisa received from her mother came from her maternal grandmother or her maternal grandfather. If Lisa has inherited her maternal grandfather's X chromosome, she is not at increased risk for having a son with hemophilia. If she has inherited her maternal grandmother's X chromosome, then it would need to be determined whether it was the X chromosome associated with hemophilia. Obtaining a blood sample from Lisa's unaffected brother would be helpful in determining whether Lisa has the same X chromosome as her unaffected brother (i.e., the X chromosome without the hemophilia mutation). Depending on the results, additional samples from Lisa's father and maternal grandfather may be needed. It is also possible that even after all these samples have been obtained, the disease-associated chromosome and Lisa's carrier status for hemophilia would not be determined because of the limitations inherent in linkage analysis (e.g., informativeness of markers, recombination).

Coordination of linkage analysis cases involves careful thought about who should be asked to provide samples; complex arrangements for sending blood samples from

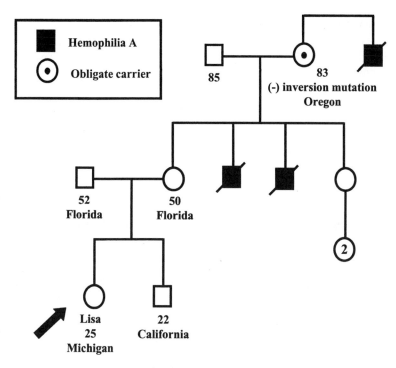

Figure 8.5 Pedigree of Hemophilia case.

family members residing in different states are sometimes necessary. If obtaining blood samples cannot be coordinated through a local genetics center, you may be able to request the testing laboratory to send test kits and pertinent paperwork to family members' physicians.

Discussing the case with the laboratory in advance is usually necessary and helps in determining the minimum number of family members needed to provide optimal information, while at the same time minimizing cost. Budgetary considerations are particularly important because even though linkage analysis involves obtaining blood samples from several family members, the cost is often borne by the individual requesting the testing. Moreover, even though a family member's sample may be critical for linkage analysis, insurance generally will not cover the cost unless clinically relevant information about that specific family member (who is not the patient) is being obtained.

It is important that patients proceeding with linkage analysis understand that optimal interpretation of the results will depend on family members being biologically related to the patient and that information about their clinical status must be accurate. The sensitive issue of potentially discovering nonpaternity through the testing needs to be addressed before linkage analysis is undertaken and similarly should be raised in direct mutation analysis cases involving multiple family members. You may need to find a way to make private inquiries about paternity with pertinent family members submitting samples for testing, since nonpaternity is unlikely to be acknowledged in the presence of other family members.

Linkage analysis often takes several weeks. If a patient is considering a pregnancy, it is best to initiate linkage analysis studies prior to conception because of the time involved in obtaining samples from family members and performing the analysis. Obtaining samples in a stepwise fashion to minimize cost is also difficult with a pregnancy underway. In addition, as mentioned above, it is possible that linkage analysis will not yield a result, in which case it will not be possible to test the pregnancy. Another consideration is the possibility that key family members needed for linkage analysis studies will not wish to participate, and such refusals may prevent the patient from proceeding with linkage analysis studies.

Arranging Genetic Testing for Prenatal Samples: Special Considerations

Making arrangements for genetic testing on prenatal samples requires special attention because the results could have significant implications for the pregnancy and because time limitations are involved. Often, in these cases, patients will be having cytogenetic studies performed to rule out a chromosome abnormality in addition to genetic testing for a specific condition. Even if the cytogenetic studies can be done through your institution, the laboratory performing the genetic testing may prefer to receive the entire sample to ensure that the sample size is adequate to complete both analyses. Depending on the genetic condition and type of testing, laboratories may also request a maternal blood sample (to rule out the possibility of maternal cell contamination of a prenatal specimen), one or more control samples, or samples from both parents (to confirm carrier status).

Clear communication with the laboratory is key: there must be precise instructions about how the specimen should be handled, and the testing priorities must be clearly defined. For example, if genetic testing for hemophilia is being performed, the fetal sex needs to be rapidly determined, since genetic testing will not be clinically indicated prenatally if the fetus is female. In such a case, cytogenetic studies generally are performed first, followed by molecular testing. In contrast, if genetic testing for cystic fibrosis is being performed, the DNA analysis for cystic fibrosis takes priority over the cytogenetic studies (since fetal sex is not a testing consideration, and the risk for having a child with cystic fibrosis exceeds the risk for having a child with a chromosome abnormality).

DNA Banking

DNA banking is the long-term storage of an individual's DNA. Genetic testing often relies on identifying a mutation in an affected family member or, if linkage analysis is being performed, obtaining blood samples from several family members. Banking samples on key family members who may not be alive when testing becomes feasible ensures that at-risk family members will have the option of testing in the future. DNA banking is an option that should be discussed when genetic testing is unavailable or is available, but limited in its detection rate and/or costly. It is particularly important to raise the option of DNA banking when samples from multiple family members are being provided to a research laboratory, since the research samples may later be inaccessible to family members. When making arrangements for DNA banking, it is important to help patients carefully specify and document to whom family members' samples can be released for future testing. DNA banking is generally not covered by insurance.

Other Considerations Associated with Genetic Testing

The limitations and implications of genetic testing need to be thoroughly discussed with patients. It is not unusual for patients to assume that because they provided a blood sample, a result will be obtained. If the genetic test is new and has a limited detection rate, consideration should be given to delaying testing until it becomes more accurate and less costly. If genetic testing does not yield a result or testing is not available, the importance of keeping in touch with a genetics center to learn about advances in testing should be emphasized.

Patients should be fully informed about genetic testing fees, since insurance may not cover the costs and the patient would be held financially responsible. At present, presymptomatic testing often is not covered by insurance, since the testing is not being performed to diagnose a current medical condition but rather one that might develop in the future. The possible impact on insurance coverage of genetic test results, particularly presymptomatic test results, is increasingly an important issue to address prior to testing.

FINDING SUPPORT GROUPS AND PATIENT RESOURCES

Given the many issues associated with a genetic condition, an important part of genetic counseling is helping patients to realize that they are not alone and providing them with information about and access to support groups and other resources. Therefore, you will need to become knowledgeable about resources at your center and those at the state and national levels. For example, knowing about special education programs in your state is critical if you work in a pediatric setting and see patients with developmental delay.

Support groups provide patients with the opportunity to meet other individuals who are dealing with similar issues and facing similar challenges. Often support group members can supply much information that is not available through any other route (e.g., where to find needed services and resources specific to the genetic condition). Support groups are not geared solely toward the individual with the genetic condition but often serve other family members and caregivers as well.

Locating national and local support groups is relatively easy thanks to directories, databases, the Internet, and your local phone book. A good starting point is the *Directory of National Genetic Voluntary Organizations and Related Resources*, originally published by the Alliance of Genetic Support Groups in 1989 and subsequently updated. It is also available on the Internet. The Alliance of Genetic Support Groups, established in 1986, is a national coalition of voluntary genetic organizations, consumers, and professionals; it serves as a bridge between individuals affected by genetic conditions and health care providers. The directory lists support groups and contact information, along with helpful information on services and educational resources available through the group. Another useful resource is the National Organization for Rare Disorders (NORD), an umbrella organization that provides educational materials, maintains a computer-accessible database for patients and professionals, and links patients with others who have the same rare disorder.

Support groups often have both patient and professional packets of information. Genetics clinics often purchase booklets for specific genetic conditions commonly seen in their clinics, but providing materials on less commonly seen genetic conditions is not always financially or logistically feasible. Even if you cannot provide patients with actual brochures, you can make a handout for patients indicating the contact information for the national support group and local chapter, if one exists. The letter sent to patients and referring physicians following a clinic visit should also include this information.

While locating support groups is relatively straightforward, determining whether the support group's focus matches the patient's needs can take some thought. For example, a patient who electively terminated a pregnancy diagnosed with a specific condition may not feel comfortable attending a support group for patients with the specific condition or a group for women who experienced spontaneous pregnancy loss. An asymptomatic patient who carries the gene for Huntington disease may feel uncomfortable attending a support group for symptomatic individuals. It is helpful to not only provide patients with the support group information but to bring up these related considerations as well.

You should take the time to become familiar with a support group's publications prior to providing them to patients. Support groups often have a variety of publications addressing different aspects of a genetic condition and intended for different audiences. Some support groups even offer publications geared toward children, teenagers, or siblings. Therefore, as you read a brochure, think about its intended audience. Keep in mind that a very well written brochure may not be appropriate for a specific patient. Care should be taken with patients who have a tentative diagnosis: if they are given brochures, they may perceive the literature itself as confirmation of the diagnosis. Questions to consider as you evaluate brochures are listed in Table 8.8.

If you have been unsuccessful in locating a support group through the resources discussed above and are a member of NSGC, you may consider putting an inquiry on the NSGC LISTSERV. Genetic counselors have also used the NSGC LISTSERV to establish if other genetic counselors are aware of individuals with a specific genetic condition who might be available to talk with other affected individuals.

Internet Resources

Patients often arrive at a clinic fortified with information about a genetic condition obtained from Internet searches. Many support groups now make their publications available on the Internet. Patients may welcome learning about a genetic condition through the Internet because of its accessibility and the ability to remain anonymous. For some genetic conditions, there are Internet sites that permit communication with other individuals dealing with the genetic condition, and questions addressed to some sites will

TABLE 8.8   Evaluating Support Group Literature

Who is the intended audience for the brochure? (e.g., patients, family members, professionals)
Would the brochure be appropriate for
   A patient with a tentative diagnosis?
   A patient with an abnormal pregnancy?
   A teenager?
Is the information accurate and up to date?
Is the information clearly presented? Are there helpful visual aids?
Is the content balanced? Too optimistic? Too discouraging?
What is the readability of the brochure?
   Is the writing technical?
   Are definitions of medical terms provided?
   Is the language appropriate for a patient with limited education or a cognitive impairment?
How is the patient likely to feel after reading this brochure?

---

receive a response from a physician or researcher. It is important to point out, however, that not all information posted on the Internet is accurate: postings come from medical and nonmedical sources, some of which are unreliable. As you review information on a site, consider the source, find out whether there is a reputable organization or scientific advisory board overseeing the site, and review the information, much as you would look over published information, before providing it to patients. Additional information about evaluating computer resources is presented in Chapter 15.

FORMULATING A PLAN FOR THE CLINIC VISIT

As you prepare for a case, it is important to determine what to present, the main points to emphasize, how information should be presented to facilitate the patient's understanding, and what visual aids can assist in this process. Care should be taken not to overwhelm the patient with information, but sufficient information must be presented to allow a clear understanding of the bottom line. In short, a key part of case preparation is envisioning how the session will unfold and planning accordingly. Initially, it may be helpful to prepare a brief outline of your main points and the order in which you anticipate presenting the information. Such an outline will help you to formulate a "game plan," but it is not to be seen as a rigid framework for the genetic counseling session. You have to be flexible enough to "go with the flow" of a session, while at the same time being sure to cover the necessary information. In addition, you need to be able to respond to both verbal and nonverbal cues to how the patient is assimilating the information and adapt the counseling as indicated.

Locating and Developing Counseling Aids

An important part of case preparation is thinking about how you can make it easier for the patient to comprehend and retain the information. Diagrams and pictures, for example, help make abstract concepts seem more concrete. In genetics clinics, you will generally find a copy of *Counseling Aids for Geneticists*, originally published by the Greenwood Genetic Center in 1984 and subsequently updated. It contains karyotype examples of common chromosome abnormalities and diagrams of different patterns of inheritance, nondisjunction, segregation of translocations and inversions, prenatal diagnostic procedures, molecular testing methodologies, and actual case examples; essentially, it provides a set of visual images to accompany your explanation of basic genetic concepts.

Rather than relying on published counseling aids, you also should think about creating your own. Showing a patient a diagram illustrating a pattern of inheritance and then reviewing the person's own pedigree can help the patient see more clearly how a condition is inherited in his or her family and how risk figures are derived. It is instructive to color-code the results of laboratory reports to assist the patient in understanding what has been inherited from each parent, and to indicate whether the disease-associated mutation has been transmitted, particularly in linkage analysis testing and other cases requiring samples from family members. When a chromosome abnormality is involved, providing the patient with a copy of the actual karyotype can enhance their understanding of a written report, which generally contains unfamiliar terminology. For chromosome rearrangements, deletions, or duplications, ideograms of the chromosomes obtained from genetics reference books or on the Internet can help the patient see what may not be evident from looking at a karyotype. Other helpful examples of educational approaches and counseling aids that you can develop are presented in Chapter 5.

Communicating with Colleagues/Negotiating Your Role in a Case

Genetic counselors often see patients as part of a team. It is therefore important to determine your role in advance and to know who will be responsible for addressing specific aspects of a case. Such preparation makes for smoother transitions in a session and less duplication of information. If a physical examination is to be performed, it is important to establish what will be addressed prior to the exam and what counseling will be deferred until afterward. Risk figures and other pertinent information should be reviewed in advance with team members to make sure that counseling provided will be consistent. It is already a challenge for patients to understand all the information being presented without having to deal with discrepancies in the information provided by different team members.

MANAGING A SESSION

An important aspect of case management is letting patients know at the start of the session:

- who they will be seeing (names and roles)
- what will happen during the clinic visit
- whether a physical examination will be performed

- whether documentation of the visit will be provided to the referring physician and/or the patient
- time limitations for the visit

Successful case management depends on being able to integrate the issues the patient wants addressed with the established components of a genetics clinic visit. At the beginning of the session, the genetic counselor/geneticist should ascertain from the patient his or her primary issues, concerns, and expectations for the clinic visit. This information could significantly impact the focus of the session and alter the approach to the genetics clinic visit. When diagnosis is the focus, it is important to let patients know at the outset that additional evaluations and follow-up visits may be needed and that it is possible that a diagnosis will not be made.

Managing the flow of a session is also an important part of case management and is addressed in the section that follows. If you are independently seeing a patient before being joined by other team members and have obtained information the others should know, you can sometimes convey this information while the patient is present (e.g., "Ms. Brown was telling me about her son Jason's recent surgeries.") If issues have come up that you feel are best communicated to your team without the patient present, however, you will need to initiate a break in the counseling session. This can be accomplished by telling the patient that you would like to go over all the information that he or she has provided with your team in the staff-room. This effectively communicates to the other team members that you have important information to share and lets the patient know what you are doing when you leave the room. Managing a case also entails bringing a session to a proper close, reiterating key points, and making sure that the recommendations and plan for follow-up are clearly communicated.

Handling Time Constraints

Your patient's bladder is full, she has a scheduled appointment for prenatal testing, and you still need to discuss the genetic conditions you just learned about when taking her family history.

The patient scheduled to see you at 10:00 A.M. has arrived at 11:00 A.M.

Your patient keeps bringing up new issues and you have yet to complete the family history, but unless a geneticist comes in to perform the physical exam, the whole morning's schedule will be off.

Your patient is confused and is having a difficult time understanding what you are saying.

Your patient is crying.

You have no time left but clearly there are issues that need to be addressed.

What do you do when the hourglass timer has emptied, and you do not have the option of turning it over to "gain" more time?

As these examples illustrate, any number of events can take a genetic counseling session down an unanticipated path. However, you must keep mindful of the time available and the possible impact of extending the session on the rest of the day's clinic schedule. When it becomes evident that more time will be needed for the patient to understand the information or have all questions and concerns addressed, you will need to do some triaging. Is it possible, given your clinic schedule, to extend the counseling session? Is another clinic appointment needed, or is it possible to meet with the patient again that same day? For example, if a patient is having prenatal testing and the information remaining to be discussed has no bearing on this decision, you can ask the patient to return to your office following the prenatal testing. While not optimal, if a session needs to be extended, particularly if there are significant issues remaining to be addressed, you may be able to have another team member see your next scheduled patient, or delay seeing that patient.

It is important to recognize that you need to have sufficient time available to effectively provide genetic counseling. In addition, patients can take in only so much information at one time. If a follow-up appointment needs to be scheduled, because a session cannot be extended, clearly communicate this to the patient: "Certain issues have come up in reviewing your family history (in meeting with you today, etc.) that we unfortunately do not have time to address. We feel these issues and your questions are important, and that is why we would like to see you back in clinic so that we have sufficient time to address them." Such an approach would be appropriate in the following case:

> You are meeting with a new patient who has recently been diagnosed with von Hippel–Lindau syndrome. While taking his family history, you learn that he has a family history of Huntington disease and thalassemia. How would you handle this session, and what would you address?

It would be appropriate here to focus on the diagnosis of von Hippel–Lindau syndrome and defer counseling about Huntington disease and thalassemia to another clinic visit. It should be clearly documented that these genetic conditions were identified and that counseling was recommended for a subsequent clinic visit. Disappointment over not having everything addressed at a single visit can be minimized if patients are informed about this possibility at the time the appointment is scheduled and reminded of the time limitations at the start of the session. As genetic testing becomes available for more conditions, the issues of "information overload" and how much can be accomplished in a single clinic visit will become even more pressing.

Dealing with time constraints is also an issue when patients need to make certain decisions—for example, whether to proceed with prenatal testing, whether to continue or terminate a pregnancy. If, given the gestational age of the pregnancy, a decision has to be made on the day of the patient's appointment, there are ways to facilitate that decision making. In addition to helping patients focus on how they have made decisions in the past and how they feel about different decision options, it is helpful to provide an environment that is conducive for arriving at a decision that best meets her needs. Locating a quiet space to talk or suggesting that the patient take a break and come back can help the patient feel in control of making a decision despite time constraints.

COMMUNICATING RESULTS

It is 4:00 P.M. on a Friday. The cytogenetics laboratory has just informed you that Mrs. Green's amniocentesis results indicate a 47,XX,+21 female karyotype, consistent with the diagnosis of Down syndrome. Should you call the patient immediately or wait? What if this result was obtained from chorionic villus sampling? Would this make a difference in when to call? What if these results had arrived at the same time on a Tuesday? Should you call the patient at work or call her at home in the evening? When do you call, and what should you say?

There should always be a plan for communicating test results, as the scenario above illustrates. At the clinic visit, you should discuss how normal and abnormal results will be communicated, whether the patient will be returning to clinic to receive results, and if results can be phoned to the patient at work. The results of genetic testing may be confusing to patients, and subject to misinterpretation, which is why it is optimal to meet in person to discuss the results. Test results may alter risk assessment and may have significant implications for the patient and family members. Patients with chronic conditions commonly have already undergone extensive workups and lived through years of speculation. Test results may replace this living with uncertainty or the lack of an explanation with the reality of a diagnosis, to which the patient now needs to adjust.

Before providing results to a patient, you should review the report to make sure it is accurate. Do you have the report for the right patient, and is the patient's name spelled correctly? Was the specific test you or the referring physician requested actually performed? Is the interpretation consistent with the patient's test result? Are there any errors in the report? Do you need to contact the laboratory to clarify any part of the report? In other words, don't look at the test results for the first time when you are providing them to the patient. Especially if you are giving the patient a copy of the results, it is important to make sure that you understand the content of the report before conveying it to the patient. Try to anticipate ways of facilitating the patient's understanding of the results. Some of the suggestions presented in the counseling aids section of this chapter may be helpful. A letter clearly documenting and explaining genetic test results should be sent to the patient and the referring physician, especially if the patient is unable to return to clinic to receive the result.

Presymptomatic testing protocols, such as the one established for Huntington disease, require that patients return to clinic to receive their test results (International Huntington Association and World Federation of Neurology Research Group on Huntington's Chorea, 1994). Many clinics offering genetic testing for cancer or other adult-onset genetic conditions have similarly adopted this requirement. Scheduling a return clinic visit to receive test results means that the patient will be prepared to expect results, "good" or "bad," on a specific day. While one might be tempted to give normal results by phone, research has shown that patients receiving "good" news, such as negative presymptomatic Huntington disease test results, can have an adverse reaction: they may feel guilty that other family members are affected, or they may regret decisions made earlier based on the assumption that they would eventually be affected (Huggins et al., 1992).

It is important to keep in mind the timing of communicating a result. For example, if you receive an abnormal prenatal result on a Friday afternoon, you may consider waiting to phone the patient until Sunday night or Monday, at which time you can offer an immediate appointment. Some patients find it difficult to learn of an abnormal result and then have to wait through the weekend before having a chance to meet and discuss the results. Other patients benefit from processing the information over the weekend.

Results may be communicated by phone when genetic testing is performed to confirm an already well-established diagnosis. Because of the time frame involved and implications for a pregnancy, prenatal results are typically communicated by phone, with a follow-up clinic appointment scheduled if results are abnormal. When prenatal testing involves both cytogenetic studies and genetic testing for a specific condition, you may consider delaying communication of results until all studies are completed. Otherwise, a patient might be told that she is having a chromosomally normal boy or girl, only to learn a couple of days later, when the genetic testing is completed, that the fetus is affected with a genetic condition. When prenatal results are abnormal, it is important to discuss their significance with the patient as soon as possible. This is also important to enable a timely decision to be made about continuing or terminating the pregnancy. You should keep the call communicating the results brief and supportive, without attempting to provide the counseling you would initiate in person.

If the patient has requested that results be phoned to a work number, it is important to also establish how the patient will be located and how the call should be identified. It may be difficult for the patient to predict work circumstances in advance, so be prepared to listen for cues: if there is hesitation in the patient's voice, suggest calling back at a better time or inquire whether he or she can call you back from a more private location.

Patients should be informed about the normal time frame for receiving results. If a patient has been told to expect a result within a certain time period and the laboratory notifies you of a delay, generally it is best to share this information with the patient, who might worry if the specified waiting passes without word from you. Since patients may assume that a delay in results means that the result is abnormal, it is important to explain the reason for the delay. One can also anticipate delays around holidays, when laboratories may be closed or have fewer workers, and so inform patients.

COMMUNICATING WITH OTHER SPECIALISTS AND RESEARCHERS

Patients who are diagnosed or are being seen to rule out a genetic condition will often need referrals to other specialists (cardiologists, neurosurgeons, etc.) for evaluation. The genetic counselor is often an important link between the patient, the genetics clinic, and other specialists, making sure that pertinent information is communicated and follow-up initiated. Sometimes, it is necessary to educate specialists about the specific genetic condition. For example, if you suspect a genetic condition in a pregnancy for which there are associated ultrasound abnormalities, sharing this information with the ultrasound staff can help them do a more focused assessment. If an autopsy is being performed and you suspect a specific genetic condition, providing the pathologist with information, or a genetic reference book that clearly describes the clinical features, will be particularly helpful. Providing specialists with pertinent test results, records, and information about the genetic condition enhances patient care.

Letting patients know that there are different counseling resources available to them is important. It is not unusual for issues to arise in genetic counseling sessions that are better addressed by a therapist trained in long-term therapy. Although genetic counselors are certainly able to address psychosocial issues, most do not work in a setting where they can meet with a patient regularly to provide ongoing counseling. If your hospital has clergy and therapists on staff, contact them to ascertain their availability to meet with your patients. For example, a patient terminating a pregnancy may benefit from addressing religious concerns with hospital clergy. In addition to suggesting that patients can obtain recommendations for therapists from their friends, clergy, or physician, therapists on staff at your hospital may be helpful in identifying practitioners in other communities.

Genetic counselors and support groups both play a major role in providing patients with information about research studies. Researchers are often looking for patients to include in their studies, and patients are often interested in contributing to research in the hopes that advances can be made in understanding and treating a particular genetic condition. Patient involvement in research can be as simple as consenting to be listed in a registry of individuals with the genetic condition, or completing surveys. Other research may entail undergoing evaluations or providing blood or other samples from themselves and family members, along with medical record documentation. It is important that patients be informed about research in a manner that does not pressure them to participate. Since the genetic counselor will often be the direct link between the patient and the researcher, it is important that you take the time to understand the focus of the research, what the patient is being asked to provide, and what will be provided to the patient.

CASE DOCUMENTATION

As you prepare a case, obtain information from patients, coordinate testing, and follow up, it is important to document what you have done so that any member of your team reviewing that chart will know the status of the case. In other words, you should clearly document what you would want other members of your team to know about the patient and make it possible for them to have sufficient information to manage the case, especially in your absence. Recording names and phone numbers of individuals you have contacted as well as the dates is important and will make it easier for you or others to follow up if there is a question as to what has transpired or if information you requested is not received.

Considering all there is to do for a single case, it goes without saying that you will need to develop an efficient system to keep track of the status of your "active" cases. A tracking system is important because genetic studies often take a month or longer to complete, by which time you will already be involved in counseling other patients. While you may wish to keep in your office charts of patients whose cases are ongoing, as a visual prompt that something remains to be done, this will generally not be possible or desirable, given the potential for office clutter! One alternative is to consult with your colleagues about computer database programs to track results or evaluations that are pending. The establishment of a comprehensive patient database is also useful for conducting case conferences, completing clinic surveys and reports, and planning potential research studies.

Letter Writing

An important part of case management is writing letters to the referring physician and/or the patient which summarize the genetic counseling that was provided. Such letters not only are important for documenting patient care, they also serve to facilitate communication. A whole chapter (Chapter 9) is devoted to the critical topic of documentation.

IS A GENETICS CASE EVER COMPLETE?

It is not unusual for several years to pass between an initial clinic visit and a return visit. A case that you may have considered complete may later reopen as a result of advances in the understanding of a genetic condition. Given how rapidly genetics is evolving, it is important to emphasize to patients the need to keep in touch with a genetics center. While ideally, one would like to be able to inform patients by letter when advances pertinent to their genetic conditions are made, logistically, this is next to impossible. Encouraging patients to join support groups is another way to keep them updated, since support groups often have publications through which they disseminate information about advances.

SUMMARY

Case management is initiated when the patient schedules an appointment and continues long after the patient has left the clinic. It is a complex task that has many important components and details to oversee. As this chapter illustrates, genetic counselors play a pivotal role in case management. The ability to successfully manage a case will directly impact the quality of the care, education, and counseling provided. Developing the skills to anticipate, evaluate, prioritize, coordinate, and triage will be critical for successful case management. Your case management skills will continue to evolve throughout your clinical experiences, and each case will provide a new opportunity to implement and refine these skills.

REFERENCES

Alliance of Genetic Support Groups (1998) *Directory of National Genetic Voluntary Organizations and Related Resources*, 4th ed. Washington, DC: AGSG.
  Phone: (800) 336-4363
  URL: *http://www.geneticalliance.org*

Antonarakis SE, Rossiter JP, Young M (1995) Factor VIII gene inversions in severe hemophilia A: Results of an international consortium study. *Blood* 86:2206–2212.

Greenwood Genetic Center (1995) *Counseling Aids for Geneticists*, 3rd ed. Greenwood, SC: Jacobs Press.

Hayden MR (1981) *Huntington's Chorea*. New York: Springer-Verlag.

HELIX. University of Washington, Children's Hospital and Medical Center, CH-94, PO Box 5371, Seattle, WA 98105-5371.

Phone: (206) 527-5742

URL: *http://healthlinks.washington.edu/helix/*

Huggins M, Bloch M, Wiggins S. Adam S, Suchowersky O, Trew M, Klimek ML, Greenberg CR, Eleff M, Thompson LP, Knight J, MacLeod P, Girard K, Theilmann J, Hedrick A, Hayden MR (1992) Predictive testing for Huntington disease in Canada: Adverse effects and unexpected results in those receiving a decreased risk. *Am J Med Genet* 42:508–515.

International Huntington Association and World Federation of Neurology Research Group on Huntington's Chorea (1994) Guidelines for the molecular genetics predictive test in Huntington's disease. *Neurology* 44:1533–1536.

Jones KL (ed) (1997) *Smith's Recognizable Patterns of Human Malformation*, 5th ed. Philadelphia: Saunders.

McKusick VA (1994) *Mendelian Inheritance in Man, Catalogs of Human Genes and Genetic Disorders*, 11th ed. Baltimore: Johns Hopkins University Press.

Online Mendelian Inheritance in Man, OMIM (^{TM}). Center for Medical Genetics, Johns Hopkins University (Baltimore, MD) and National Center for Biotechnology Information, National Library of Medicine (Bethesda, MD), 1997.

URL: *http://www.ncbi.nlm.nih.gov/omim/*

MEDLINE/PubMed.

URL: *http://www.ncbi.nlm.nih.gov/PubMed/*

National Organization for Rare Disorders, Inc. (NORD) PO Box 8923, New Fairfield, CT 06812-8923.

Phone: (800) 999-6673

URL: *http://www.pcnet.com/~orphan/*

National Society of Genetic Counselors (1995) *Molecular Genetics Laboratory Questionnaire*. Wallingford, PA: NSGC.

Rimoin DL, Connor JM, Pyeritz RE (eds) (1997) *Emery and Rimoin's Principles and Practice of Medical Genetics*, 3rd ed. New York: Churchill Livingston.

Scriver CR, Beaudet AL, Sly WS, Valle D (eds) (1995) *The Metabolic and Molecular Bases of Inherited Disease*, 7th ed. New York: McGraw-Hill.

Thompson MW, McInnes RR, Willard HF (eds) (1991) *Thompson & Thompson Genetics in Medicine*, 5th ed. Philadelphia: Saunders.

9

Medical Documentation

Debra Lochner Doyle

As the art of medicine has evolved, so too has the art of medical documentation. The first medical records were journals kept by practitioners to record detailed descriptions of a patient's ailments, treatments, and outcome. Such notes aided in monitoring the patient's progress following varied treatments, as well as furnished information for ongoing explorations in the field of medicine.

Medical documentation is now an integral part of overall medical practice. Information obtained through the course of a clinical interaction and transcribed into notes is no longer deemed the sole property of the clinician. While hospitals and health care providers are responsible for generating and maintaining medical records, the information contained in the records is increasingly viewed as the property of the patient. Many parties such as consulting health and social service practitioners, hospital administrators, peer review organization staff, insurers (third-party payers), and self-insured employers, as well as the patient, may have access to these records. In addition, medical documentation is often critical to the resolution of disputes that arise over health care treatment.

For all the importance placed on medical documentation, one finds relatively little guidance in the medical literature concerning what to document, when, where, how, and why. These aspects of medical documentation along with suggestions and examples, are addressed in this chapter. Additionally, some comments regarding disclosure of medical information and medical record retention are provided. The many issues that surround the increasing use of electronic record keeping are beyond the scope of this chapter, however, the reader can assume that the *information* contained in an electronic patient record is no different from the information contained in a written medical record.

A Guide to Genetic Counseling, Edited by Diane L. Baker, Jane L. Schuette, and Wendy R. Uhlmann.
ISBN 0-471-18541-8 (cloth), 0-471-18867-0 (paper).   Copyright © 1998, Wiley-Liss, Inc.

TYPES OF MEDICAL DOCUMENTATION

The term "medical record" typically includes both inpatient, hospital-based records and outpatient clinic charts, and the guidelines for medical documentation offered here pertain to records of both types. Most genetics or specialty clinics maintain outpatient charts, sometimes known as clinic charts or shadow files. As in other medical specialties, documentation of the genetics evaluation and counseling typically includes the chart note and letter to the referring physician. More unique to genetics is a third form of documentation, a letter to the patient or family summarizing the visit.

The several types of medical documentation typically found in the charts of patients who have been seen for genetic counseling are presented in Table 9.1. The chart note and the letters of correspondence (letters to the physician and patient) are usually considered "official" documentation, especially if the chart note refers to the correspondence. In addition to the official documentation, many genetics charts include standardized forms such as intake forms for collecting demographic information and pedigree forms. These forms help to ensure that pertinent information is collected. Standardized clinical and laboratory record forms also guarantee that documented information is preserved as a single record, and that desired information is presented in a particular way. Standardized forms are limited, however, in that not every clinical or laboratory circumstance can be anticipated.

A **chart note** is a summary note that describes a clinical encounter. It includes a statement about the problem or nature of the referral, past medical history, history of the present illness, family history, social history, results of the physical examination, an assessment or diagnosis, the information or counseling provided, and the recommendations for medical management, treatment, and referral. The chart note is always signed and dated (Kettenbach, 1995).

SOAP is an acronym for a charting method in which each letter stands for a particular section of the chart note. It is discussed here as an example of one of many formats used for charting patient notes. "S" stands for *subjective* and includes the

---

TABLE 9.1   Genetics Clinic Chart Contents

Patient intake form
Pedigree
Summary chart note(s), outpatient and/or inpatient
Laboratory report(s): cytogenetics, DNA studies, other labs
Copies of formal correspondence
   Letter to/from referring physician
   Patient letter
Radiology report(s)
Pathology report(s)
Signed consent forms
Documentation of phone calls
Photographs
Outside records
Copies of medical records obtained from other pertinent family members
Insurance and billing information

Problem: Jacob is a 16-year-old who returns to genetics clinic for continued evaluation and counseling regarding his diagnosis of NF 1. Jacob was last seen in clinic in July, 1996. His diagnosis was based on the presence of multiple café-au-lait macules, axillary and inguinal freckling, and a pseudoarthrosis of the left ulna.

Subjective: At Jacob's last clinic evaluation, he had a new complaint of headaches. These were limited to the frontal regions and were occurring with a frequency of approximately one every 2–3 days. Since Jacob's last visit, he had a brain MRI. The results were reported by the patient's mother to have been within normal limits. Since his last visit, his headaches have increased somewhat in intensity. He is treating them successfully with Tylenol. Otherwise, his symptomatology has not changed.

Jacob was evaluated by an ophthalmologist several months ago, per the patient's mother. No abnormalities were noted; his peripheral vision was within normal limits.

Review of systems is negative. Jacob has not noticed any change in the number or size of the palpable neurofibromas. He has not noticed any pain or numbness of his extremities and he has not noted any change in balance.

Jacob is entering his junior year of high school. He swims and is on the track team. His grades are B's and C's.

Objective: On physical examination, the patient presents as alert, active, and is a cooperative young man who is in no acute distress. His blood pressure is 120/60. His height is 183 cm, which is the 75th to 90th percentile, and his weight is 75.4 kg, which is the 50th to 70th percentile. The patient's HEENT exam is within normal limits. His back has normal curvature. Chest is clear to auscultation and percussion. There is a normal S1 and S2 without murmurs. Abdominal exam is benign. Genitalia is that of a Tanner V male. There are 2 neurofibromas palpable in the inguinal region of the right groin. Lower extremities are symmetrical. The left forearm is shortened and clubbed with a surgical scar present proximally. Numerous café-au-lait macules are present on the trunk and extremities. Axillary freckling is present bilaterally. Neurologic examination is nonfocal. The patient has normal strength and sensation. Tandem gait maneuvers are performed normally and deep tendon reflexes are 2/2 in upper and lower extremities bilaterally.

Assessment: Impression:

    1. Neurofibromatosis type 1.
    2. Left congenital pseudoarthrosis of the left ulna, status postsurgical correction and fusion.
    3. History of headaches.

The patient's neurofibromatosis is stable. There are no new manifestations and no new associated problems. The patient's growth and development are within normal limits.

Plan: We request that the patient return to the genetics clinic in one year's time for continued evaluation for potential complications of neurofibromatosis. Continued use of Tylenol for headaches was recommended. Any change in frequency or in intensity should be reported to the patient's primary care physician and to the genetics clinic. Per the patient's request, we will refer the patient to orthopedics and reevaluation of the patient's left pseudoarthrosis.

Signed:
Date:

Figure 9.1 *Sample SOAP note.*

pertinent information that the patient gives the provider (e.g., "The patient reports no leaking of amniotic fluid, bleeding, or illness during her pregnancy."). "O" is for *objective*, which incorporates the measurements and observations of the provider (e.g., "Mrs. Smith walks with a wide-based gait and has difficulties with speech."). "A" stands for *assessment*, a listing of identified risk factors and a review of the information conveyed to the patient concerning each. "P" refers to the treatment *plan*, which is a description of the recommended treatment or follow-up plan. A sample SOAP note is included in Figure 9.1. Note how the sections include the various elements of the medical history, social history, and other information as mentioned above. The SOAP format, however, does not include certain information that is critical to medical record documentation, namely, the source of the patient referral and the duration of the patient interaction. These factors are discussed in greater detail later in the chapter (see below "Governmental and Fiduciary Requirements").

The **letter to the referring provider**, like the chart note, summarizes the clinical encounter. This is a professional standard that facilitates communication between health care providers because it serves as written documentation about the genetic counseling session to be included in the patient's medical records maintained by the referring provider. This letter can also be a valuable educational tool for the referring physician and staff, since it includes information about the diagnosis, its pathology, risks for inheritance, and possible treatment options. Such documentation can be especially useful for rare conditions, since the primary care provider is unlikely to have encountered similar cases.

The **patient letter** represents another important form of formal correspondence and has long been regarded as a vital tool of the genetic counseling process. The patient letter summarizes important information from the genetic visit, the diagnostic issues considered, and the counseling provided. It documents the pertinent aspects of the patient's medical and family history, and provides a mechanism for sharing information with family members.

THE IMPORTANCE OF MEDICAL DOCUMENTATION

The most important reasons for recording medical information are as follows:

- to ensure the best possible care for the individual and family
- to document the events of an inpatient or outpatient visit
- to facilitate communication among health care providers

While these objectives relate to the enhancement of individual patient care, there are other uses of medical documentation that relate to enhancing care for patients in the future. For example, the information can be used for quality assurance and improvement efforts, and for research.

First and foremost, medical documentation is essential to providing the best possible care by identifying the services or treatments provided to or discussed with a client. Health care is typically delivered by multiple providers, and the majority of genetic counseling clients are referred by an outside health care provider. Therefore, good documentation serves to improve communication between all the health and social service providers working with a patient (Smith, 1993). The permanent nature of medical records may contribute, now or in the future, to establishing or confirming a diagnosis, and to determining an accurate risk assessment. In addition, medical documentation can serve to assist genetic counselors in recalling a previous counseling session as they prepare for future consultations with the same client or family. The institution continues to improve its overall health care service when records are selected as part of the hospital or clinics' quality assurance and quality improvement exercises. While these benefits may not be immediately realized by a patient or family, they do hold the promise of long-term improved care to patients in general.

RECOMMENDATIONS FOR MEDICAL DOCUMENTATION

All documentation should be objective and factual, since the medical record is intended to be a tool for the delivery of health care services, documenting clinically relevant information and supporting the assertion that services were performed within the accepted standards of care. The events being depicted should be recorded in chronological order, and the notation should then be dated and signed, or countersigned.

In general, it is preferable for documentation to be as brief as possible. Only pertinent and relevant information that relates directly to the client's health care is entered into the record. Long narrative paragraphs should be avoided but may be necessary under exceptional circumstances. For example, genetic counseling sessions often include detailed review and assessment of risks associated with multiple factors identified in a family or personal health history. In these circumstances, care should be taken to record all the pertinent information succinctly. It is imperative, however, that the medical documentation be complete, because there is oftentimes a perception that anything not included in the medical documentation didn't happen: "Partial record keeping implies partial health care" (Kapp, 1993).

Records should be neat and legible, and standard medical abbreviations should be used appropriately. Medical abbreviations help convey complex terms or concepts in a manner that is clear and brief. For example, ROM is standard for "range of motion." However, many abbreviations have more than one meaning, depending on context in the medical record. For example, AMA could mean "against medical advice" or "advanced maternal age." If the meaning of an abbreviation is not clear within the context of the note, avoid its use.

All entries in the medical record should be value neutral (Mangels, 1990; Fiesta, 1993). Labels that may be perceived as derogatory should not be used to describe a patient's appearance, condition, or behavior. Some examples of value neutral terms are compared with connotative terms in Table 9.2. The medical record is not an appropriate forum for airing grievances or differences of opinions. Editorializing about the facility, the patient, the family, or other health care providers should be avoided.

TABLE 9.2   Examples of Value-Free Language

Connotative	Value Free
The patient's cousin **suffered** from a cleft lip.	The patient's cousin had a cleft lip.
Ms. Jones is a 22-year-old **unwed mother** of a child with Down syndrome.	Ms. Jones is 22 years old and is the **sole provider** caring for a child with Down syndrome
Mrs. Smith is a **substance abuser** and is pregnant with a singleton at approximately 20 weeks gestation.	Mrs. Smith admits to continued **use of** alcohol and cocaine during her pregnancy. She is currently 20 weeks gestation.

Source: Adapted from Baker et al. (unpublished).

The source of medical information should be documented (e.g., "according to medical records from . . . Hospital"). If the source of information is not an official document, this should be made clear. Commonly used phrases such as "the patient denied" and "the patient reported" are helpful because they convey both that the patient was asked about an issue and that a specific reply was given.

Whenever relevant, refer in the medical record to booklets, literature, World Wide Web sites, or instructional sheets that were used in counseling or given to the patient. Once referenced, these items also become part of the medical documentation.

Inevitably in the course of a genetic counseling visit, names and other identifying information about the client's family members will be revealed. It is important to balance the need to document pertinent genetic, medical, and social information about family members with the need to protect the privacy of the family members. If it is determined that the identifying information is relevant to the health care of the client and as such must be documented, the record should reflect where the information was obtained and plans for verifying the accuracy of the information. For example,

> Mr. Barnes reports that his 35-year-old brother was diagnosed earlier this year with hypercholesterolemia following a minor heart attack. Mr. Barnes has indicated that he will speak with his brother, Bernard Barnes, concerning a request to release his medical records to our office.

If the client is unable or unwilling to assist in the verification of a significant reported history, identifying names may not be useful and should be omitted from the record. The genetic counselor may consider using only initials under these circumstances. Likewise, unrelated individuals (e.g., the friend who accompanies your client to the counseling session) should not be referenced by name in the medical documentation. The genetic counselor should also carefully consider whether identifying information should be included in the patient letter, since these letters are often shared with other family members, friends, and health care providers. It is best to discuss with the client in advance what family information will and will not appear in the patient letter. Table 9.3 summarizes these general recommendations for medical documentation.

While the suggestions provided apply to all forms of medical documentation, some additional considerations are worth noting. For instance, in drafting the patient letter, care should be taken to use clear and easily understood language. In addition,

TABLE 9.3   The "Do's and Don'ts" of Medical Documentation

	Do
Write legibly	Reference source of information
Write brief, succinct, and complete notes	Reference resources
Record events in a chronological fashion	Make *obvious* additions or corrections
Use value-free language	Sign and date all documentation (include time when appropriate)
Be objective and factual	Obtain countersignature when necessary
Use abbreviations sparingly	Protect the privacy of the client and his or her family
	Don't
Delay in documenting events	Obliterate anything from a record
Record nonpertinent identifying information	Alter another provider's documentation
Record disparaging or inflammatory remarks	Justify or explain actions/mistakes
Assign abbreviations to demeaning phrases	Reference incident reports (unusual occurrence reports)
Editorialize	

medical jargon should be avoided whenever possible and defined when it is used. Explanations of medical terminology should be provided in lay phrases; in some instances parenthetical phrases or notes are sufficient (e.g., "During the examination, Millie was noted to have hypotelorism (closely set eyes.") (Baker et al., unpublished). It should be clear from the outline for developing a letter to the patient or family represented in Figure 9.2 that the patient letter discusses the condition in considerable depth compared to the letter to the referring provider.

When Should Medical Documentation Occur?

Medical documentation should occur with every patient visit and in a timely fashion (George 1991). Medical records are typically viewed as accurate and reliable because they are recorded as soon as possible following the clinical encounter. All individuals, including health care providers, experience a fading of memory over time. Therefore, it is essential that documentation occur soon after the consultation. In fact, Medicare and the Joint Commission on the Accreditation of Health Care Organizations (JCAHCO) have established standards for record keeping that include specific time limitations for recording or updating medical documentation. For example, Medicare requires that medical documentation occur within 15 days from hospital discharge (Kapp, 1993).

This is not to say that a genetic counselor is without recourse should additions, clarifications, or corrections to the chart note be desired (Fiesta 1991). After all, it is critical that the documentation be accurate and truthful. Changes to documentation, whether they occur in an outpatient clinical chart or as part of the inpatient hospital-based medical record, should be obvious and legible. A single line should be drawn through the error and the correction written above or below the line. In addition, the correction should be dated, including the time if pertinent, and signed or countersigned.

[Genetics Clinic Letterhead]

Date
Family name
Address   re: patient name
                                                                           hospital registration number

Dear (*Patient, Parent, or Guardian*),

I.   **Introduction** (Generally only 1–3 sentences)
     A. Date of visit
     B. Clinic
     C. Reason for referral or visit (include name of referring provider)
     D. Purpose of the letter

II. **Body**
     A. Description of significant family, medical, pregnancy, and developmental histories and pertinent tests results*
     B. Review of physical exam and/or diagnosis*
          Findings on physical exam/basis for diagnosis
          Interpretation of significant (+) or (–) family history
     C. Natural history of condition
          Description of clinical features and prognosis
          Incidence of carrier and disease frequency in general population
     D. Explanation of inheritance
          Brief description of genes and chromosomes
          Suspected or established inheritance pattern(s)
     E. Summary of risk assessment
          Recurrence risk
          Risks to other family members
          Baseline risk of birth defects, if relevant
     F. Outline of reproductive options/prenatal diagnosis
     G. Recommendations for medical management, diagnostic workup, carrier testing, etc.
          Clear, unambiguous statement of recommendations
          Review of referrals made
          Laboratory testing undertaken or under consideration
     H. Significant psychosocial issues and concerns
     I. Recommendation of nonmedical resources (including contact names, telephone numbers, etc.)
          Support services
          Patient literature
          Educational resources

III. **Closing**
     A. Plan for reporting test results
     B. Schedule of return visit to genetics clinic
     C. Invitation to recontact genetics clinic
     D. Other closing remarks

IV. **Signatures**
     Genetic counseling student
     Staff genetic counselor
     Fellow
     Attending geneticist
     cc: local physician/others as requested by family

*Reference the source of this information (i.e., patient/parent report, physical examination, consultation report, medical institution/laboratory from which records were obtained, etc.).

FIGURE 9.2 Guidelines for writing genetic counseling letters (Baker et al., unpublished)

Since all documentation should be chronological, if additional insights or alterations are warranted, the documentation should also indicate why the additions were made, perhaps out of chronological order:

> In further considering Mrs. Jones' reproductive options, it was suggested that artificial insemination might be beneficial. Therefore, Mrs. Jones was contacted and this alternative was reviewed. [date/sign].

White-outs, erasures, or complete obliteration of any material (achieved, e.g., by marker pen to blacken a section) can create a sense of secrecy and could be called into question. Under no circumstances should one health care practitioner alter or obscure another health care provider's documentation (Gilbert et al., 1994; Hirsh, 1994; West, 1995).

Who Records in the Chart?

Every health care provider who interacts with a patient has a responsibility to record on the patient's chart the pertinent information concerning the interaction. In some institutions, however, not all categories of health care providers are permitted to record in hospital records. In most institutions, authorization is required before a health care provider can write in a patient's record. Most genetic counselors who work in hospital settings are able to write in patient charts, but you should inquire within your own institution about the guidelines for recording, signing, and countersigning patient chart notes. Genetic counseling students should have a supervisor countersign their medical documentation. Countersigning an entry implies that another person reviewed the entry and approved the care given; it does not imply that the countersigner performed the service. For this reason, all entries that are countersigned should clearly delineate who performed the services and should be carefully reviewed by the countersigner to ensure the accuracy and completeness of the information presented. It must be emphasized that by providing a signature, the countersigner assumes responsibility for anything written in the notes covered by the signature. For example if the chart note states that a specified health care provider will arrange for a follow-up ultrasound appointment, the countersigner is responsible for ensuring that this appointment is scheduled.

DISCLOSURE OF INFORMATION CONTAINED IN MEDICAL RECORDS

> *It has been said that the concept of medical confidentiality is as old*
> *as the practice of medicine.*
> *—Norrie, 1984*

The issues of privacy and confidentiality are certainly not new to genetic counselors. Yet, with all the discourse surrounding privacy and confidentiality in the field of genetics, few hard and fast recommendations are available.

It is worth noting that requests for access to medical records are most frequently initiated from patients themselves, as opposed to other family members or third-party payers. By signing a specific form, furthermore, clients frequently authorize the release of their medical information at the time of applying for insurance or receiving treatment (Smith and Jones, 1991). Nonetheless, when offices receive appropriately authorized requests for medical records, they are required to provide this access in a timely fashion (Fordham, 1993). Authorization is generally considered adequate when the patient has signed the request form; when the patient is deceased, an immediate family member or first-degree relative can appropriately sign for the release of the records.

On certain occasions it is legally permitted, if not mandated, to divulge medical information without the patient's consent (McCunney, 1996). This circumstance most frequently arises in public health scenarios with respect to duty to report infectious disease or other reportable conditions (e.g., HIV/AIDS, birth defects in newborns, gunshot wounds). In such cases health care practitioners are obligated by state statutes to report the conditions to the appropriate state agencies and do not need to obtain patient or parental permission beforehand.

Special care must be taken to safeguard confidentiality when patients are recognized celebrities. This is not to imply that celebrities deserve more protection than noncelebrities; it must be recognized, however, that celebrities may be targets of intense media attention that strives to uncover personal and confidential information. The obvious policy is that "a patient's medical records are confidential and will not be released except under proper patient authorization or as required by law" (Roach, 1991). Such a policy confers protection to *all* clients. Additional strategies for consideration include omitting the celebrity name from the record and instead using a code name corresponding to a master list maintained by a medical records director. This action is disallowed in some states, however, so legal counsel should be consulted before such coding is applied. Another strategy entails placing the medical records of celebrity patients in a secure file accessible to designated personnel only. None of these approaches completely guarantees that no breach of confidentiality will occur. Therefore, periodic in-services or trainings to remind all staff about the need for privacy should be instituted as a means to enhance the success of privacy procedures.

Perhaps the best guidance for the genetic counselor concerning disclosure of information can be found within the National Society of Genetic Counselors (NSGC) Code of Ethics, which explicitly states that "the primary concern of genetic counselors is the interests of their clients. Therefore, genetic counselors strive to . . . maintain as confidential any information received from clients, unless released by the client" (NSGC, 1992). In fact, many professional societies have formulated codes of conduct to help health care providers make informed decisions about requests for medical information. These professional codes of conduct serve as respected frameworks for health care providers in a variety of situations.

In addition, some federal and state regulations can also offer guidance to clinicians. These include the Americans with Disabilities Act (ADA), the Occupational Safety and Health Act, which established the Occupational Safety and Health Administration (OSHA), and the Uniform Health Care Information Act. The ADA allows release from the medical record recommendations for work restrictions and corresponding accommodations but does not entitle the employer to receive specific diagnostic information. OSHA requirements "preserve confidentiality of medical records by specifically stating that [such] requirements are not intended to affect existing legal and ethical obligations" (Hahn, 1996). In 1985, the National Conference of Commissioners on Uniform State Laws approved the Uniform Health Care Information Act, which recognizes a right on the part of patients to consent to the release of their health care information, or revoke such consent. In addition, it offers a mechanism for clients to review, and in some circumstances correct, medical documentation they dispute. While few states have adopted the act as originally drafted, several states have enacted similar legislation.

The accountability of third-party payers with regard to preserving confidential information is less clear. There are relatively few regulations in place that address the obligations of insurers in handling confidential and sensitive medical information.

Finally, it is worth considering some of the advantages that electronic record keeping systems may offer. On August 21, 1996, President Clinton signed into law the Health Insurance Portability and Accountability Act (HIPAA). Section F of HIPAA mandates the establishment of an electronic patient records system, as well as privacy rules for these electronic records. HIPAA is sure to expedite the development of national standards for electronic patient records, and it is not unthinkable that these national standards will also significantly impact policies pertaining to written documentation. Clearly, as these policies are formulated, there will be a need to monitor evolving guidelines and the potential for misuse of health information (McAuliffe, 1996; Brannigan, 1992).

Governmental and Fiduciary Requirements

Medical information also is recorded to comply with regulatory requirements and to secure eligibility for legal protections. Health care practitioners have a fiduciary responsibility to compile and maintain accurate medical records. Federal regulations require providers participating in the Medicare program, including hospitals and other institutions, to ensure that medical records are "accurate, promptly completed, filed, retained and are easily accessible" (42 CFR, 428.24) (Stratton, 1994). The regulations pertaining to Medicare are developed by the Health Care Financing Administration (HCFA), which also dictates the federal program requirements for Medicaid. The policies developed through HCFA have historically been adopted by third-party payers. Therefore, while relatively few genetics patients are recipients of Medicare, regulations or policies pertaining to Medicare are oftentimes held as the standard for all health care providers and payers.

The Joint Commission on the Accreditation of Health Care Organizations (JCAHCO), the National Commission of Quality Assurance (NCQA), and some state facility licensing statutes require that institutions generate and maintain medical records for licensure, reimbursement, and accreditation purposes. Of note, the JCAHCO and HCFA requirement that entries into medical records be signed is being challenged and may be revised as the use of electronic records becomes more widespread (Brandt, 1994).

The medical record should provide documentation that corroborates the Current Procedural Terminology (CPT) billing codes used. The Evaluation and Management (E/M) CPT coding system is a mechanism for describing the type and duration of services provided. E/M CPT codes were devised based on the recommendations of a consensus panel of physicians, allied health care providers, Medicare carriers, private insurers, consumers, and other interested parties, and are published annually by the American Medical Association. The E/M CPT codes also define levels of complexity and distinguish between new and established patients, and inpatient and outpatient consultations. The E/M CPT code components of the patient encounter for billing purposes by E/M CPT codes are as follows:

- obtaining a history
- performing an examination
- facilitating medical decision making
- counseling
- coordinating patient care

Other aspects of the encounter that are considered include the duration (time spent), complexity (high, medium, or low), and the level of risk (also ranked as high, medium or low).

An institution can use E/M codes even if every component is not a part of an individual case. However, these codes convey the level of complexity and the level of risk associated with a case that the health care provider can rightfully claim (CPT, 1997). With few exceptions genetic counseling sessions are billed using E/M codes, and it is therefore important to document in the medical record the various components of a genetic counseling session, including those listed above.

DOCUMENTATION THAT IS SUBJECT TO EXTERNAL REVIEW

What constitutes bona fide medical documentation? The answer to this question varies depending on who is doing the asking. If the question is being asked by a patient, a medical provider, or an attorney, any and all recorded information may be deemed medical documentation. In the broadest sense, virtually all information that is recorded in the regular course of health care delivery could be viewed as "medical documentation." For these purposes, there is no difference between an outpatient clinic chart, or shadow file, and an inpatient, hospital-based medical record. The value of the contents and the potential uses are equal irrespective of where the documentation is housed.

In the strictest sense however, only the chart note, and in some circumstances the letter to the referring physician (or family), would be considered "medical documentation" and would be requested by a third-party payer in the course of an audit. If, however, other documentation is referenced within the chart note or the letter, it too becomes part of the accepted medical documentation. For example, if a letter is generated to the referring physician in lieu of a clinical chart note, and the letter to the referring physician includes the information that the patient supplied medical records pertaining to her mother's diagnosis of breast cancer, which were used for the risk assessment, the letter along with the mother's medical records could be considered the "medical documentation" for this case.

In some circumstances even formal correspondence between physicians is excluded as part of the medical documentation being reviewed. This practice is justified because auditors (known as medical abstractionists) hired by third-party payers or governmental agencies to conduct either a fiscal audit or performance audit find it unrealistic and impractical to hunt through every written document in search of a particular piece of information. For this reason, typically only the chart note, in some cases supplemented by formal correspondence to a referring physician, are reviewed and considered in an audit.

THE MEDICAL RECORDS AUDIT

Third-party payers, including Medicare and Medicaid, utilize medical records when they conduct institutional or agency audits. During a fiscal audit, a sample of medical bills will be compiled and the corresponding medical records collected for review. An analysis of the chart note in Figure 9.3 will demonstrate what occurs during a record review for purposes of a fiscal audit.

The first problem with the sample chart note is the lack of reference to a referring social or health care provider. In the absence of such information, an auditor would assume that the visit was the result of a self-referral. There is only one E/M CPT code that can be used when the client is self-referred, and that is *Office Visit*. This code is typically reimbursed at a lower level than other E/M codes, such as *Consultation* or *Counseling and Risk Reduction*, but the latter codes require that the patient be referred by another clinician. It is clear that a detailed family history was obtained, risk assessment was performed, and patient education occurred. The provider did a nice job of referencing the educational materials used as well as laboratory results obtained, so these documents would also be included in the auditor's review. The documentation does not indicate that a physical examination was conducted during the visit. There is also no reference to time spent with the patient, which becomes a key billing component if counseling or coordination of care make up over 50% of the encounter. Given the available documentation, the medical records abstractionist would likely classify this case as being of medium complexity and low risk. Thus, if a provider, for billing purposes, used the code for high complexity and high risk, the existing medical documentation would fail to support the claim.

It is less common for nongovernmental agencies to conduct audits; however, as with Medicaid/Medicare audits, the rate of nongovernmental audits has increased in recent years. Health care providers and agencies that receive Medicaid or Medicare funds are required to participate in external quality review programs by professional review organizations (PROs), sometimes called peer review organizations or professional standards review organizations. The PROs were created in 1972 by the Social Security Act Amendment specifically for this purpose. The PROs use medical documentation to determine whether appropriate practice parameters were adhered to

Melinda Correll is 33 years of age and is seen today with her husband for genetic counseling.

Family History: The family history is shown below in the pedigree. Ms. Correll has had three miscarriages at 10 weeks, 6 weeks, and 8 weeks gestation, respectively. The couple has a 4-year-old child who is reportedly in good health and achieving his developmental milestones appropriately. Chromosome analysis was done on one of the products of conception and was reportedly 46,XX. The last miscarriage occurred in December 1996. Peripheral blood samples for both Mr. and Ms. Correll were obtained at that time and sent to [name of lab]. Ms. Correll's chromosomes are 46,XX and Mr. Correll's are 46,XY,t(6;18)(q21;q23). Mr. Correll appears to have a balanced translocation between chromosome number 6 and 18. There is no family history of mental retardation, stillbirths, neonatal deaths, infertility, or other birth defects with the exception of a paternal cousin who had a cleft lip. This individual died at 24 years of age as a result of a motor vehicle accident but was reportedly in good health until the time of his accidental death. Both Mr. Correll's parents died in their mid 80s due to heart attacks. Mr. Correll is unaware of any pregnancy losses experienced by his parents, and his only sibling, a 43-year-old brother, is unmarried.

The recurrence risks associated with a balanced translocation between chromosomes 6 and 18 were discussed at length. Illustrations from the *Counseling Aids for Genetics* manual were used to demonstrate the approximate 50% risk of unbalanced gametes, the other 50% of the gametes resulting in either normal or balanced karyotypes. Mr. and Ms. Correll were informed that empirical risks for unbalanced conceptions are significantly less than the 50% relative risk. Prenatal diagnostic procedures were described including amniocentesis and chrorionic villus sampling. The benefits, risks, and limitations of each were described in full. Additional reproductive options explored included artificial insemination by donor sperm; however, both Mr. and Ms. Correll rejected this option.

Mr. and Ms. Correll have indicated a desire to pursue another pregnancy and would be interested in proceeding with amniocentesis. They will contact our offices when a pregnancy is achieved so that follow-up counseling can be initiated and the appropriate tests arranged. We have encouraged this couple to contact us if they should have additional questions or concerns.

Date:
Signature:

FIGURE 9.3 A sample medical record for critical review.

(e.g., that a pregnant patient was offered maternal serum α-fetoprotein or multiple marker screening, or that a pediatric patient received age-appropriate well-child visits). Medical records can also be reviewed by some PROs to investigate for fraudulent or abusive billing practices. For example, a third-party payer such as Blue Cross may question billing patterns of one institution that are not consistent with billing patterns of another comparable institution. To investigate, the payer may hire a PRO to conduct a medical/fiscal audit.

LEGAL ISSUES

It is important to recognize that patients' rights are preserved in good medical documentation. It is also worth noting that the best medical documentation cannot and should not replace good communication between health care practitioners and their clients (Berry, 1992; Tammelleo, 1995).

Most clients receive appropriate care and do not initiate lawsuits, but in the 1980s, medical malpractice suits began to proliferate. As a result, some health care practitioners were influenced by fear of lawsuits and began to practice what was later coined "defensive medicine"—ordering more tests, treating patients with more conservative approaches, refusing to see some clients for fear of litigation—as well as documenting more meticulously (Dollar, 1993). Information that historically had been absent in most medical documentation became commonplace; such new components include reviews of alternative forms of treatment and references to materials provided to, or used with, the client as part of the informed consent process (Nisonson, 1991). Meticulous documentation is a noteworthy goal, given the benefits of accurate and complete documentation as reviewed earlier.

Some institutions have developed policies and procedures that allow for "charting by exception," which can be useful in situations for which certain procedures are routine. For example, if it is customary for pediatric genetics patients with metabolic disorders to be seen yearly for a follow-up evaluation, and after the visit parents are given an information sheet (how to contact the clinic staff in the interim, when the next visit should occur, etc.), the medical record could contain a simple "follow-up" check box. A mark in the box tells the reviewer that such standard follow-up instructions were issued, and no further documentation was required. However, failure to check off such a box might support a patient's claim that no such information was ever provided. In the 1993 case of *Palinkas* v. *Bennett* (620 N.E.2d 775) the Supreme Judicial Court of Massachusetts found that Dr. Bennett could testify regarding his routine practice and, upon hearing the physician's testimony, concluded that his actions constituted a habit. So even in the light of missing documentation, the testimonial evidence of routine practice may be viewed as fact. Documentary evidence no doubt carries greater weight (Hershey, 1994).

RETENTION OF MEDICAL RECORDS

> *Ideally, patient records should be retained forever.*
> —*Harris and Thal, 1992*

The parameters of record retention have long been debated. There are, of course, numerous practical as well as legal issues in this area. A major issue is that of storage, especially for a large facility that may have many records on clients who have not been seen for many years. Nonetheless, failure to maintain medical records can have serious adverse consequences for patients, as well as for the health care provider or clinic. Geneticists and their clients know all too well the frustration of not being able to verify a significant medical history in a long-dead family member whose records are vitally important in accurately assessing a recurrence risk. However, even with the increased use of electronic record keeping as well as microfiche, many institutions find it impractical to maintain records indefinitely.

Many states have statutes that regulate the preservation and retention of medical records. Sometimes these laws also dictate how much can be charged for reproducing a medical record upon request. This is done to prevent institutions from using copying charges to create a barrier to some providers or patients seeking access to information. There are also federal regulations that concern minimum record retention practices under certain circumstances. For example, the federal antidumping law, passed as part of the Consolidated Omnibus Reconciliation Act of 1986, requires all hospitals that participate in Medicare to maintain medical and other records for a minimum of 5 years for individuals transferred to, or from, another hospital (Sickon, 1992). It is generally recommended that medical records containing financial information be retained for a minimum of 3 years for income tax audit purposes.

Legal statutes of limitations must also be considered in determining the length of time that records should be retained. These regulations are established to ensure that malpractice claims are brought within a reasonable time after the events in question occurred. These statutes, however, generally apply to adults; minors (under the age of 18) may have more time to file a claim. For example, in 1982 the California Appellate Court suspended the statute of limitations in a suit against an obstetrician who had failed to diagnose Down syndrome in a fetus (*Calle* v. *Kezirian*, 145 Cal. App. 3d 189). The parents brought an action against the obstetrician when they learned at a later date that their child's condition could have been diagnosed through amniocentesis. It may therefore be advisable to retain obstetric or pediatric records until the minor has reached the age of 18.

Genetics record retention policies should be developed with legal advice about federal and state regulations. The American Hospital Association, the American Medical Record Association, and many professional societies recommend retaining original or reproduced (i.e., microfiche or electronic copies) records for a period of 10 years after the most recent adult patient care entry. They recommend retaining pediatric records longer and even advocate that certain portions of the medical record be retained permanently. Hospitals and other medical facilities should weigh all these many factors when establishing record retention policies and procedures. Furthermore, there are no published guidelines concerning if, or how, to maintain records of telephone inquiries or patient intake forms for patients who failed to keep or canceled their appointments. Clinics tend to establish individual protocols for dealing with information obtained under these circumstances and are encouraged to seek guidance on the issue from their hospital attorney.

Once policies have been established and the records have reached the end of their lifetime, the next issue is how to discard them in a responsible fashion. Medical records that contain private and confidential information should be incinerated or shredded.

SUMMARY

Medical documentation assists the patient by optimizing health care and facilitating effective communication between providers. Medical documentation allows providers, institutions, third-party payers, and employers to comply with state and federal regulations. It also assists providers in supporting or defending their practice, should clients initiate legal claims against them or their institution. Understanding the varied uses of the information can be valuable in helping to identify what, where, when, and how to document.

Genetic counselors should consider and periodically reevaluate the various tips recommended for documenting, such as writing brief and succinct notes, keeping comments objective and factual, keeping in mind the necessary CPT components, and signing and dating all documentation. Furthermore, genetics clinics should develop and implement policies that will confer the greatest degree of privacy and confidentiality to their clients' medical information.

REFERENCES

Baker D, Schuette J, Uhlmann W *Guidelines for Writing Genetic Counseling Letters*. Unpublished.

Berry R (1992) Patient fights for right to records. *J Am Dent Assoc*, 123 (7): 238.

Brandt M (1994) New rules for the CPR (computer-based patient record): No more signing on the dotted line. American Health Information Management Association, Chicago. *Healthcare Inform* 11(10):30–32, 34.

Brannigan VM (1992) Protecting the privacy of patient information in clinical networks: Regulatory effectiveness analysis. *Acad Sci* 670:190–201.

CPT (1997) *Physician's Current Procedural Terminology*, 4th ed. Chicago: American Medical Association, pp. 9–39.

Dollar CJ (1993) Promoting better health care: Policy arguments for concurrent quality assurance and attorney-client hospital incident report privileges. *Health Matrix* 3(1):259–308.

Fiesta J (1991) If it wasn't charted, it wasn't done! *Nurs Manage* Aug 22(8):17.

Fiesta J (1993) Charting—One national standard, one form. *Nurs Manage* 24(6):22–24.

Fordham H (1993) Judicial commission & risk management: commission routinely receives questions, complaints about access to medical records. *Mich Med* 92(11):45–48.

George JE (1991) Law and the emergency nurse: Poor emergency department record keeping may hamper legal defense. *J Emergency Nurs* 17(3):167.

Gilbert JL, Whitworth RL, Ollanik SA, Hare FH Jr, James L (1994) Evidence destruction—Legal consequences of spoilation of records, Gilbert and Associates, P.C. Arvada, Colorado. *Leg Med (US)*: 181–200.

Hahn JR (1996) Medical records: Information gathering and confidentiality. *Benefits Q* Fourth Quarter 48–56.

Harris MG, Thal LS (1992) Retention of patient records. *J Am Optom Assoc* 63(6):430–435.

Hershey N (1994) Evidentiary ruling favorable to physician. *Palinkas* v. *Bennett. Hosp Law Newsl* 11(10):4–6.

Hirsh BD (1994) When medical records are altered or missing. *J Cardiovasc Manage* 5(6):13–14, 16.

Kapp M (1993) General guidelines for recording patient care: Discharge planning update. Department of Community Health, Wright State University, Dayton, OH 13(3):21–22.

Kettenbach G (1995) *Writing SOAP NOTES*, 2nd ed. Philadelphia: F.A. Davis.

Mangels L (1990) Chart notes from a malpractice insurer's hell. *Med Econ* Nov. 12.

McAuliffe DM (1996) Nurse attorney notes. Documentation—An Rx for success. *Fla Nurse* 44(5):13.

McCunney RJ (1996) Preserving confidentiality in occupational medical practice. *Am Fam Physician* 53(5):175–160.

National Society of Genetic Counselors (1992) *Code of Ethics.* Wallingford, PA: NSGC.

Nisonson, I (1991) The medical record. *Bull Am Coll Surg* 76(9):24–26.

Norrie, (1984) Medical confidence: Conflicts of duties. *Med. Sci Law* 24:26.

Roach WH Jr (1991) Legal review: Coping with celebrity patients. *Top Health Rec Manage* 12(2):67–72.

Sickon AC (1992) Legal review: The medical records implications of state and federal anti-dumping provisions. *Top Health Rec Manage* 12(4):83–90.

Smith AW, Jones A (1991) Computerizing medical records: Legal and administrative changes necessary. *Healthspan* 8(11):3–6.

Smith J (1993) Good medical records key to prevention, defense against malpractice claims. *Mich Med* 92(6):14–15.

Stratton WT (1994) Necessity of physician's signature on hospital medical records. *Kans Med* 95(3):57.

Tammelleo AD (1995) Good charting—bad communications: Recipe for disaster. Case in point: *Critchfield* v. *McNamara* 532 N.W. 2d 287—NE 1995) *Regan Rep Nurs Law* 36(2):2.

West JC (1995) Punitive damages allowable for record alteration. *Moskovitz* v. *Mt. Sinai Medical Center, Sister of Charity Health Care Systems, Inc., Cincinnati. J Healthcare Risk Manage* 15(1):43–45.

10

Ethical and Legal Issues

Susan Schmerler

A basic knowledge of ethical theory and principles and an understanding of the law are important in a profession that deals with problems at the cutting edge of science. In this chapter, we look at ethical and legal issues that impact genetic counseling. It is to some extent artificial to address these areas separately, since for each section of text there is a relationship, a common source, and much overlap to the material in other sections. Because many ethical and legal concepts are complex and easily misunderstood, terms are defined throughout the chapter.

It has become traditional in ethics and legal literature to use the feminine pronoun. We observe that tradition here.

ETHICAL ISSUES

Why Study Ethics?

Theoretical ethics does not occupy us on a day-to-day basis. On a personal level, for an individual whose behavior is usually moral, or socially acceptable, the choices she makes are automatic. At times in our personal and professional lives, however, we are unsure how to respond to a situation. There are also situations in which an otherwise moral person may not have the motivation to do the morally right thing. The guidance of rules at these times helps in making the moral choice. An understanding of the source of the rules of conduct may be of value for complicated situations.

A Guide to Genetic Counseling, Edited by Diane L. Baker, Jane L. Schuette, and Wendy R. Uhlmann. ISBN 0-471-18541-8 (cloth), 0-471-18867-0 (paper).   Copyright © 1998, Wiley-Liss, Inc.

What Is Moral Behavior?

Society as a community of people has a common code of conduct, an agreed-on view of what is acceptable behavior and what behavior is not acceptable. This is what is called morality. There are, for example, certain kinds of behavior that everyone agrees are immoral, hence not acceptable. In certain situations, however, society can justify such basically repugnant actions as killing, imprisoning, and deceiving. For example, killing in self-defense is generally deemed justifiable. The sum of this "agreed on" conduct makes up a common morality. The goal of moral behavior is to decrease the harms suffered by members of society. Moral convictions, then, provide a standard by which to evaluate our own and other people's conduct and character.

What Is Ethics?

The science of ethics is applied to help us understand the basic themes underlying and governing society's moral behavior. Theories are derived to account for, organize, and explain the themes that are identified, and to place them within a framework that allows rules of conduct to be developed. Ethics, then, is the establishment of a set of guidelines for morally acceptable conduct within a theoretical framework. The particular theory offered usually allows for continuing review of the basis for or justification of any conduct.

Ethical theories are discussed by means of terms that have uniformly accepted definitions. The **principles** offered within a theory are the source of the guidelines for behavior. For example, the principle of nonmaleficence is the source of the guideline "do no harm" among health care professionals. From principles, values are drawn and rules developed. **Values** are qualities considered good, or priorities, and are desirable and important. **Rules** are specific, and they must be observed at all times. They can be looked at as controls to prohibit or prevent harm. For example, directly causing any of the following harms is prohibited: killing, causing pain, disabling, depriving of freedom, and in certain nontrivial situations, depriving of pleasure. To prevent harm, we must follow the rules to keep promises, to obey the law, to do our duty, and to refrain from cheating and practicing other forms of deception. Rules promote and protect basic human interests, both individual and societal. Not obeying a rule usually results in punishment.

Behaviors that are encouraged are said to be the embodiment of **ideals**, or goals to which we should aspire. Not achieving an ideal, or even not attempting to achieve it, is not punishable conduct. **Duties** are defined by a person's role in society. They can be, for example, social or professional, and they include behaviors that are required of the person by that role. **Virtues** are characteristics of an individual that are morally desirable (e.g., candor, faithfulness, integrity). **Rights** are justified claims that individuals or groups can make on others or on society. Along with rights must come responsibilities on the part of the individual and obligations on the part of others. For example, my right to move about freely obligates you not to unjustifiably block my way. I also have a responsibility to move about only in my own space or in public spaces, and to not invade your space.

In summary:

- Principles are sources, or guides, for values, rules, duties, and rights.
- Values are priorities that are thought to be important and desirable.
- Rules are specific statements of what should (or should not) be done.
- Ideals are goals toward which we aspire.
- Duties are behaviors that are defined by our role in society.
- Virtues are morally and socially desirable characteristics.
- Rights are justified claims.

Depending on which theory is proposed, the same behavior can be a principle, a value, or an obligation. Trust is a good example. Trust is considered by some to be the overriding principle in medicine. It is also a value derived from the principle of respect for autonomy. Trust is the basis of the provider–patient relationship and an obligation derived from the fiduciary (i.e., the confidential or trusting) nature of the doctor–patient relationship.

Ethical dilemmas occur when equally strong arguments exist to justify the application of more than one theory or principle to a situation. To help clarify what appears to be a complicated problem, ethical theories can be applied to organize our thinking about the various choices in a difficult predicament. This is called moral reasoning. Ethical theories provide a structure for case analysis. Disagreements that involve the facts of a case can be separated from those that involve either a difference in the weighting or ranking of the benefits and harms found in the case or an equal weighing of different principles as defined in the case. We will discuss later how the process of moral reasoning can be applied to genetic counseling cases and can be used to derive an ethically justifiable course of action with respect to a particular situation.

Ethical Theories

Theorists have developed many strategies in an attempt to understand moral behavior. Each theory has had its supporters and detractors. There is no one ideal theory. The various theories are not necessarily mutually exclusive. Some look to more than one theory to provide a comprehensive framework for understanding and directing behavior.

Several theories have had an impact on modern medicine. **Consequence-based utilitarianism** is prominent in medical ethics. The primary focus of this theory is the promotion of happiness. Actions that maximize good and promote the greatest amount of happiness over pain are "right" or acceptable actions. An ethical dilemma is resolved by looking at the consequences of doing or of not doing an action. **Virtue ethics** and **principle-based ethics** have also had a great influence on medicine. The virtue-based theory of ethics focuses on the character traits or virtues a good person should have. Since a person with such traits will naturally act in a morally acceptable way, there should be no need to dictate conduct in a particular situation or to establish general moral rules to determine acceptable conduct. An ethical dilemma is resolved by asking how a virtuous person would act in the particular situation. Principle-based ethics emphasizes the role of moral reasoning and analysis in ethical decision making. The core principles of autonomy, beneficence, nonmaleficence, and justice clarify moral duties and obligations. An ethical dilemma is addressed by weighing competing principles, duties, and values. When applied in medicine, these theories are clinician-oriented. A more recent contribution to theoretical ethics is the **ethic of care**. The focus of care ethics is the maintenance and enhancement of caring, while conserving the traditional values of other ethical theories. Care ethics is focused on the humanistic virtues—the characteristics that are valued in interactive, intimate relationships. Ethical dilemmas are addressed by promoting respect for equality while at the same time recognizing and valuing differences.

THEORETICAL INFLUENCES ON GENETIC COUNSELING

The profession of genetic counseling has developed within the medical model and has been influenced by the ethical and moral positions of medicine. It has also been shaped by the individuals who have chosen to practice in the field and view themselves in the context of their relationships. We can gain a greater understanding of the ethical standards that have thus emerged in genetic counseling by more closely examining two theories: the ethic of care and principle-based ethics.

Ethic of Care

Humanistic virtues, those that are valued in relationships, are the basis of care ethics. These include sympathy, compassion, fidelity, discernment, and love. The source of this ethic comes from natural human caring. The ethic of care is based on interpersonal relations, with mutual interdependence and emotional response emphasized. "Care" in this context is the care for, emotional commitment to, and the willingness to act on behalf of those with whom one has a significant relationship. It involves insight into and understanding of someone else's circumstances, needs, and feelings, and a responsiveness to that person's needs as she defines them for herself.

Care ethics is often termed feminine because it emphasizes receptivity, relatedness, and responsiveness, as opposed to logic. The elements of care include relations, attention, compassion, fulfilling needs of others, and helping others to grow as caring individuals. Attachment is the standard for an ethic of care. Reasoning within the ethic of care framework depends on an understanding as opposed to a knowing, with a focus on relationships. It necessitates attention to context and interrelatedness.

The ethic of care perspective is especially meaningful for the role of a genetic counselor. It is a bilateral theory, involving the relationship of two individuals, with the focus on the relationship.

Principle-Based Ethics

The study of what is morally right and wrong in the practice of medicine has evolved into the field of biomedical ethics. Broadly, it has meant a study of the ethics of the life sciences and health care. In a more narrow sense, medical ethics is the code of conduct followed by the medical profession. Beginning with the Hippocratic school, the ethics of medicine has been influenced by virtue- or character-based ethics. In modern times, guidelines derived from principle-based ethics have been used in medical case analysis. The principles of beneficence, nonmaleficence, autonomy, and justice are used to frame the guides of this ethical theory.

The role of a health care provider (e.g., physician or counselor) defines the duties of that practitioner. The specific duties come from the employer, the profession, and/or the expectations of society. From these duties are derived the rights and obligations of the professional. The obligations described below have been organized under one principle only. This is done for practical purposes. Clearly, the same rules can be supported by different principles, and the same obligations can be derived from more than one principle.

Beneficence   Beneficence is the promotion of personal well-being in others. The type of conduct derived from this principle is positive: we should do good; when benefits have been balanced against harms and costs, the outcome should be a net benefit. This principle applies in a society in which one has some discretion in defining one's contributions to the general welfare. It applies specifically in case-related situations, in which the provider and the patient are assumed to have similar values and views of what constitutes a benefit.

Professionally, beneficence requires that the health care provider be a trustee of the patient's welfare. The provider–patient relationship is founded on trust or confidence. The provider is an advocate for the patient, acting in good faith for the benefit of that patient. The fiduciary relationship thus constituted requires honesty and fidelity.

Fidelity requires the provider not to withdraw from a patient's care without notice to the patient, to submerge her own self-interests if they are in conflict with the patient's interests, and to put the patient's health care interests first. The virtues of candor, loyalty, and integrity are derived from this principle.

Beneficence can be interpreted as paternalism when, for example, the professional's perception of her duty of beneficence toward the patient and her definition of the limits of the patient's autonomy do not agree with views of the patient. The refusal by the professional to acquiesce in the patient's wishes, choices, or actions can result in the neglect and violation of the patient's autonomy. In the extreme, paternalism is the opposite of the principle of respect for autonomy. In an emergency, however, paternalism is considered to be justified.

Nonmaleficence   Nonmaleficence, which involves restrictions on behavior as opposed to actions that promote behavior, is framed in negative terms: we should not inflict harm. The rule to do no harm applies to all people in our society. "Thou shalt not kill" is a moral rule reflecting the principle of nonmaleficence.

Nonmaleficence can be seen as an obligation encompassed by or derived from the principle of beneficence. The ethical issues raised by the use of people as subjects in research or for testing experimental therapies are not uncommon. Medical research is both therapeutic and nontherapeutic. People who might be involved in research include groups of healthy adults and children, ill adults and children, people living in controlled situations (e.g., prisoners, members of the armed services), and mentally incompetent adults and children. The use of a person as an object of study or as a means to an end emphasizes the researcher's obligation to do no harm. There is a duty to remain the protector of the life and health of the research subject. International guidelines have been established to protect human research subjects. The Nuremberg Code and the Declaration of Helsinki (Grodin et al., 1993) address ethical issues of research. Voluntary consent prior to participation in medical research has been made an absolute essential.

Autonomy Respect for autonomy is based on the recognition of the intrinsic value of each individual, that person's capacities, and his or her point of view. It represents the individual's personal rule of self, remaining free from controlling interferences that prevent the making of meaningful choices. In medicine, respect for autonomy is especially evident in the recognition of the patient's right to decide what will be done to her body. There are values and obligations derived from this principle, and there are applications of the principle that are relevant both to the health care provider and to the patient. We turn now to some of the important obligations derived from the respect for autonomy that are relevant to genetic counseling, truth telling, confidentiality, and informed consent.

Truth Telling Respect for others requires telling the truth. Both lying and inadequate disclosure show disrespect for other people and threaten any fiduciary relationship. Adherence to the rule of veracity is essential for fostering trust in the medical setting. Truth telling is related to the obligations of fidelity and promise keeping that are inherent in the medical relationship. Truth telling also pertains to the management of information that could affect a person's understanding or decision making. Shielding a patient from the truth is a form of paternalism. Valid consent depends on truthful communication.

The patient also has an obligation to be truthful; she is expected to cooperate in her own care by providing an honest and complete history. She has a duty to inform the health care provider of any important issues and to make it known whether she has a clear understanding of the material presented by the provider.

Confidentiality Confidentiality relates not only to the communication between people but to the fact that a relationship exists between them. In a fiduciary relationship, any communication that is intended to be kept secret is classified as confidential. A relationship with a genetic counselor or other health care provider is a confidential relation. It is one of trust. The patient relies on the discretion of the counselor.

It is a goal of society to maintain the public health. It is in the public interest for people to seek and obtain health care. Much of medicine involves the individual's having to reveal personal, intimate information. Without access to such information, a provider often cannot effectively help a patient. Thus the patient must be able to trust the provider to respect her privacy and the confidentiality of their relationship, which in turn implies the obligation not to share any privileged information without the patient's consent. The patient does not have to ask to have the information kept confidential, because confidentiality is the standard of care. Any breach is a violation of the duty of fidelity. If a practitioner has any standard exceptions to complete confidentiality in her practice, it is her duty to disclose those exceptions to the patient before accepting any confidences.

There is a difference between confidentiality and privacy. Privacy relates to limited or restricted access to an individual. This relates to the person herself, to objects intimately connected with her, and to information about her. A person controls the relationships she has to some extent by controlling what personal information she shares with others. She determines what she wants others to know by allowing them access through a "for your ears only" exchange. The relationship does not have to be professional. Sharing information that has been given in this way, as a secret, without the person's consent, is a violation of her privacy.

Confidentiality can be waived by the patient. For example, a waiver has occurred when a patient signs a health insurance form that includes a release for disclosure to that third party. A waiver also is implied when the patient knowingly shares confidences in front of a third party who is not a member of the health care team. A patient who later contemplates bringing a malpractice suit in which her medical information is an issue will find that she had waived confidentiality in respect to that information.

There are exceptions to the expectation of confidentiality. A state can require by statute that certain information be disclosed. Mandatory reporting usually is restricted to situations in which the value of confidentiality conflicts with the goal of protecting the public health, as in cases of child abuse, or the requirement to report birth defects, specific communicable diseases, or the results of newborn screening. The area of discretionary exception to confidentiality is very controversial. Because of its importance to genetic counseling, we will look closely at the arguments put forward on both sides of the issue in a later section.

Informed Consent   Informed consent has been obtained when a patient with substantial understanding, and in the absence of control by others, intentionally authorizes a professional to do something. Informed consent is often considered to be the goal of the patient–provider interaction and is an important part of the process of shared health care decision making. A patient's autonomy is manifested in her right to make her own health care decisions, including declining treatment and making use of experimental therapies, and in her privacy and bodily integrity. This obligation is especially relevant to research involving human subjects.

There are five elements to informed consent that must be met for the consent to be truly informed. The threshold element is **competence**, or the capacity to make a rational choice. To be judged capable of making a rational choice, a patient must be rational, as well as able to communicate a choice, and able to understand the information provided and the consequences of the choices available. A physician is considered to have the ability to make the judgment as to a person's capacity.

"Competency" is a legal term. A person is presumed to be competent in making decisions regarding her health care at 18 years of age. Minors can be emancipated as determined by their state statutes to make some or all of their own decisions. A person who is competent to authorize her own treatment is also competent to refuse treatment. A court will determine the competency of an individual if the question is raised. There is a continuum of competency. The criterion that is applied depends on the context of the task. For example, an individual may be considered competent to sign a will while at the same time not competent to make health care decisions.

There are two information elements to informed consent. The first is the **amount and accuracy** of the information provided to the patient. Disclosure of the possible benefits and risks of an intervention (e.g., chorionic villus sampling) is an obligation of the professional. Disclosure also includes the obligation to discuss the available alternatives (amniocentesis, triple screen, targeted ultrasound scan). The rule of truth telling, or veracity, is essential to this element of informed consent. The second necessary information element, **patient's understanding**, presents a myriad of barriers to informed consent. Patients may be fearful, sick, or uneducated; they may hold unscientific beliefs; they may be in denial; or they may not speak the same language as the provider. It is the obligation of the provider to identify these barriers and endeavor to overcome them.

Consent also involves **voluntariness** and **authorization** by the patient. As used here, voluntariness means the absence of control by others. The absence of control must be substantial: that is, the patient's authorization has to be an active agreement, an agreement reached not simply by yielding to or complying with a suggestion from the provider.

There are exceptions to the need for informed consent. For more common, low risk interventions where the risks and benefits are obvious, explicit agreement may not be required, and consent is implied. In emergency situations it is commonly assumed that patients are unable to make decisions about their care or to participate in it because of pain and fear, lack of understanding of the danger, or an unconscious state. An objective, "reasonable person" standard is applied in these situations. That is, if a reasonable person under the same or similar circumstances would consent to treatment, then consent is presumed.

Justice   The principle of justice is sometimes explained as giving all patients their due, or treating "like" people in a "like" manner. In health care systems, four values are recognized. **Equality** is the provision of equal care for all, while **liberty** is the freedom of choice, for both the provider and the consumer. **Excellence** is the provision of the best possible care for everyone, while **efficiency** is a broad category that includes the containment of health care costs.

The application of justice involves the intersection of ethics and public policy. This is best illustrated through distributive justice. On one level is the distribution of resources. When resources, such as donor organs, are scarce, we want to know who has access to those resources and how is such qualification determined. At this level, justice implies equal access to services. On a second level is the distribution of risks, which is well illustrated by population-based genetic screening tests. The benefits and burdens of false positive and false negative test results are considered when establishing cutoffs for distinguishing normal versus abnormal screening results. This distributes the benefits and burdens of the testing most equitably.

CODES OF ETHICS: THE NSGC CODE

In general, the code of ethics of a profession presents the moral obligations deduced from the kinds of activity in which the members of the profession are engaged. Traditionally it consists of a rational and systematic ordering of the principles, rules, duties, and virtues characteristic of the profession and the intrinsic achievement of the ends to which the profession is dedicated. A code can express duties and goals in rules that all members are required to obey and/or in ideals toward which all members are encouraged to strive. An enforceable code lists the duties that are required, with penalties for failure to perform them. A professional who wants to behave ethically should be able to use the code of ethics of that profession as a guide.

The code of ethics written for the members of the National Society of Genetic Counselors (1992) presents ideals for the profession of genetic counseling. These are goals towards which the practitioner strives. Such a document does not put forth rules to follow but offers guides for the pursuit of the ideals.

The NSGC Code of Ethics was written from the perspective of the ethic of care, which is defined by interpersonal relationships (Benkendorf et al., 1992). The major relationships established by genetic counselors were used to organize the values, principles, and beliefs that are defined by the profession of genetic counseling. These relationships were identified as being with oneself, the client, colleagues and society. Chapter 11 presents a complete discussion of the NSGC Code of Ethics.

CASE ANALYSIS

When a dilemma or an ethical conflict arises in the course of a clinical interaction, it is not always necessary to resolve it immediately. It is helpful to sort out the facts and separate the real dilemmas from disagreements over the use of language or problems with communication. When such issues have been clarified, ethical principles can be applied and an acceptable resolution achieved. The steps listed in this section are a combination of the approaches of several theorists. They are offered as a guide to the process of analyzing a case. The case analysis that follows demonstrates how ethical principles can be used to study a particular situation. Case analysis is applicable to both clinical and laboratory work.

A genetic counselor, Anne, receives a phone call from a friend and colleague, Carlos, requesting confirmation of a diagnosis in a young man, Chris. Carlos has a client, Diane, who is related to Chris. If the suspected diagnosis is confirmed, Diane will be at risk for being a gene carrier. Chris lives in Anne's geographical area. It is most likely that he has been seen by Anne. Anne is told that time is of the essence. Diane has not been able to get a release from her relative, and she is now in her second trimester of pregnancy. The cast of characters is represented in Figure 10.1.

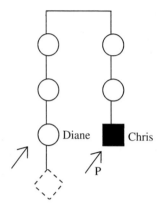

FIGURE 10.1   *The relationship between clients Diane and Chris. Diane is Carlos's client. Chris is Anne's client. Chris has an X-linked disorder.*

1. The first step is to gather all the facts, that is, to hear the story of the case. Before beginning the analysis, be sure that there is no information you want about the case, including social, medical, or genetic facts available to date.

Diane was referred to Carlos because of her family history (i.e., because of Chris's problems). Carlos has surmised from the description of those problems that Chris has an X-linked recessive disorder. Chris and Diane are second cousins, their maternal grandmothers being sisters. Anne recognized the description of Chris. Anne does in fact know Chris and his diagnosis. When Anne requested a signed release of information, she was told that Diane had already asked for one but had been refused. Chris's family's response was inconsistent with Anne's knowledge of the family's understanding and concern for other family members. Anne's dilemma: Should she release the diagnosis?

Further information would be helpful in sorting out the issues. For example, how long has Diane been aware of the possible risk? Would knowing her cousin's diagnosis make a difference to Diane's management of her pregnancy? Has the fetal sex been identified on ultrasound scan? We also need to know whether timely, accurate carrier testing and/or prenatal testing is available for Diane, possibly without involving her cousin. These are a sample of the questions that can be asked in this case.

2. The cast of characters of the story has to be defined: Who are the interested parties? Identify the role of each in the case. Who are the decision makers? Who has a stake in the outcome of the case? Who has what information (patient's wishes, family history) about the case? What is the patient's capacity to take part in any decisions that will be made?

The responsibilities of each party should be clear, as should the rights and obligations of each. Any outside influences need to be identified, such as hospital policies or legal implications.

In our scenario the cast includes genetic counselors Anne and Carlos, and clients Chris and Diane. We should not, however, overlook the interests of the fetus, the father of the fetus, Anne's institution, and society in general. If Chris were a child, his guardian would have to be included in our list. We know, however, that Chris is an adult.

Focusing on the main characters, we can state that Diane has an interest in obtaining information so that she can make informed decisions after weighing the risks and benefits of testing. She also has a responsibility to bring the family history to the attention of her doctor in a timely fashion and to pursue a release of information from her family. Carlos has an interest in and duty to provide the best care for his client, and to use his resources to gain important information for his client. He also has a responsibility to know the standard of practice regarding confidentiality and a duty to encourage a colleague to act ethically.

Chris has a privacy interest. Not only is his medical information confidential, the fact he was seen by Anne is confidential. There may be a question of whether he has a duty to his family to share his diagnosis.

Anne has a duty to respect the privacy and confidentiality of Chris. She also has a duty to conform to her profession's code of ethics. If a risk exists, does she have a duty to breach confidentiality and warn Diane? Anne may have a legal duty not to breach confidentiality if her state recognizes a genetic counselor–client privilege— that is, if the confidentiality of the client's information is accorded legal status. (The question of a duty to warn is discussed in more detail later in the chapter.)

3. Defining the problem (asking what decisions are to be made) is important for separating the problems of definition or communication from the actual ethical dilemma(s). The way people define a moral problem, the situations they interpret as moral conflicts, and the values they use in their resolution are all functions of social conditioning.

The primary problem in this case is the weighing of the benefits (mainly to Diane) of having medical information against the harms (mainly to Chris) that might be caused by breaching confidentiality. Breaching (or being asked to breach) a professional code is also a harm to the counselor, Anne.

4. Identifying the principles involved in the conflict will help clarify where the various obligations lie. The content of the obligations and any interobligational conflicts should then become apparent. You can then examine the priorities of the parties involved.

Anne has to weigh the principles of beneficence (helping Diane in the decision-making process and helping Carlos fulfill his responsibilities and nonmaleficence (avoiding unnecessary pregnancy risks and anxiety) against respect for autonomy (Chris's), and nonmaleficence also defined as undermining Chris's trust in the profession).

5. In resolving conflicts, you will find it helpful to use all your available resources, including personnel such as colleagues, religious leaders, hospital ethics committees, and an attorney if necessary. In gathering support information, you should also refer to other cases presented in the literature, institutional policies, professional practice guidelines, and the NSGC Code of Ethics.

6. In determining a course of action, apply the principles of the profession in your assessment of the burdens and benefits of all alternatives. Any action you decide to take must be ethically justified on the basis of a good result, primarily for your client and secondarily for any others who may be affected.

7. It is important to evaluate the course of action you pursued. Did it resolve the dilemma presented? Is any follow-up action necessary? Can another such conflict be avoided in the future? How will the individuals involved deal with the outcome?

Anne's solution to the dilemma will be shaped by how she weighs her various duties. If she feels that Chris's confidentiality should be respected at all costs, she will refuse to provide any information without a signed release. This position can be defended by calculating the theoretical risk of harm to Diane's fetus and by interpreting Diane's statements as reluctance to pursue the release of information from her cousin. If Anne's goal is to protect Chris's confidentiality while providing some help to Diane, she can contact Chris before responding to Carlos. Anne can ask Chris if he would be open to sharing his diagnosis. On the other hand, if Anne has made unsuccessful attempts to obtain consent for releasing the information and believes that the benefits to Diane and the fetus outweigh the harms to Chris, she may elect to provide Chris's diagnosis without a release.

LEGAL ISSUES

The material in the remainder of the chapter is presented to alert you to legal matters of interest to genetic counselors. The discussion is not meant to provide specific legal advice. If such advice becomes necessary, the services of a qualified attorney should be sought.

Relation Between Ethics and Law

As discussed above, moral values result from a combination of cultural factors such as shared values, institutions, and traditions. The rights and obligations derived from this "community conscience" are the moral minimums that are expected of all members of society. Immoral, unacceptable conduct is deterred, prevented, or punished by sanctions imposed by society. Setting boundaries for behavior is the function of the law.

We have already stated that "rights" are justified claims that an individual or a group can make on others or on society. Moral rights—for example, the right to move about freely—are derived from the community conscience. They encompass the right to make and act on one's own choice, to control one's own actions and affairs. They also include the right to certain social benefits and goods, or entitlements. Education is an example of a modern entitlement. Having a right presumes the existence of an obligation on the part of someone else to act in a particular manner. My right to move about freely (e.g., to enter a health facility) obligates others not to block the facility entrance (i.e., to unjustifiably restrict my movements).

Rights can be violated or infringed. A violation is unjustified (blocking the entrance to a health care facility), while infringement is a justified overriding of a right (requiring children to be immunized before they enroll in public school).

A legal right is defined by what is legally allowed. Rights are conferred, granted, given, awarded, bestowed, or gained through the law. These rights are not absolute or final. They can be revoked or infringed. An individual has, holds, and owns these rights. A set of rules (i.e., our legal system) has been developed so that people can exercise their rights. The legal system has been established as a mechanism for identifying and settling disputes about rights and enforcing rights by applying rules. It also allows for judging among claims to rights that may seem to conflict.

Sources of Law

The laws that impact health care services can be found on the federal and state levels. On the federal level, we have the Constitution, federal statutes, and case law. The Constitution is the supreme law of the land. It protects certain individual liberties, such as the freedom of speech, from federal and state government abuses. It limits the actions the government can take and mandates certain specific governmental obligations (e.g., to convene Congress at least once a year). Any federal or state law that is in conflict with the Constitution is subject to be declared unconstitutional by the U.S. Supreme Court.

Many federal laws have an impact on health care in general and genetics in particular. The federal Privacy Act of 1974 limits the disclosure of information obtained by employees of federal services, federal agencies, and government contractors. The Rehabilitation Act of 1973 and the civil rights statutes dating from the 1960s prohibit the infringement of a person's rights by private entities involved in employment, housing, or public accommodations on the basis of race or sex, for example. The Americans with Disabilities Act of 1990 adds disabilities to the characteristics covered by the civil rights statutes and extends the Rehabilitation Act to cover private businesses.

Federal laws are enacted by Congress, while the Supreme Court interprets the law through the cases it hears. Supreme Court decisions on issues regarding the federal constitution are binding on all state and federal courts. For example, through a series of cases heard by the Supreme Court regarding a woman's right to make reproductive decisions for herself, the fundamental right to privacy, to make those decisions without undue government interference has emerged (see, e.g., *Roe* v. *Wade*, 1973). It also has become clear, however, that the government is not obligated to ensure that the individual can act on those decisions, by for example, providing facilities or funding. [The decisions in *Maher* v. *Roe* (1977) and *Harris* v. *McRae* (1980) address this issue.]

Within the federal judicial system, a body of case law has developed from the decisions addressing federal law and federal constitutional rights. Once a court decision has been reached, a precedent has been created for future decisions by that court and by lower courts.

For areas regulated by both the federal and state governments, a state may be stricter than the federal government but not more lenient. For example, it is possible to have both federal and state genetic privacy acts. The federal statute may protect genetic information from misuse by public agencies and private enterprises dealing with the public, while a state might elect to expand that protection to cover insurance companies. However, a state could not allow the use of genetic information within the state in a manner proscribed by the federal statute.

States also have constitutions, and the state legislatures enact laws. State court judges look first to statutes and the case law developed in their respective states. When no state precedents on an issue are found, judges may consider approaches developed in other states.

Some areas of law, such as public health, insurance, and professional licensing, are left solely to the states to regulate. State laws that infringe on fundamental rights such as reproductive decisions may be upheld only if they are shown to meet a very high standard of importance to the state.

A common problem for professionals is the lack of statutory consistency or uniformity from state to state. A professional can be licensed in one state and not another. Privileges that are recognized in one state may not exist in another, and the limits of confidentiality obligations on health care professionals may differ. The federal government does not usually address these inconsistencies.

The law is constantly developing and changing, through both the enactment of laws and their interpretation. A professional should be familiar with the laws in her state that apply to her work and should be aware of changes as they occur.

Medical Decision Making by Minors

At the age of 18 years, in the absence of cognitive impairment, a U.S. citizen is considered legally competent. He or she can make personal health care decisions and enter into contracts. For adults who cannot decide for themselves, family members are the first choice as surrogate decision makers. A court can appoint a surrogate when necessary. For someone who had been competent but is now comatose or mentally ill, a judge may apply substituted judgment using advance directives or past conversations. For the individual who has never been competent (e.g., an infant or an intellectually impaired adult), a proxy acts as an advocate and uses her own judgment as to the individual's best interests.

Since parents have direct sovereignty over minor children, they are entitlted to make decisions on behalf of their offspring. This practice is derived from the notions of family privacy and bonding, parental autonomy, and the legal responsibility of parents for the care and support of their children. Parents are presumed to have the best interests of their child at heart. This presumption can be rebutted by evidence of neglect or abuse. A parent may consent to therapy or decline it, but may not refuse lifesaving therapy for a child. Court-ordered blood transfusions are a familiar example.

Most children lack the capacity to make appropriate health care decisions for themselves. The participation of a minor in her own health care decisions changes with age and circumstances. As a child acquires the capacity for moral reasoning and the cognitive skills to recognize cause and effect, and develops a sense of the future, she is entitled to participate more in health care decisions. In some states the ages at which a minor may consent to various treatments or procedures are defined by statutes. A minor is recognized by most states as emancipated (i.e., accorded adult status) under circumstances such as marriage, pregnancy, or financial independence. The status of a mature minor is recognized and applied to adolescents who though not yet 18 years old are able to demonstrate to a judge that they possess a certain level of maturity and understanding of the consequences of medical decisions. This status usually is invoked in situations in which the state has an interest in encouraging the adolescent to seek medical care that she might not pursue if her parents were to be informed, as with the treatment of venereal disease. The President's Commission for the Study of Ethical Problems in Medicine and Biomedical and Behavioral Research (1983) recommended that to the extent possible, individuals who are recognized to lack decision-making capacity still be consulted about their preferences out of respect for them as individuals.

The Law and Genetic Counseling

A medical malpractice suit is a public accusation of wrongdoing that has not yet, and may not ever, be proven. Usually such an action is brought by an individual against the provider. Malpractice issues for genetic counselors could be expected to be addressed most often in terms of the laws that govern medical malpractice: contract, battery, and negligence.

The nature of the provider–patient relationship is fiduciary and can be considered contractual. All transactions between the parties in the relationship require a high degree of good faith and fairness. One party (the patient) places her confidence and trust in the other (the provider). She incorporates the representations of the provider into her thinking and behavior. Certain conditions (e.g., the maintenance of confidentiality with respect to disclosures made by the patient) are understood or implied. There should be no failure on the part of either party in regard to these understandings. Some rules of law that may apply include the following:

Neither party should exert pressure or influence on the other.

Neither should take selfish advantage of the trust of the other.

Confidences shared should not be used to benefit one or disadvantage the other without full disclosure and consent.

When a provider agrees to take care of a patient, an obligation to properly attend to the patient is established. This obligation lasts as long as both parties agree to the relationship, or until the relationship is affirmatively terminated by either party. For a specialist, this duty ends when the consultation or testing is completed. A provider who totally neglects a patient or fails to give the patient any care or attention while a need for continuing care exists may be considered to have abandoned that patient.

Such abandonment also constitutes negligence. Abandonment can be a breach of a contract with the patient to provide professional services, but this is not a common basis for suits in medicine.

A second basis for malpractice is battery. Battery is unwanted, unconsented touching. The legal prohibition of this type of behavior is reflected in the necessity for consent. It is not necessary that there be physical injury, financial loss, or an unsuccessful treatment. The injury can be to personal dignity, creating the feeling of having been insulted or violated.

Most malpractice cases are the result of real or perceived negligence. Suits in genetics usually are based on claims of injury to a patient because a provider either did not use or misused certain information or a technique. Examples of misuse of information include failure to recognize a genetic disorder in a child, a misdiagnosis, the ordering of a wrong test, or the misinterpretation of a laboratory result. Failure to act may also be the basis of negligence. Not taking a family history, not ordering the appropriate tests, and not identifying a high risk situation are examples of failure to act. Failure to provide complete and accurate counseling has been the basis of negligence claims. The provision of incorrect recurrence risks for hearing loss were the basis of *Turpin* v. *Sortini* (1981), while *Schroeder* v. *Perkel* (1981) claimed a failure to diagnose a child's cystic fibrosis.

If a person thinks she has been injured by a battery or by negligence, she can initiate a malpractice suit. Malpractice is a tort action. A tort is an injury or a wrong done by a private person, as opposed to one done by the government or a government agent. The injury can be to the person herself or to her property, as the result of a negligent act or intentional misconduct. Only injuries for which compensation in money damages can be recovered are included.

A tort action has four elements that must be proven by the person who brings the suit (the plaintiff):

1. There was a provider–patient relationship and a duty owed by the defendant (the person being sued) to the plaintiff.
2. The defendant breached that duty either by not doing what should have been done (failure to act) or by doing something improper (deviation from the standard of care).
3. The breach of duty was the direct cause of the harm suffered by the plaintiff.
4. An actual injury (physical, financial, or emotional) resulted, which can be compensated for by a ruling of the court.

The first element, **duty**, is shown by looking at the standard of care of the profession. The substance of duty is established through the practice of the profession. A professional who represents herself and provides services as a specialist must possess and apply the knowledge, skill, and care that a reasonable, well-qualified specialist in that field would use in a similar case or circumstance. An appropriate (not necessarily the highest) level of skill and knowledge can be found through national certification requirements, practice guidelines, and codes of ethics. These provide an objective, uniform measure of the standard of care for that professional.

The plaintiff then has to show that the provider did not conform to those standards (i.e., she *breached her duty*) either in what she did or what she did not do. The **causation** element is usually most difficult to prove. The birth of a child with defects is often the damage claimed by parents and is said to have resulted in two specific torts, wrongful birth and wrongful life.

The tort of wrongful birth is a specific negligence tort; it can be brought by parents when a child is born with a disorder or a defect. The defendants could be a physician and/or other health care providers, the hospital or university that employs them, and/or support services such as a laboratory. The parents have to show that if they had known there was an increased risk, they would have avoided the birth of an affected child. The duty that was owed to them comprised accurate genetic counseling and an explanation of risks and available tests. By not providing the plaintiffs with the information that was needed, the defendant breached her duty and deprived this set of parents of their right to make an informed decision. A breach could involve the failure to use due care in performing and interpreting tests, to diagnose a genetic condition or ascertain the genetic nature of a condition, or to inform the plaintiffs of accurate risks and available tests. The parents then claim that the breach of duty they have demonstrated is the cause of the birth of the affected child, since they allege that if they had been properly warned, they would have avoided the child's birth.

A claim can be brought by a child against health care providers if it is desired to assert that the child should not have been born. In such wrongful life suits, an affected child claims there was a duty owed to his or her parents by the health care providers to inform them of possible defects in future children. As in the wrongful birth suit, it is claimed that failure to meet that duty deprived the parents of the right to make informed decisions, resulting in the birth of a child with defects, and possible pain and suffering. The child has to assert that it would have been better not to have been born at all than to be born with defects and with pain and suffering. These claims are not usually successful. The courts usually do not see life itself as an injury.

Fraud is another example of a tort that may involve genetic counseling. A pattern of unfounded statements of reassurance are fraudulent, as is the making of promises that are known to be false. If a patient claims to have been intentionally misled by a provider, she will further assert that having relied on the provider's false counsel, she was prevented from making an informed decision.

Law suits are brought for different reasons. Some are used to punish the wrongdoer, while others are used to challenge existing laws. The remedy sought is a compensation for actual injuries.

The remedies for malpractice depend on the injury claimed. A contract can be enforced by the court. Money damages can be awarded for breach of contract or tort claims. Administrative sanctions can include suspending a license or funding, or assessing fines.

No one likes to be sued. It is devastating, insulting, embarrassing. It uses up valuable time and resources. You can do everything according to the standards of the profession. You can follow every suggestion here. There is still no guarantee that you will never be sued or that you will prevail if you are named in a suit. You can, however, minimize the possibility of finding yourself in court by observing the following recommendations, both as an individual and with your colleagues as a profession.

1. *What Can Be Done as a Profession?* As mentioned above, courts look to the standards of the profession to ascertain the duties and obligations of its practitioners. A profession can actively set its own standards by developing practice guidelines and by offering certification and recertification. If the individuals who practice the specialty neglect to act in these matters, standards may be set by legislators and other health care providers.

The profession of genetic counseling has taken an initiative to assure the quality of its practitioners by establishing the American Board of Genetic Counseling. This entity administers a national certification examination. The profession's national organization, the National Society of Genetic Counselors, Inc. (NSGC), also sponsors and supports the development of practice guidelines. It provides two vehicles (*Perspectives in Genetic Counseling* and the *Journal of Genetic Counseling*) for the publication and disbursement of these guidelines. The Code of Ethics, accepted through ratification by the NSGC membership, is another source of the profession's standards. When more than one group of professionals (such as M.D. and M.S. genetic counselors) perform the same service, each is expected to meet at least a common minimum standard.

2. *What Can Be Done as Professionals?* It is most important to be aware of and comply with the standards of practice of the profession. The ultimate goal is to provide the highest quality of care, not because doing so will keep you out of court, but because it is an ethical principle of the profession. Reasonable errors in judgment should not lead to malpractice claims. Other considerations include the following.

(a) *Effective communication is paramount.* Communication, both with the client and with colleagues or team members, should be based on realistic expectations. Informed consent is a goal of the dialogue that involves the disclosure of the information the client needs to make an appropriate decision. A signed consent is a record of that communication process. It does not replace the process, but is written evidence that the communication took place. In addition to a formal consent form, it is helpful to attach to the client chart a note stating what was said, that the client understood the material, and that all her questions were answered. This may seem obvious, but it cannot be said too often: open communication is a key element in preventing malpractice claims.

(b) *Keep accurate and complete records.* It is essential to document what was done and why it was done. Documentation provides a record of the information given to you by the client and by you to the client. Significant telephone calls as well as important face-to-face conversations should be included. If possible, notes should be typed. At a minimum, they should be written in ink. Each page in the chart needs a signature. Knowing what not to write is as important as knowing what to include in the chart. Abbreviations and acronyms that have multiple interpretations should be avoided. Gratuitous remarks about a client (personality or appearance) are not useful. Open criticism of another professional's judgment should not be included. Careful documentation can be used to support your position that professional standards of care were met. Chapter 7 addresses these issues in more detail.

(c) *Preserve patient confidentiality.* When you are asked to release information from a client's chart, only information generated by your office should be included. Records you used to form your opinion may be included. Do not provide information on the telephone unless you are absolutely certain the person you are speaking with has the client's permission to hear it. Middle names and social security numbers are good identifiers in this situation.

(d) *Do not practice without malpractice insurance.* Confirm your insurance coverage with your employer. Insurance policies can be purchased through your professional organization if an employer does not provide coverage.

(e) *Be prepared to terminate an unsatisfactory relationship.* Barring any statutory obligations, a professional is free not to enter into a relationship with a particular client. If you need to sever your relationship with a patient, it is prudent to take affirmative action to notify the patient, suggest substitute care, and document why the relationship was severed.

AREAS OF PRACTICE RAISING ETHICAL AND LEGAL QUESTIONS

Genetics is at the cutting edge of science, ethics, and law. The possible applications of genetic information have ramifications that go far beyond medical care. Ethical, legal, and social considerations should be in the forefront of all aspects of genetics and have been incorporated as an essential part of the Human Genome Project (Andrews et al., 1994).

The following topics are a selected few of the subjects that are much debated in genetics. They represent some of the ethical dilemmas in genetic counseling and are used here to stimulate continuing discussion by genetic counselors. The pros and cons found in the literature for each topic are presented.

Sex Selection

The desire to have offspring of a chosen gender was not created by the availability of prenatal techniques. Cultural, economic, and political pressures throughout history have resulted in the development of an astonishing array of pre- or periconception techniques used by parents to influence the gender of their offspring (Jones, 1992). Today, medical techniques for the preconception, preimplantation, and postimplantation identification of sex are available. Because the techniques applied following implantation are inevitably associated with abortion, they stimulate the most discussion.

Arguments in support of sex selection emphasize social and economic benefits. Beside the identification of a fetus with a sex-linked disorder (e.g., hemophilia, Duchenne muscular dystrophy), positive factors are cited that include benefits to the family by enhancing parent–child compatibility, reducing neglect and abuse based on sex, and reducing the number of unwanted children. Benefits to society include reducing the birthrate and secondarily improving parent–child relations (Warren, 1985). Legal arguments emphasize the constitutional protection of procreative liberty and privacy (Robertson, 1990). As long as the reason for termination is not questioned in any other context, how can such questioning after prenatal diagnosis be justified?

Those who argue for the prohibition of sex selection find that the harms outweigh the benefits. Deleterious effects on parents, besides the costs in time and money, include the guilt and physical effects of repeated pregnancies and terminations. A failed attempt at gender selection could lead to rejection of the child of the undesired sex and increased burdens on any children of the desired sex. Society is seen to be harmed through the reinforcement of sex stereotypes and the devaluation of the human worth. A slippery-slope argument suggests that sex selection is a first step toward more intrusive genetic engineering. A shift in the use of technology from a physician–patient model to a service provider–consumer model could lead to market exploitation of the technology (Jones, 1992).

Can we solve this dilemma in a way that does not limit women's rights or involve termination? The request for sex selection, although limited, is a symptom of a greater social and cultural problem of the unequal value assigned to people on the basis of gender. The suggestions for solutions that address the issue in terms of legal prohibition, professional coercion, or moral sanctions serve to reinforce in part and ignore in part this basic problem. The elimination of the economic and cultural basis of the need for sex selection will make it less desired.

For genetic counselors, sex selection highlights the necessity of balancing the need to be culturally sensitive with the need to follow our ethical obligations to ourselves. Although Fletcher and Wertz (1987) found that M.D. and Ph.D. geneticists were supportive of prenatal diagnosis for sex selection, only 32% would do the testing in their centers. Because the purpose of prenatal diagnosis for sex identification is to trigger the abortion of the undesired-sex fetus, many institutions do not accept it as an indication for testing.

Presymptomatic Testing in Children

The advances in molecular genetics that have led to presymptomatic or predisposition diagnostic testing raise questions regarding intrafamilial relationships, disclosure, and confidentiality. When the testing is applied to the care of children, these issues are magnified. Such testing has the potential for great benefits and great harms.

In assessing the risks and benefits of testing in the process of proposing guidelines for the testing of children, Wertz et al. (1994) categorized genetic tests in terms of their utility. Testing can offer immediate medical benefits for the child, as in the case of newborn screening for treatable disorders (e.g., phenylketonuria), or it can be of benefit to the older child when making reproductive decisions (e.g., carrier testing for fragile X syndrome, Charcot–Marie–Tooth disease). Some testing that does not offer benefits to a child is nevertheless requested by the parents or the child, or done to benefit another family member.

Presymptomatic testing for children may be considered beneficial for several reasons. Psychologically, such tests benefit parents by removing worries due to uncertainty, by avoiding resentment from children later in life, and by facilitating the planning necessary to provide support for an affected child. Social benefits include taking advantage of available technology and helping to prepare for the future. And presymptomatic testing offers such medical benefits as encouraging vigilance in health care or eliminating the need for intensive medical surveillance. Parents consider it their role, responsibility, and right to decide whether and when to test children (Mitchie et al., 1996).

The harms that may result when children are tested include disturbance of parent–child or sibling–sibling relationships, damage to a child's self-esteem, unwarranted anxiety related to the anticipation of symptoms, and the removal of the child's right to decide whether to be tested as an adult.

At present, the testing of children cannot be justified for disorders in which symptoms are rare in childhood and for which no treatment is available (e.g., Huntington disease). A committee of the British Clinical Genetics Society (Working Party, 1994) recommended that there be a general presumption against testing when there is no direct health benefit to the child. The American Society of Human Genetics (ASHG) and the American College of Medical Genetics (ACMG) (1995) presented a joint report suggesting points to consider when parents raise the issue of childhood testing. The report emphasized the importance of medical, social, and psychological issues. The NSGC, while endorsing the ASHG/ACMG statement, developed its own resolution addressing the issue (1995). The positions of these organizations concurred on, among other things, the acknowledged benefits and risks of such testing. They differ in that the NSGC addresses prenatal testing. Some of the discussion on these statements can be found in *Perspectives in Genetic Counseling* (vol. 17, no. 4, and vol. 18, no. 1). Although there does not seem to be a consensus regarding the time at which to test children for disorders for which medical intervention is available (e.g., familial polyposis), it is agreed that testing should be considered by the time clinical surveillance would be initiated.

The Use of Genetic Information for Discrimination

Genetic discrimination occurs when genetic information is used to differentially treat individuals, deny them normal privileges, or to treat them unfairly. The issue is particularly important in the areas of employment and insurance coverage, when a person's privacy rights are in conflict with the rights of the employer or insurance company to have the information needed for determining the extent of coverage.

Federal law (the Americans with Disabilities Act) prohibits discrimination in employment on the basis of disability. Disability is broadly defined by the ADA as a physical or mental impairment that substantially limits one or more life activities. A person who has a record of having such an impairment or is regarded in this light is defined by the act as disabled even when there is no apparent incapacity. Included in the definition of disability, then, are people with a genetic predisposition and those who are asymptomatic carriers of a late-onset disorder.

Employers and insurers are permitted to consider risks in making underwriting decisions and to refuse to cover preexisting conditions. This can lead to a restriction of job mobility for people with genetic conditions. Other questions are raised as well: What is the reliability of genetic tests in predicting future health status? How is "preexisting" to be defined? Who decides which tests are appropriate? and Who will pay for them?

Insurance companies are regulated by the states. Until recently, state laws tended to focus on testing rather than on the information itself. Hudson et al. (1995) presented recommendations for the protection of individuals against genetic discrimination. However, insurance companies have a strong financial incentive to avoid risk. In assessing rates, insurance companies use information on current health status, family history, and medical records. Not being truthful on an application for insurance about a known future risk is fraud and grounds for denying or canceling coverage.

Duty to Warn Third Parties

The responsibility of health care providers to third parties is a developing and controversial area. Genetic information can be relevant to those who are related to a genetic counselor's client. A duty to warn third parties can be derived from the counselor's responsibility to society.

It can be argued that the impact of a client's genetic information on the risks for her relatives and their children puts this information into a category that is different from other medical information. This difference is thought to limit the provider's duty of confidentiality. We can consider the issue from the point of view of those third-party relatives who seek the information and of those who may have risks of which they are unaware.

The President's Commission for the Study of Ethical Problems in Medicine and Biomedical Research (1983) supports the position that the duty to prevent harm may at times limit the professional's duty of confidentiality. They outline conditions that should be met before a patient's confidentiality is breached. First, reasonable efforts to convince the patient to consent must have failed. Second, there should be a high probability that harm (risk of occurrence) will occur if the information is not shared and that the information will actually be used to avert that harm. Third, the risk must be to an identified third party, and the disorder must be a serious one. Finally, only the genetic information needed to prevent that harm (i.e., needed for diagnosis or treatment) should be shared. The commission strongly advised that any professional considering breaching a patient's confidentiality have the circumstances of the case reviewed by an appropriate third party. There is no definition given of "reasonable," "high probability," or "serious" disorder.

In developing a code of ethics for the Council of Regional Networks (CORN), Baumiller et al. (1996) incorporated the presidential commission's recommendations. In their Code of Ethical Principles for Genetics Professionals, they recommended that in the infrequent circumstance that there will be great harm, a provider can override the patient's confidentiality after first informing the patient. Again, "great" harm is not defined, and we do not know how that is related to the seriousness of a disorder. From this point of view, it would not be acceptable for a genetic counselor to contact relatives without consent to inform them that they are at risk for an untreatable disorder or for a disorder for which the testing is common, public knowledge.

The reasons for *not* sharing a patient's genetic information without consent are well presented by Fost (1992). He points out that genetic counseling situations do not usually involve imminent life-threatening risks to third parties. He also emphasizes that the doctor has a special relationship with the patient but none with her relatives, who are strangers to the provider. Genetic information involves areas of life that are very personal, such as one's identity and reproductive fitness. Privacy concerns should be of essential importance. Fost points out that there are no legal mandates for reporting genetic information to third parties, and telling at-risk relatives is not a recognized obligation or duty. Any harm suffered by that third person was not intentionally inflicted by the patient, but is the result of an act of nature. Disclosure can cause financial harm in the form of higher insurance rates, loss of employment opportunities, and possible loss of insurance, as well as emotional harm. The possible consequences of disclosure and nondisclosure should be carefully weighed.

Both Suter (1993) and Pelias (1991) approach disclosure to third parties from a legal perspective. These authors agree that at this time there is no legal mandate to warn third parties of a risk that may be posed by the genetic information of a client. They emphasize that a genetic professional has the primary duty to respect the privacy, autonomy, and confidentiality of the client with whom she has direct contact.

The responsibility to a party who is not your client is emphasized in the special situation of monozygotic twins. Heimler and Zanko (1995) confronted this issue when a request for predictive testing for Huntington disease was made by one of a pair of monozygotic twins. The authors required the third party (the other twin) to participate in the testing situation. This position stimulated some debate in the genetic counseling literature (Hodge, 1995; Reich, 1996).

Encouraging open communication among family members may help to circumvent conflicts. It is useful to offer to some clients a written summary or an article about the condition in question that can be shared with relatives. Some families may benefit from group genetic counseling sessions.

Unexpected Findings

The story of finding nonpaternity during family genetic testing is a familiar one. (An example of such a discovery is included in one of the genetic counseling cases discussed in Chapter 16.) Unexpected information can occur in a variety of testing and professional situations. For example, an unexpected chromosome abnormality or variant can be detected, an ambiguous ultrasound finding may be observed, or a conflict of interest can come to light. At times, test results are open to interpretation or subject to controversy.

When deciding how to approach the unexpected findings, we have to weigh the benefits and harms of nondisclosure with those of disclosure. The first considerations include the relevance of the information to the client's situation and the consequences of the finding(s).

The arguments for disclosing findings are based on the principle of autonomy. The information belongs to the individual. Duty to the client requires truth telling and full disclosure. The client may learn of the findings in a different context from some other provider. Beneficence would require that we avoid undermining the client's trust in the profession, which could happen if she learns a genetic counselor was not totally honest with her.

The reasons for nondisclosure of an unexpected finding should be compelling. Arguments for *not* disclosing incidental findings include respect for autonomy. A person has a right not to know, and disclosure may violate that right. Because issues such as nonpaternity or gene carrier status can be emotionally charged, the principle of nonmaleficence may prevail in some situations. However, facts relevant to medical decision making cannot be justifiably withheld.

The President's Commission (1983) recommends that the client be advised before testing of the possibility that unexpected information may be revealed. Details of the institutional policy on disclosure (if one exists) should be included in the discussion.

Duty to Recontact

A major difference between genetic testing and diagnosis and other areas of health care is the volume of information that new genetic technologies are continuously adding to our fund of knowledge. This creates a continuous need to examine the impact of the rapid advances in genetics on professional practice. New applications for tests and easier access to testing could possibly lead to the recognition of new rights and privileges for clients. We have to address the implications of these changes to our duties to clients with whom we were formerly actively involved.

Usually the duty a professional owes a client lasts as long as there is a need for those services—for example, the duration of an illness. For the specialist, the duty to the client usually ends with the consultation. Part of the duty is the full disclosure of known facts and information regarding the genetic condition.

Pelias (1991) suggests that there may be a continuing obligation to recontact clients when new information becomes available that would have an impact on that client's decision making. This expanded duty to disclose could be based on the recognized duty of physicians to recontact patients when new information regarding past medication or therapy is discovered. For the genetic counselor who deals in information, this possible new duty may apply to information regarding changes in diagnostic availability and new interpretations of prior test results.

As in all situations, documentation by the consultant that includes a request for the client to keep in touch with the genetics clinic, especially if individual circumstances change or develop, is helpful. The counselor may also include in client letters statements about the potential future availability of testing and technological advances and information about genetic support groups, especially those providing members with periodic updates with respect to scientific and medical developments. However, these efforts may not relieve the provider of her duty to recontact that client. A communication in writing by means of a first-class letter to the last known address is evidence of a good-faith effort to notify a former client. It is not clear how far back in time the duty to recontact applies.

The argument that there is no duty to recontact involves the client's responsibility for herself and her own health care. As an adult, she should keep up to date regarding any testing or services available to her. She should contact the health care provider on a regular basis if her concerns are ongoing. The administrative nightmare that a duty to recontact may represent to some practitioners is often included in the argument. Respect for the client's autonomy can also be invoked. Not every client wants to keep thinking about the reasons that brought her to genetic counseling. Some need to "put the past behind them," while others have preferred to change providers. By initiating contact with the client after a period of time, we may be violating her right of privacy and her right "not to know."

CONCLUSIONS

The goal of this chapter was to introduce some of the ethical and legal issues that impact genetic counseling. The subjects addressed were a sampling of those that many practitioners confront at some time. Hopefully, a sense of the depth of these issues will stimulate ongoing discussion, research and further study.

REFERENCES

Americans with Disabilities Act of 1990 42 U.S.C. S12101, 12201-12213 (Supp. V 1994).

Andrews LB (1992) Torts and the double helix: Malpractice liability for failure to warn of genetic risks. *Houston Law Rev* 29:149.

Andres LB, Fullarton JE, Holtzman NA, Motulsky AG (eds) (1994) *Assessing Genetic Risks. Implications for Health & Social Policy.* Washington DC: National Academy Press.

ASHG/ACMG Report (1995) Points to consider: Ethical, legal and psychological implications of genetic testing in children and adolescents. *Am J Hum Genet* 57:1233–1241.

Baier AC (1986) Extending the limits of moral thinking. *J Philos* 133(10):538–545.

Bartels DM, Leroy BS, Caplan AL (eds) (1993) *Prescribing Our Future: Ethical Challenges in Genetic Counseling.* New York: Aldine DeGruyter.

Baumiller RC, Cunningham G, Fisher N, Fox L, Henderson M, Lebel R, McGrath G, Pelias MZ, Porter I, Seydel F, Willson NR (1996) Code of Ethical Principles for Genetics Professionals: An explication. *Am J Med Genet* 65:179–183.

Beauchamp TL, Childress JF (1994) *Principles of Biomedical Ethics*, 4th ed. New York: Oxford University Press.

Benkendorf JL, Callanan NP, Grobstein R, Schmerler S, FitzGerald KT (1992) An explication of the National Society of Genetic Counselors (NSGC) Code of Ethics. *J Genet Counsel* 1 (1):31–39.

Brody H (1976) The physician–patient contract: Legal and ethical aspects. *J Legal Med* (4):25–29.

Capron AM (1979 Tort liability in genetic counseling. *Columbia Law Rev* 79:619–684.

Council on Ethical and Judicial Affairs, AMA (1995) The use of anencephalic neonates as organ donors. *JAMA* 273(20):1614–1618.

Crossley M (1996) Infants with anencephaly, the ADA, and the Child Abuse Amendments. *Issues Law Med* 11(4):379–410.

Elliot C (1992) Where ethics come from and what to do about it. *Hastings Center Rep* 22(4):28–35.

Fletcher JC, Wertz DC (1987) Ethical aspects of prenatal diagnosis: Views of U.S. medical geneticists. *Clin Perinatol* 14(2):293–312.

Fost N (1992) Ethical issues in genetics. *Med Ethics* 39(1):79–89.

Fost N (1993) Genetic diagnosis and treatment. *Am J Dis Child* 147:1190–1195.

Gert B, Berger EM, Cahill GF Jr, Clouser KD, Culver CM, Moeschler JB, Singer GHS (1996) *Morality and the New Genetics.* Boston: Jones and Bartlett.

Gilligan C (1982) *In a Different Voice: Psychological Theory and Women's Development.* Cambridge, MA: Harvard University Press.

Gilligan C (1987) Moral orientation and moral development. In: Kittay EF, Meyers DT (eds) *Women and Moral Theory.* Savage, MD: Rowman and Littlefield, pp. 19–33.

Grodin MA, Annas GJ, Glantz LH (1993) Medicine and human rights: A proposal for international action. *Hastings Center Rep* 23(4):8–12.

Harris v. McRae (1980) 448 US 297.

Heimler A, Zanko A (1995) Huntington disease: A case study describing the complexities and nuances of predictive testing of monozygotic twins. *J Genet Couns* 4(2):125–137.

Hodge SE (1995) Paternalistic and protective? *J Genet Couns* 4(4):351–352.

Hudson KL, Rothenberg KH, Andrews LB, Kahn MJE, Collins FS (1995) Genetic discrimination and health insurance: An urgent need for reform. *Science* 270:391–393.

In re Baby K (1993) 832 F. Supp 1022 (ED VA).

In re T.A.C.P. (1992) 609 So 2d 588.

Jones OD (1992) Sex selection: Regulating technology enabling the predetermination of a child's gender. *Harvard J Law Technol* 6:1–61.

Kuczewski MG (1996) Reconceiving the family: The process of consent in medical decision making. *Hastings Center Rep* 26(2):30–37.

Maher v. Roe (1977) 432 US 464.

Mahowald MB (1993) *Women and Children in Health Care: An Unequal Majority.* New York: Oxford University Press.

Maley JA (1994) *An Ethics Casebook for Genetic Counselors.* Charlottesville: University of Virginia.

Medearis DN Jr, Holmes LB (1989) The use of anencephalic infants as organ donors. *N Engl J Med* 321(6):391–393.

Mitchie S, McDonald V, Bobrow M, McKeown C, Martequ T (1996) Parents' responses to predictive genetic testing in their children: Report of a single case study. *J Med Genet* 33:313–318.

National Society of Genetic Counselors (1992) Code of Ethics of the National Society of Genetics Counselors. *J Genet Couns* 1(1):41–43.

National Society of Genetic Counselors (1995) Prenatal and childhood testing for adult-onset disorders. *Perspect Genet Couns* 17(3):5.

Natowicz MR, Alper JK, Alper JS (1992) Genetic discrimination and the law. *Am J Hum Genet* 50:465–475.

Noddings N (1984) *Caring: A Feminine Approach to Ethics and Moral Education.* Berkeley: University of California Press.

Pelias MZ (1986) Torts of wrongful birth and wrongful life: A review. *Am J Med Genet* 25:71–80.

Pelias MZ (1991) Duty to disclose in medical genetics: A legal perspective. *Am J Med Genet* 39:347–354.

Pelligrino E (1989) Character, virtue and self-interest in the ethics of the professions. *J Contemp Health Law Policy* 5:53–73.

Pelligrino E (1993) The metamorphosis of medical ethics: A 30 year retrospective. *JAMA* 269(9):1158–1162.

President's Commission for the Study of Ethical Problems in Medicine and Biomedical and Behavioral Research (1983) *Genetic Screening and Counseling: Ethical and Legal Implications.* Washington, DC: Government Printing Office, pp. 41–88.

Rehabilitation Act of 1973 (29 USCA S791 et seq).

Reich E (1996) Testing for HD in twins. *J Genet Couns* 5(1):4749.

Robertson JA (1990) The Randolph W. Thrower Symposium: Genetics and the law, procreative liberty and human genetics. *Emory Law J* 39:697–719.

Roe v. *Wade* (1973) 410 US 113.

Schroeder v. *Perkel* (1981) 87 NJ 53, 432 A2d 834.

Spital A (1991) The shortage of organs for transplantation: Where do we go from here? *N Engl J Med* 325(17):1243–1246.

Suter SM (1993) Whose genes are these anyway: Familial conflicts over access to genetic information. *Mich Law Rev* 91:1854–1908.

Troug RD (1997) Is it time to abandon brain death? *Hastings Center Rep* 27(1):29–37.

Turpin v. *Sortini* (1981) 643 P2d 954 (CA).

Warren MA (1985) *Gendercide: The Implications of Sex Selection.* Totowa, NJ: Rowman and Allanheld.

Wertz DC, Fanos JH, Reilly P (1994) Genetic testing for children and adolescents. *JAMA* 272(11):875–881.

Working Party of the Clinical Genetics Society (UK) (1994) The genetic testing of children. *J Med Genet* 31:785–797.

11

Relationally Based Professional Conduct for Genetic Counselors

Deborah Lee Eunpu

In a broad sense, professional conduct encompasses the behaviors, attitudes, and attributes that are associated with holding a relationship of trust with the public. This conduct is shaped and guided by many factors, including training, experience, and a professional code of ethics. This chapter examines how genetic counselors establish and maintain appropriate professional boundaries in their work with clients and colleagues, how the Code of Ethics of the National Society of Genetic Counselors (NSGC) can be applied in day-to-day practice, and the role of personal values in defining professional conduct. While common sense and caring are essential to professional conduct, a definable set of personal and professional values is critical, as it promotes the selection of an appropriate course of action, especially under uncommon or unanticipated circumstances.

DEVELOPMENT OF THE NSGC CODE OF ETHICS

Each professional group has the opportunity to define the responsibilities and behaviors that are expected of its members. The professional group's beliefs and values determine the context in which these behaviors are to be interpreted. A code of ethics is

A Guide to Genetic Counseling, Edited by Diane L. Baker, Jane L. Schuette, and Wendy R. Uhlmann.
ISBN 0-471-18541-8 (cloth), 0-471-18867-0 (paper).   Copyright © 1998, Wiley-Liss, Inc.

a document grounded in a profession's beliefs and values that provides guidelines and standards for professional behavior.

In 1985 the NSGC created an ad hoc committee that initiated a process leading to the development, ratification, and adoption of a code of ethics. The NSGC Code of Ethics was approved in 1991 (Anon., 1991) and was formally adopted on January 1, 1992 (National Society of Genetic Counselors, 1992). (The text of this document is presented in the chapter appendix.) By accepting membership in the National Society of Genetic Counselors, one is expected to be aware of and to abide by the Code of Ethics.

The NSGC ad hoc Committee on Ethical Codes and Principles defined a code of ethics as a "statement of beliefs and guidelines for professional behavior which reflect the responsibilities, obligations, and goals of a professional group's membership" (Benkendorf et al., 1992). The NSGC's adoption of a code of ethics "demonstrates to society that it accepts responsibility for defining professional conduct, for sensitizing its members to important ethical issues, and for affirming professional accountability" (Benkendorf et al., 1992). The code is grounded in an ethic of care, and all its parts ultimately rest on respect for self, respect for others, and the primacy of the relationship with the client. Even though individual values will vary among counselors, an awareness of one's personal ethic of caring is likely to enhance one's understanding and application of the Code of Ethics.

RELATIONALLY BASED CONCEPTS OF PROFESSIONAL CONDUCT

A genetic counselor's work is performed in the context of relationships. The NSGC Code of Ethics is grounded in this fact, and it discusses the genetic counselor's significant relationships with self, clients, colleagues, and society. The ethic of care perspective of the Code is expressed through descriptions of the responsibilities that arise from these primary relationships (Benkendorf et al., 1992). In practical terms, the Code provides guidance for what a genetic counselor should do, the appropriate nature of relationships with others, and a foundation from which to formulate answers to questions about one's professional behavior. The NSGC Code of Ethics is not intended to explicitly address every issue a genetic counselor may encounter, but to provide a general framework for professional conduct. In general, the Code provides guidance on responsibilities, dual relationships, and personal and professional values.

Responsibilities

The genetic counselor's responsibilities, broadly stated, include remaining current in knowledge about new developments in clinical genetics regarding diagnosis, management, treatment, and referral resources; providing client care that is accurate, complete, accessible, and culturally sensitive; managing, completing, and documenting clinical work; recognizing and carrying out ethical practices; and understanding professional boundaries and limits. Given the student genetic counselor's interest in, and training with regard to, helping people, the responsibilities toward clients as outlined in the Code of Ethics are likely to seem logical and easy to intuit. However, this is a complex area, and therefore this chapter focuses on one's relationships with clients, because this is the primary responsibility of genetic counselors.

Boundaries

Boundaries and dual relationships present examples of issues for which the overarching ethic of care and respect for self and others provides guidance. Several statements in Section II of the Code apply to the counselor–client relationship. Among them are the following: that we strive to respect the feelings of our clients (Section II.2), that we strive to enable clients to make informed independent decisions (Section II.3), that we refer clients to other professionals when we are unable to support them (Section II.4), and that we avoid exploiting our clients (Section II.6). To achieve these standards, the genetic counselor should develop a personal definition of appropriate boundaries with her clients. Some boundaries will be fluid and will change on a case-by-case basis, whereas others will stand throughout a variety of circumstances. For example, does the counselor wish to be known to the client on a first-name basis or by a more formal form of address? Will the counselor invite the client to have unrestricted access to her time and resources, or will she outline the circumstances under which client-initiated contact is welcomed or limited? With regard to the need for referrals, follow-ups, and return visits, will the counselor's role be to secure these services or to initiate the activities and ask the client to take responsibility for completing them? How does the counselor define the circumstances and limitations under which she will be involved with clients? In this context, appropriate boundaries are defined as limits in the extent or quality of the relationship between genetic counselor and client which are established by the genetic counselor for the optimal benefit of the client.

The boundaries one establishes will shape the form and content of one's relationship with clients. Two important areas to consider with regard to establishing appropriate boundaries are the structural elements in genetic counseling and the extent to which personal information about the genetic counselor is shared.

Structural elements in genetic counseling are best described as noncontent (i.e., nonmedical and nongenetic) issues related to the arrangement of appointments, staffing, access, organization, and location of services. These are the elements that must mesh properly to make a clinical encounter flow smoothly from first contact through follow-up. Although an individual genetic counselor may not have control over a number of these elements, she is likely to have primary responsibility for some aspects, such as maintaining contact with clients.

A commonly encountered structural element is that of establishing how a client may have access to the genetic counselor. As with our colleagues in other areas of health care, established hours, access to message-taking services, and protected time during the workday, are necessary to support the genetic counseling role. Under certain circumstances a counselor might call a client from home in the evening; however, she generally would not give that client her home phone number to use at will.

Boundaries set by a counselor about when and where work is done are necessary in maintaining professional relationships with clients. Allowing the client to have unlimited access to the genetic counselor may convey an inaccurate idea about the counselor's role or create the impression that the genetic counselor is a friend rather than a professional, which may result in unrealistic expectations of the genetic counselor based on the client's ideas of what a friend is and does. When the client is led to have unrealistic assumptions about the role of the genetic counselor, there is the potential for unfounded expectations and confusion that can persist and interfere with the already complex tasks of genetic counseling. Alternatively, clients given unlimited access to the genetic counselor may conclude that they are not competent to make decisions or cope without assistance. In a short-term relationship like genetic counseling, validating client autonomy and capacity for making decisions is part of one's work with clients, beginning with the first contact and continuing through all subsequent interactions.

It is quite common for the genetic counselor to be questioned by a client about her personal life or experiences. Such questions (Do you have children?, Have you been pregnant?, and Have you had an amnio/CVS/etc.?) require a thoughtful response. A student may be strongly tempted to answer by sharing personal information. The genetic counseling student may feel that answering such questions in the affirmative will create a strong client–counselor bond or raise her level of expertise in the client's opinion. However, this is not necessarily the case. The genetic counselor needs to determine what response is most likely to benefit the client and the process of genetic counseling. The disclosure of personal information by the genetic counselor may arouse a response of empathy and concern from the client, which in turn may cause the client to prematurely stop evaluating his own feelings and decisions (Kessler, 1992). The act of answering such questions may be interpreted by the client as evidence that the genetic counselor's experience allows her to know more about how to make the type of decisions called for than the client herself. Therefore, in most situations, this type of client question is best handled by an exploration of the motivation for the question, not by providing personal information.

There are numerous possible client motivations for asking personal questions about the genetic counselor. These include the following:

- Client's desire to feel less unique and isolated (e.g., "Am I the only person this has ever happened to?")
- Client's belief or attitude that she cannot be fully understood unless there is a shared experience
- Client's desire to know what the counselor would do if she were in the client's situation, to make a decision based on the wisdom of others
- Client's desire to delay whatever comes next, be it a procedure, leaving, or having a distressful feeling
- Client's desire to have a personal connection with the genetic counselor instead of a professional–client relationship
- Client's desire or need for additional information about a diagnosis, procedure, or decision.

The degree to which a client may be aware of her motivations for asking personal questions is variable. Asking a client, "Tell me something about why this information is important to you?" gives the client an opportunity to discover more about how she is experiencing the situation and, therefore, may lead to greater self-knowledge and improved self-esteem. The short-term nature of genetic counseling makes it important that the client be encouraged to stay focused on her own experience of the situation, albeit with some guidance and support from the genetic counselor.

More permeability in the boundaries between genetic counselor and client may be warranted in some circumstances. Consider the client who participates in a support group after counseling and then asks her genetic counselor to also become involved. This new relationship is different from the prior counselor–client relationship in that the parties are now collaborators. With each change in the relationship, it is the genetic counselor's responsibility to determine the appropriateness of the change, to set a tone, and to provide an environment that fosters readjustment of the boundaries as appropriate for the new situation.

In some situations, the client may have strong or painful feelings about past experiences in genetic counseling, and these issues must be addressed before a new working relationship can take shape. Even when the client's experiences have been with another genetic counselor or center, it is useful to address the client's feeling about such encounters before undertaking new work.

Another boundary issue is that of an invitation to attend a family celebration or rite such as a graduation, funeral, or wedding. The intimacy of the genetic counselor's work with clients presents many reasons and opportunities for clients and counselors to feel close to each other and to desire ways to share significant events happening in the client's life, whether losses or accomplishments. In considering whether to participate in such an event, the genetic counselor needs to give careful thought to his role with the family, the reasons why the family wants his participation, and his reasons for wanting to be present for an important family event. Perhaps the primary issue for the genetic counselor to consider is the personal needs associated with attending the event. This does not mean that the genetic counselor cannot participate in a family's event because of feelings or needs, but that it is appropriate to attend to such needs for support or consolation through resources other than the family's. This is a good demonstration of the importance of self-care as a condition for interacting appropriately with clients.

Dual Relationships

Section II of the NSGC Code of Ethics also guides the genetic counselor's actions in cases of dual relationships with clients. Dual relationships are inherently difficult in that they may obscure the roles, expectations, responsibilities, and objectivity of either the genetic counselor or the client. Dual relationships, as discussed below, are to be avoided if at all possible.

Friends or Colleagues as Clients   Consider the situation in which a genetic counselor is asked to provide professional services to a client who is also a friend or colleague. Among the inherent risks are the following:

- The client may experience an undesirable loss of privacy.
- Genetic counseling will be complicated by the need to renegotiate former roles of both parties.
- The client may receive abnormal results, necessitating an even more involved relationship with the counselor.
- The client may not receive optimal care as a result of loss of objectivity by the genetic counselor.

How one deals with this type of dual relationship is in part determined by the community in which one practices. In a large metropolitan area, the genetic counselor may have the option of referring the friend to another genetic counselor or medical center, where she will receive the needed services without disrupting the original relationship with the counselor/friend.

In a setting in which the genetic counselor is the sole provider of services, the option of referring elsewhere may not be open. Nonetheless, the same attention must be given to the dual nature of the relationship to protect the interests of the colleague/friend. It is preferable to discuss the dual nature of one's relationship with the client directly at the beginning of the genetic counseling process and to set limits about what will be discussed inside versus outside the office. Potential problems should be anticipated and examined. The genetic counselor and the colleague/friend may then cocreate a mutually acceptable way to address their new relationship and the relevant issues. Another alternative might be to have only the geneticist see this particular client. By attending to the potential problematic aspects of dual relationships directly, one fosters the client's ability to make informed decisions while maintaining an optimal level of privacy.

Potential of Gain from Association with a Client   Another aspect of dual relationships arises when the genetic counselor is in the actual or perceived position of benefiting as a consequence of a client's action. The genetic counselor may gain financially through an investment in a particular test or procedure for which a client is a candidate. Or the counselor may benefit intellectually or professionally from a client's participation in research protocols with which the genetic counselor is associated. Whenever there are potential conflicts of interest, the genetic counselor should address them and seek consultation to develop a plan to minimize the risk to the client or to herself.

Another possible conflict between a genetic counselor's and a client's interests is exemplified by the family who desires to end a genetic evaluation before it is completed, while the genetic counselor recognizes her own interest in continuing the relationship. The dual relationship consists of the genetic counselor's primary professional responsibilities to client autonomy and her intellectual, medical, or research interests in reaching a diagnostic outcome. The genetic counselor should recognize this discrepancy and direct her energies toward understanding and supporting the client's goals. If the client understands the information provided and still chooses to terminate an evaluation, then the genetic counselor's responsibility is to explore the reasons for this decision, to confirm that it is not based on incorrect information, to provide the client with the best possible interpretation of the evaluation results to date, to inform the client how the evaluation can be continued in the future (specifying under what conditions and limitations), and to support the client in the decision to withdraw from the evaluation.

Interactions with the media provide another example of potential conflicts introduced by dual roles. The press views the almost daily developments in clinical genetics as exciting story opportunities. A genetic counselor is frequently the individual asked to provide vignettes about client experiences or to arrange interviews with clients with specific conditions. The genetic counselor may wish to promote the media's access to accurate information; however, the client's best interests should receive primary consideration. The genetic counselor should not exploit clients or their experiences for personal gain or the gain of others. The counselor is balancing two potentially conflicting goals: to provide the media with access to personal stories, and thereby help inform the general public, and to protect the confidentiality of clients. Not all media requests represent such conflicts. Often clients will indicate their willingness to be contacted by the media out of a belief that they can help to educate others through sharing their story and experience. It is important to obtain a clear understanding of the media's request and purpose and to work with your institution's media relations department, which is skilled in mediating such interactions. Additionally, many local, regional, and national genetic support groups welcome media interest and have already identified and trained certain members to interact with the press.

Relationships with Multiple Members of the Same Family   In genetic counseling, one is often involved with multiple members of the same family, either simultaneously or at different points in time. When a family demonstrates respect for its members and tolerance for each member's right and ability to make decisions about genetic testing, the likelihood of conflict as a result of seeing the same provider is low. However, this is not always the case in families.

Consider a family in which one member (the consultand) has had a genetic evaluation and has made it known that she does not want to share the results or her medical records with her family. Later, a relative who knows that the evaluation has been performed sees the same genetic counselor for services. The genetic counselor has now been placed in the situation of a dual relationship with the family. To disclose any information about the consultand without her consent would be a violation of confidentiality. Yet, not to disclose genetic risks within the family may have a consequence for the relative (e.g., the relative may have a risk for developing a disorder or having an affected child).

There is no single best way to handle a circumstance like this. The genetic counselor must identify the dual relationship, assess the potential for effect on various family members, and develop a plan for proceeding. That plan should consider the values and goals stated in Section II of the Code of Ethics (genetic counselors and their clients) by seeking to balance the obligation to maintain confidentiality with the need to enable each client to make informed decisions.

The existence of intrafamilial conflicts around disclosure of testing results is not uncommon. Therefore, in the context of genetic counseling, it is helpful to consider the entire family as the patient. To deal with the family effectively, one must understand that the family is a system with dynamics and rules of its own. Most of the time the family is healthy and functional. In the context of this chapter, one might say that the family has its own code of ethics to which members adhere for the benefit of the individual as well as the entire family. Generally, the family is capable of flexibility and change as it is challenged by new experiences—the definition of a healthy family (Nichols and Schwartz, 1991; Kaslow and Celano, 1995). However, not all families function well, and many families are severely stressed by the events that bring them to genetic counseling.

If there are power struggles within families, members may try to involve the genetic counselor in attempting to accomplish individual goals. In families in which communication is not occurring directly between two parties who respectively have and want information, attempts may be made to enlist the genetic counselor as a go-between. For example, a client wanting to convince another family member to proceed with genetic testing may seek to engage the genetic counselor in supporting, or even facilitating, this outcome. In this case, the counselor needs to determine how to work in the context of this family's structure. We cannot change families, and we cannot always assume the roles they have in mind for us. Any given family's cultural norm regarding the importance of sharing information for the benefit of the entire family may be different from what the genetic counselor is familiar with or values. Alternatively, a family may observe gender-based rules that define how one is expected to share information. In any of these situations, a careful consideration of the family from different perspectives will help the genetic counselor formulate an approach that is sensitive to the needs and values of the family, yet mindful of the counselor's own professional boundaries.

The Genetic Counselor's Ability to Relate Appropriately with Self and Others

The ability to integrate the Code of Ethics into everyday practice is not easily or quickly learned. Didactic experiences as well as supervision that is directed at development of the professional person of the genetic counselor will be provided in one's training (American Board of Genetic Counseling, 1996; Fine et al., 1996). Both personal and external factors contribute to the development of the genetic counselor's professional conduct.

Personal Factors Each developing genetic counselor brings to her training not only knowledge but an assortment of attitudes and beliefs about ethical decision making and moral behavior that derive from gender, family of origin, cultural or ethnic group, spiritual community, and prior personal experiences. Values are also shaped by emotional and physical health, as well as by one's position in the life cycle.

The first step toward the integration of personal values with professional actions requires that the developing genetic counselor become aware of her personal values and their origins and meanings. This is to ensure that clients' needs can be attended to with the least interference from the counselor's individual perspective. Since each genetic counseling student will be at a different level of personal value awareness, both didactic and evaluative experiences are useful in developing insight around these issues. Didactic components will introduce the student to the principles of bioethics and the NSGC Code of Ethics. Evaluative components will offer opportunities to explore her personal values and to determine where they came from and how they may enhance or complicate her ability to apply the NSGC Code of Ethics in practice. The evaluative component may take place in class discussion, while working with one's supervisor, or through writing. The goal is to use whatever techniques are useful to increase awareness of one's personal values.

External Factors A second set of factors, those that are external to the genetic counselor, will also influence the development of professional behavior. External factors are those pertaining to the environment and circumstances within which genetic counseling occurs. They include the setting (typically a medical institution), the model for care provision (most often the medical model), the time available for a clinical interaction (limited), the involvement of other providers during this particular visit (e.g., ultrasound evaluation or other procedures), and the availability of return visits (dependent on transportation options, client's distance from the medical center, and other circumstances). How one manages these external contingencies will shape how one provides genetic counseling. Questions about external factors that bear exploration include, but are not limited to, the following:

- Are there aspects of client scheduling or clinic visits that can be modified to better serve clients?
- What is the genetic counselor's role in relation to the other health professionals who are involved in serving the client?
- How do professional practice standards of genetic counseling shape the genetic counselor's behavior?
- What institutional or community resources are available to assist with cases presenting particular needs?

The first step in managing external factors is to develop an understanding of what they are and to identify the expected normative behavior for genetic counselors in this setting. This understanding serves as a scaffolding on which the genetic counselor then applies the Code of Ethics to everyday practice. As with personal factors, exploration of external factors will provide the developing genetic counselor with important information that may influence and guide the definition of appropriate professional conduct ultimately arrived at.

DEVELOPING A PERSONAL PROFESSIONAL CONDUCT CODE

Once one has achieved awareness of the personal and external factors that may influence how one works in genetic counseling, this awareness must be integrated with actual case material. Case-based factors can be as varied as the clients one works with, so the best starting point is to have an open, observing perspective. This requires a high level of flexibility from the genetic counselor, as illustrated in the following examples.

Clients with the Same Diagnosis at Different Stages of Life

One of the most common examples of the flexibility required of a genetic counselor relates to the stages at which she works with a family or client. The issues at prenatal, newborn, pediatric, or adult diagnosis differ vastly and require that the genetic counselor interact in different ways with clients even though the condition may be identical in all.

The genetic counselor decides what information to share and how to impart it based on an assessment of the client's needs, ability to understand the information cognitively, and resources for coping with the information emotionally. For example, the family who has just received an abnormal prenatal test result at 18 weeks of gestation is likely to have very different concerns and needs from those of parents who have been given the same diagnosis for their newborn child. To support the clients' sense of autonomy and their ability to make critical decisions in both these cases, the genetic counselor must structure the delivery of information for the particular circumstance. While it is important to provide the correct clinical picture, it is also important to infuse that picture with an appropriate measure of hope. The tempo is individualized to the diagnosis and to the client's needs.

The genetic counselor who works in a specialty clinic is particularly likely to see clients who are dealing with the same diagnosis but at different life stages. Three consecutive sessions may involve a couple requesting information on which to base a decision about whether to continue a pregnancy that has received an abnormal prenatal diagnosis, a family whose child was recently diagnosed and for whom having a perspective of hopefulness is necessary to cope with extensive medical treatments, and a family whose members are interested in carrier testing but have no direct experience with the diagnosis. Each of these sessions calls for a different approach, and the genetic counselor's approach to finding the correct balance will be rooted in his interpretation and application of appropriate professional conduct.

Clinical Service Versus Research Protocols

Work with clients will differ depending on whether one is providing clinical services or access to research protocols. Just as the genetic counselor who deals with the same clients in different settings is responsible for setting the tone and for negotiating appropriate boundaries, the genetic counselor must also provide clients with guidance about the differences between clinical and research activities. In this regard, the genetic counselor is working to minimize the potential conflicts of a dual relationship, to protect the client from exploitation, and to serve the client's needs. In this situation, one may experience conflicting pressures from a colleague who has a research interest as the fundamental agenda. In evaluating such conflicts, the best interests of the client should take precedence. It is sometimes difficult to maintain a respectful relationship with colleagues while promoting the well-being of clients, but the achievement of such a difficult balance is upheld by the Code of Ethics.

Working with Clients Who Exhibit Difficult Behaviors

"Difficult client" is one of the misnomers in clinical practice. There are no difficult clients, but rather clients with difficult behaviors. These are most likely clients who are very needy, angry, anxious, or hurt. Once one acknowledges that clients are not difficult but are exhibiting behaviors related to their needs or feelings, the way to work with such clients becomes clearer.

The first step in determining the cause of difficult behaviors is to get the best possible assessment of what the client is experiencing. When a client's anger is directed at the genetic counselor, is the client really angry at the hospital because of a lost test result? Underneath that anger, does the client feel sadness, grief, or loss because she even had to consider such testing? Going even deeper, is the client feeling isolated and damaged as the result of a loss in self-esteem?

When clients present with behaviors that make it difficult to move forward with the business of genetic counseling, it is the genetic counselor's responsibility to identify as clearly as possible the cause of the behaviors and to find ways to work with the client. It may be necessary to obtain consultation on the case, and ultimately the client may be referred to someone else, in the hope that a more constructive relationship can be achieved. Blaming clients for needs and feelings that result in disruptive behaviors without trying to understand their basis can interfere with the client's ability to benefit from the counseling relationship.

Encounters in Social Settings

The genetic counselor may encounter clients in social settings that have nothing to do with the professional activities that initiated the relationship. As described previously, the genetic counselor is responsible for redefining the appropriate boundaries in response to new circumstances and the client's cues. If, in a social encounter, the client does not acknowledge the presence of the genetic counselor, it is best to avoid approaching the client. The primary motivation for not approaching is to maintain the client's apparent wish for privacy about their relationship with your program. If the client approaches the genetic counselor, then obviously the counselor should acknowledge the client. Under most circumstances, it is inappropriate to engage in a discussion of the client's case, but if the client appears distressed, or initiates a discussion about her clinical circumstances, some brief comments may be made, followed by asking whether the client would like to call the counselor/clinic for more discussion. Asking how the client wishes to be acknowledged in a social setting will help guide the genetic counselor's subsequent interactions.

Awkwardness also can arise for a genetic counselor when a fellow guest at a social gathering learns of the counselor's area of expertise, volunteers information about her genetic circumstances, and questions the counselor about her risks. In this situation, one is unlikely to have sufficient clinical or psychosocial information about the individual to make an informed response. Nor would a response in such a setting be appropriate. The better course is to acknowledge the validity of the question and encourage the individual to contact a genetics clinic. A simple explanation of your need to have appropriate resources, sufficient time, and undivided attention in order to address the questions should put your response in perspective.

THE GENETIC COUNSELOR'S CONDUCT IN RELATIONSHIPS WITH COLLEAGUES AND SOCIETY

The NSGC Code of Ethics also considers how genetic counselors relate to colleagues and society. These relationships are secondary to those with clients, but are still important, especially since collaboration with colleagues is usually necessary to provide genetic counseling.

Relationships with Colleagues

In comparison to relationships with clients, relationships with colleagues are characterized by more equality in terms of power, and therefore concerns about the vulnerability of one party or the other are less important than issues of mutual responsibility and respect. Conflicts may arise when there are competing opinions of what constitutes adequate client care. For example, what should one's response be if a client has received less than optimal care from a colleague? Or, what if a genetic counselor realizes that a planned course of treatment or testing described by a colleague may not be adequate or appropriate? In each of these situations, the well-being of the client comes first. However, one is also obligated to address the differences of opinions with colleagues with sensitivity and respect.

Another potentially difficult situation arises when a collaborating colleague fails to demonstrate the values of mutual respect and cooperation for colleagues. This collaborator may be a sole source for clinical diagnosis or a research protocol, therefore making it difficult to avoid continued contact. As with clients who demonstrate difficult behaviors, one should first assess the situation to try to gain some understanding of the colleague's behavior. It may be necessary to obtain consultation from a peer who can provide objective feedback and interpretation, or to seek outside consultation, such as through an office of human resources. With a clearer understanding of the factors motivating apparently obstructive behaviors, the genetic counselor can speak with the collaborator about how she would like to see the relationship change. Such discussion should be based in the genetic counselor's experience of the behavior—the feelings and responses it engenders and their detrimental effect on the work on behalf of clients. If no change can be achieved, the genetic counselor has several choices:

- Identify another source for the same services.
- If the conflicts are based on personal differences, consider having someone else from your program interact with the individual.
- Continue working with the individual, being sure to obtain adequate consultation and support while doing so.

Relationships with Society

Genetic counselors perform their work within the context of society at large. When there are discrepancies between what is happening in this broader arena and the needs of clients, the genetic counselor needs to be aware of the conflicts and be prepared to act as an advocate for client needs. In current practice, this means giving consideration to the ramifications of health care reform on the quality and accessibility of genetics services, externally imposed limitations on reproductive decision making, and the possibility that testing protocols will become available to clients without benefit of adequate pretest education and counseling.

Not every genetic counselor will have the time or interest to become active in working for changes of attitudes, policies, and laws that have the potential to benefit genetic counseling clients. However, at a minimum, a genetic counselor should be knowledgeable about societal issues that have an effect on clients and should know the experts and organizations who work with these issues. Thus the genetic counselor must stay well informed about trends in societal developments that could affect genetic counseling practices or clients.

CONCLUSIONS

Professional conduct for genetic counselors is described in general by the Code of Ethics of the National Society of Genetic Counselors. The fundamental values underlying professional conduct are those of caring and respect for self, clients, colleagues, and society. The context of one's professional behavior is relational. The development of a personal code of professional behavior is a career-long process, which requires that one understand the nature, origins, and potential ramifications of one's knowledge, attitudes, and beliefs about ethical decision making. One can then develop a professional approach that integrates personal awareness, the National Society of Genetic Counselors Code of Ethics, and clinical experience.

REFERENCES

American Board of Genetic Counseling (1996) *Requirements for Graduate Programs in Genetic Counseling Seeking Accreditation by the American Board of Genetic Counseling.* Rockville, MD: ABGC.

Anon. (1991) Code of Ethics Approved. *Perspect Genet Couns* 13(3):8.

Benkendorf JL, Callanan NP, Grobstein R, Schmerler S, FitzGerald KT (1992) An explication of the National Society of Genetic Counselors (NSGC) Code of Ethics. *J Genet Couns* 1(1):31–40.

Fine BA, Baker DL, Fiddler MB, and ABGC Consensus Development Consortium (1996) Practice-based competencies for accreditation of and training in graduate programs in genetic counseling. *J Genet Couns* 5(3):113–121.

Kaslow NJ, Celano MP (1995) The family therapies. In: Gurman AS, Messer SB (eds), *Essential Psychotherapies Theory and Practice.* New York: Guilford Press, pp. 343–402.

Kessler S (1992) Psychological aspects of genetic counseling. VIII. Suffering and counter-transference. *J Genet Couns* 1(4):303–308.

National Society of Genetic Counselors (1992) National Society of Genetic Counselors Code of Ethics. *J Genet Couns* 1(1):41–43.

Nichols MP, Schwartz RC (1991) *Family Therapy Concepts and Methods,* 2nd ed. Boston: Allyn & Bacon, pp. 77–133.

The Code of Ethics of the National Society of Genetic Counselors

PREAMBLE

Genetic counselors are health professionals with specialized education, training, and experience in medical genetics and counseling. The National Society of Genetic Counselors (NSGC) is an organization that furthers the professional interests of genetic counselors, promotes a network for communication within the profession, and deals with issues relevant to human genetics. With the establishment of this code of ethics the NSGC affirms the ethical responsibilities of its members and provides them with guidance in their relationships with self, clients, colleagues, and society. NSGC members are expected to be aware of the ethical implications of their professional actions and to adhere to the guidelines and principles set forth in this code.

INTRODUCTION

A code of ethics is a document which attempts to clarify and guide the conduct of a professional so that the goals and values of the profession might best be served. The NSGC Code of Ethics is based upon relationships. The relationships outlined in this code describe who genetic counselors are for themselves, their clients, their colleagues, and society. Each major section of this code begins with an explanation of one of these relationships, along with some of its values and characteristics. Although certain values are found in more than one relationship, these common values result in different guidelines within each relationship. No set of guidelines can provide all the assistance needed in every situation, especially when different relationships appear to conflict. Therefore, when considered appropriate for this code, specific guidelines for prioritizing the relationships have been stated. In other areas, some ambiguity remains, allowing for the experience of genetic counselors to provide the proper balance in responding to difficult situations.

SECTION I: GENETIC COUNSELORS THEMSELVES

Genetic counselors value competence, integrity, dignity, and self-respect in themselves as well as in each other. Therefore, in order to be the best possible human resource to themselves, their clients, their colleagues, and society, genetic counselors strive to:

1. Seek out and acquire all relevant information required for any given situation.
2. Continue their education and training.
3. Keep abreast of current standards of practice.
4. Recognize the limits of their own knowledge, expertise, and therefore competence in any given situation.
5. Be responsible for their own physical and emotional health as it impacts on their professional performance.

SECTION II: GENETIC COUNSELORS AND THEIR CLIENTS

The counselor–client relationship is based on values of care and respect for the client's autonomy, individuality, welfare, and freedom. The primary concern of genetic counselors is the interests of their clients. Therefore, genetic counselors strive to:

1. Equally serve all who seek services.
2. Respect their clients' beliefs, cultural traditions, inclinations, circumstances, and feelings.
3. Enable their clients to make informed independent decisions, free of coercion, by providing or illuminating the necessary facts and clarifying the alternatives and anticipated consequences.
4. Refer clients to other competent professionals when they are unable to support the clients.
5. Maintain as confidential any information received from clients, unless released by the client.
6. Avoid the exploitation of their clients for personal advantage, profit, or interest.

SECTION III: GENETIC COUNSELORS AND THEIR COLLEAGUES

The genetic counselors' relationships with other genetic counselors, genetic counseling students, and health professionals are based on mutual respect, caring, cooperation, support, and a shared loyalty to their professions and goals. Therefore, genetic counselors strive to:

1. Foster and protect their relationships with other genetic counselors and genetic counseling students by establishing mechanisms for peer support.
2. Encourage ethical behavior of colleagues.
3. Recognize the traditions, practices, and areas of competence of other health professionals and cooperate with them in providing the highest quality of service.
4. Work with their colleagues to reach consensus when issues arise about the role responsibilities of various team members so that clients receive the best possible services.

SECTION IV: GENETIC COUNSELORS AND SOCIETY

The relationships of genetic counselors to society include interest and participation in activities that have the purpose of promoting the well-being of society. Therefore, genetic counselors strive to:

1. Keep abreast of societal developments that may endanger the physical and psychological health of individuals.
2. Participate in activities necessary to bring about socially responsible change.
3. Serve as a source of reliable information and expert opinion for policymakers and public officials.
4. Keep the public informed and educated about the impact on society of new technological and scientific advances and the possible changes in society that may result from the application of these findings.
5. Prevent discrimination on the basis of race, sex, sexual orientation, age, religion, genetic status, or socio-economic status.
6. Adhere to laws and regulations of society. However, when such laws are in conflict with the principles of the profession, genetic counselors work toward change that will benefit the public interest.

Adopted in January 1992 by the National Society of Genetic Counselors, Inc.

12

Student Supervision

Patricia McCarthy and Bonnie S. LeRoy

Supervision is an essential component of genetic counseling education, as it is in the preparation for any health care profession. It serves two primary purposes: promoting the professional development of student supervisees and ensuring the continued provision of quality client services. This chapter describes the supervision process in a systematic manner. We have drawn on extensive literature from the mental health counseling and psychotherapy fields, and on our combined 35 years of experience as counselor educators and clinical supervisors.

The sections that follow define supervision and present the goals for desired outcomes. The different roles and responsibilities of the supervisor and student are described, and supervision objectives and techniques are delineated. Suggestions for assessment and evaluation are provided. Ethical and legal issues are raised, and critical issues inherent in the supervision process are also discussed. Examples of supervision issues, provided by genetic counseling supervisors and students, are included to illustrate major concepts. The emphasis of this chapter is on practical suggestions for approaching the supervision process and improving the effectiveness of the supervision relationship.

The ultimate outcome of effective supervision is a student who progresses toward becoming a fully competent genetic counseling professional. The American Board of Genetic Counseling (ABGC) has established proficiency criteria for the profession by identifying practice-based competencies that an entry-level genetic counseling professional is expected to demonstrate (Fine et al., 1996). Graduate programs that are accredited by the ABGC must provide clinical experiences that assure that students develop these competencies. Thus it is crucial that the supervision goals and the

A Guide to Genetic Counseling, Edited by Diane L. Baker, Jane L. Schuette, and Wendy R. Uhlmann.
ISBN 0-471-18541-8 (cloth), 0-471-18867-0 (paper).   Copyright © 1998, Wiley-Liss, Inc.

student evaluation be tailored to specifically address student progress in the areas of the practice-based competencies.

The practice-based competencies established by the ABGC were developed through a consensus approach using actual practice experiences to reflect the standards expected of the profession (Fiddler et al., 1996). They are categorized into four domains, and each encompasses the practice skills necessary for the student to attain expertise within the domain. The influence of Bloom's taxonomy, discussed in more detail later in this chapter, can be seen in these competencies, since they are concrete and measurable and therefore adaptable for setting goals for supervision and student evaluation. The practice-based competencies are presented as an appendix to Chapter 1 of this text and are included in this chapter in the clinical evaluation example given in Appendix 12.2.

In the assessment of outcomes achieved, as well as in the development of specific goals for outcomes expected but not achieved, student progress within these competencies needs to be addressed through the supervision relationship and student evaluation. The approach that is most effective is if supervisor and student work together as a team. Supervision, like genetic counseling, involves many skills that require time to master. This chapter provides some guidance and practical suggestions for the development of effective relationships in the supervisory setting so that the supervisor and student may learn and grow together.

DEFINITIONS OF SUPERVISION

A number of definitions of student supervision exist in the counseling and psychotherapy literature. Bernard and Goodyear (1992) defined supervision as "a means of transmitting the skills, knowledge, and attitudes of a particular profession to the next generation in that profession. It also is an essential means of ensuring that clients receive a certain minimum quality of care while trainees work with them to gain their skills." Hart (1982) defined supervision as "an on-going educational process in which one person in the role of supervisor helps another person in the role of student acquire appropriate professional behavior through an examination of the student's professional activities."

Implicit in these definitions are three dimensions: (1) an ongoing relationship between the supervisor and student that consists of reciprocal influence and support, (2) activities that promote socialization into the profession, and (3) a focus on the development of professional behaviors and skills for the benefit of clientele.

The supervision relationship evolves over time and generally ends after a specified time period. Supervisor and student need to develop trust and rapport, and they must have agreed-on goals to maintain a viable relationship. According to Stout (1987), it is important that the supervisor regularly ask herself, "Whose needs are primarily being met? Is what I'm doing within the realm of supervision?" Irrelevant, excessive self-disclosure and excessive probing into the student's personal life should be avoided. A regular time for supervision meetings should be set and adhered to, and supervision meetings should take place at a specific professional site.

A distinction has been made between clinical and administrative supervision (Bernard and Goodyear, 1992). Clinical supervision involves directly facilitating the development of an individual's genetic counseling skills, while administrative supervision concerns accountability and involves monitoring activities to promote quality care for clients. Clinical supervision is concerned with student professional and personal development, while administrative supervision involves the coordination of overall clinical services and evaluation mechanisms (policies, protocol, documentation, and overall accountability). Most of the information presented in this chapter pertains to clinical supervision.

Clinical supervision can be divided into individual supervision and peer supervision. Individual supervision typically involves a supervisor who is an experienced genetic counselor and a student with less genetic counseling experience than the supervisor. Peer supervision consists of individuals with comparable levels of experience, who usually conduct supervision without a designated supervisor. As discussed in more detail in Chapter 6, Psychosocial Counseling, peer supervision can serve the practicing genetic counselor by further enhancing skills and professional growth. Individual clinical supervision is the focus of this chapter because it is the primary method of student training in the genetic counseling field.

GENETIC COUNSELORS AS CLINICAL SUPERVISORS: RESPONSIBILITIES

The Association for Counselor Education and Supervision (ACES), which has written standards for counseling supervisor responsibilities (ACES, 1995), defines a supervisor's responsibilities as follows: "to teach the inexperienced and to foster their professional development; to serve as consultants to experienced counselors; and to assist at all levels in the effective provision of counseling services." The genetic counselor supervisor has several responsibilities, including overseeing case preparation, evaluating performance in conducting cases, reviewing written documentation, and discussing the professional development of the student. The supervisor needs to be versatile, shifting among the roles of consultant, teacher, counselor, and evaluator described later in this chapter. Supervisor responsibilities also include being available to the student, providing clear and timely feedback, offering support and guidance as needed, setting expectations for the student, and following up to be sure goals are met.

The skills that a genetic counselor uses to counsel patients are distinct from those used to supervise students, and this difference is important for supervisors and students to appreciate. The primary theoretical approach to genetic counseling involves nondirectiveness with the intent of facilitating and supporting client autonomous decision making. Genetic counselors are careful to assess each client's perceptions and belief systems and to work within his or her framework of values. Supervisors teaching in the clinical setting strive to help students work within the accepted framework of values of the profession and to separate their own values from those of the client. A survey of practicing genetic counselors found that one of the behaviors most frequently reported in the client encounter is information giving (Bartels et al., 1997). Yet an effective supervisor often intentionally provides little or no information but, rather, helps the students learn how to obtain it on their own and become creative in solving problems. Thus, the supervisor must employ an approach different from that which would typically occur with a client in the genetic counseling encounter.

Being an effective genetic counselor does not always make one an effective supervisor. It is a myth that supervision is identical to the activity being supervised (McCarthy et al., 1988). In other words, supervision skills extend beyond the skills used in genetic counseling, and there should be a continual effort to develop them.

THE GOALS OF SUPERVISION

To determine the goals for supervision, consider these questions: As a result of this rotation, what should the student know? What should she be able to do? How will she feel about her professional development? These questions are representative of the global aims or goals inherent in the supervision relationship. Two different models that present a useful framework for skill development from which specific genetic counseling supervision goals can be established and evaluated are The Discrimination Model and Bloom's Taxonomy.

The Discrimination Model developed by Bernard and Goodyear (1992) classifies student skills into four categories based on whether the skills primarily involve overt behaviors, feelings or thoughts. Examples that are specific to genetic counseling are illustrated.

- *Process Skills* These are "doing skills" that consist of the actual techniques and strategies used in a genetic counseling session:
  - obtaining appropriate family, medical, and social histories
  - using open-ended questions to help clients express concerns
- *Professional Skills* These are "doing skills" that involved adherence to professional standards of behavior:
  - coming to genetic counseling sessions adequately prepared
  - documenting genetic counseling session data accurately in client charts
- *Personalization Skills* These are "feeling skills" that pertain to the internal, subjective reactions students have toward their clients and the genetic counseling and supervision relationships:
  - recognizing how client loss triggers a student's own grief reaction
  - seeking feedback during supervision
- *Conceptualization Skills* These are "thinking skills" that involve the cognitive processes of case analysis and client conceptualization:
  - identifying appropriate case approach based on client data
  - anticipating client reaction to genetic information

The second model that provides a useful framework for developing supervision goals is a modification of Bloom's Taxonomy (1956) and is presented in Table 12.1. Developed for use in educational settings, this taxonomy consists of a hierarchy of learning processes from basic memorization of facts to highly sophisticated critical thinking. The novice student is more likely to have memorized basic genetic information but

TABLE 12.1   The Application of Bloom's Taxonomy for Developing Genetic Counseling Goals

Beginning Level Skills

Know
List, repeat, memorize, recall, state. Recall or recognize appropriate facts, concepts, principles.

EXAMPLES

1. State the mode of inheritance for neurofibromatosis.
2. List the etiology, symptoms, and prognosis for the major genetic conditions dealt with in the metabolic diseases clinic.

Comprehend
Translate, interpret, extrapolate from data, charts, graphs, etc.

EXAMPLES

1. Interpret a pedigree to identify a family at high risk for cancer.
2. Translate information presented in scientific articles into language understandable to the client.

Intermediate Level Skills

Apply
Illustrate, demonstrate, use, give an example. Use facts, concepts, and principles in a hypothetical or real situation.

EXAMPLES

1. Illustrate for a client the difference between dominant and recessive inheritance in a way that the client can understand.
2. Provide information to clients in easy-to-understand terms.

Analyze
Compare and contrast information, deduce, dissect, break down.

EXAMPLES

1. Compare and contrast the possible psychosocial impact of the diagnosis of Down syndrome in a first pregnancy for a 25-year-old versus a 41-year- woman who comes to clinic with a long history of infertility.
2. Determine from available data, the most appropriate follow up plan for a case.

Advanced Level Skills

Synthesize
Integrate information, consolidate, build, join.

EXAMPLES

1. Consolidate data from a client's family history and from the blood test results to identify what information is important to communicate.
2. Integrate and convey information from several sources that is important for a particular client.

Evaluate
Critique information, assess, judge, debate, determine the worth of. . . .

EXAMPLES

1. Assess client likely reaction to your proposed counseling plan and modify as needed.
2. Assess the extent to which clients understand the information presented in a genetic counseling session.

Source: Genetic counseling examples based on Bloom BS (ed) (1956).

tends to lack skills at application, critical analysis, and evaluation of cases. One would expect more skills representative of the higher levels of the hierarchy to be demonstrated as a student gains experience. The Bloom's Taxonomy model can be used by both supervisor and student to assess the skill level of the student in various clinical settings and to establish specific goals for attaining or improving upon skills.

Supervisor: I have a student who is significantly older than me and came to our clinic with some related experience. However, the student is a novice in the area being supervised and has some difficulty with being a beginner at anything. How do I handle this?

Response: This issue can be addressed with a concrete approach by reviewing with the student Bloom's Taxonomy (see Table 12.1) and the practiced-based competencies defined by the ABGC. The actual skill level of the student can be documented, and specific goals for achieving more advanced skills can be established.

SUPERVISION DISCUSSION TOPICS

The primary discussion topics in clinical supervision can be categorized as client-related issues, issues pertaining to the counselor, and issues pertinent to the supervision relationship (McCarthy et al., 1994). These topics involve issues that will often also be discussed in the evaluation of the student and can be used as a reference when structuring supervision sessions.

Client-Focused Topics

- Client diagnosis
- Assessment of family and medical history
- Interpretation of test results
- Determining genetic counseling approach (e.g., type of information to provide and how to present this information)
- Conceptualizing client dynamics (e.g., client indecision)
- Planning student responsibilities for genetic counseling session
- Setting short-term and long-term goals
- Assessing client progress (e.g., client decision making)
- Identifying ethical/legal issues
- Recognizing boundary issues in counseling

Counselor-Focused Topics

- Genetic counselor/client dynamics
- Professional development
- Clinical skills progression
- Impact of personal issues on provision of genetic counseling

Supervision Relationship Topics

- Relationship between supervisor and student
- Boundary issues in supervision (e.g., confidentiality)
- Administrative issues

SUPERVISOR AND STUDENT ROLES

Several authors have defined the types of roles adopted by supervisors and students over the course of their relationship. The following is an adaptation to genetic counseling of the roles discussed by Bernard (1988). Four roles are described: consultant, teacher, counselor, and evaluator. It is important to remember that supervisors usually have had little or no formal training in teaching and clinical supervision. They typically use, and overuse, the role in which they are most comfortable, or fall back on examples of supervision approaches they experienced as a student, such as the medical model of teaching. The following examples demonstrate how supervisors and students can be more versatile in their use of roles and thus enhance the effectiveness of the supervision experience.

Consultant Role

Consultation is the principal activity through which supervision is typically conducted (Boyd, 1978). When the supervisor is a consultant and the student is a consultee, the two interact collaboratively. Objectives are mutually agreed on, the supervisor encourages the student to self-evaluate, and the focus of supervision is on the student's clients. Consulting activities include brainstorming possible strategies and interventions, generating options for the client, and discussing client needs. The consultant (supervisor) acts as a facilitator who works with the consultee (student) to determine effective planning and action. The consultee has a great deal of personal responsibility for goal setting, evaluation, and problem solving. Students who are more advanced usually prefer the consultant role.

Teacher Role

When the supervisor is in the teacher role and the student is in the classical student role, the primary interaction is instruction. The teacher is a resource person who shares information, skills, and strategies, which the student is seeking. The focus is on the development of the student as a genetic counselor. Teaching activities include demonstrating, explaining, and interpreting events from genetic counseling sessions, and identifying appropriate interventions. The supervisor exerts a fair amount of power over this interaction, as she is the "expert" providing at least some of the answers. Beginning students who are eager for models of what to do usually prefer the supervisor to assume the teaching role.

Counselor Role

When the supervisor is in a counselor role and the student is in a client role, the primary interaction is one of exploration with the goal of promoting self-awareness and growth. Here the focus in on the student as a person. The supervisor assists the student in recognizing developmental tasks and becoming aware of personal issues that may impact on her work with clients. For example, most beginning mental health counselors struggle with issues of competence and confidence (Loganbill et al., 1983). A supervisor in the counselor role can introduce a discussion of these dynamics and help to normalize them (e.g., "Most students feel some anxiety over their skills. Is that what you are feeling?"). The supervisor plays a role in helping the student identify feelings and defenses and in helping her understand how these inner experiences impact the genetic counseling relationship. However, the supervisor is not responsible for helping the student change these defenses or reactions. This distinction maintains an important boundary in that the supervisor does not provide psychotherapy for the student. If personal counseling or psychotherapy appears to be appropriate, the supervisor should discuss with the student a referral to a therapist. The counselor–client roles tend to be adopted infrequently in the typical genetic counseling supervision relationship.

Evaluator Role

When the supervisor is an evaluator and the student is an evaluatee, the primary interaction is critiquing and feedback giving, and the focus is on accountability. The supervisor acts as a gatekeeper assessing the effectiveness of the student and the service the student is providing. In this interaction, elements of administrative supervision often blend with clinical supervision. Evaluation activities include formal and informal assessments, goal setting, and giving and receiving feedback. Three aspects of the clinical experience are generally included in the evaluation: the work of the student, the genetic counseling services provided, and the supervision relationship itself. The process of evaluation is covered in more depth later in this chapter.

Examples of Supervisor–Student Roles

There is no one role that is the absolute "right" role to assume in clinical supervision. Supervisors and students move in and out of these roles depending on their needs. The following supervisor–student interactions illustrate how the different roles described above can be effectively applied to the same scenario.

Scenario   The student had just completed a genetic counseling session in which the client started to cry as she heard the results of her genetic testing. The student had responded by repeating extremely complicated information about the test results in a rapid, highly technical fashion, at which point the supervisor elected to take over the session. Later, during a supervision discussion, the case was reviewed and the issue was discussed.

Consultant Role

Consultee: I'd like to talk about some other ways I could have responded when the client started to cry.

Consultant: One approach I've tried is waiting a few moments before saying anything. What other approaches have you thought of trying?

Teacher Role

Student: Is what I did OK? If not, what should I have done instead?

Teacher: This is what I think you should have said to empathize with the client. . . .

Counselor Role

Client: It's not usually like me to ignore the client's feelings. I don't know what was going on.

Counselor: It seems that you felt threatened by her show of emotion. . . . Let's discuss this.

Evaluator Role

Evaluatee: I don't think repeating all the information was the best way to respond when she cried.

Evaluator: What genetic counseling goal were you trying to meet by repeating this information? How effective was your timing and pacing in helping the client to understand the information?

It is important to note that these roles are not always discrete; elements of more than one role may be evident in any particular supervisor–student interaction. Furthermore, a mix of roles may be optimal. For example, in the preceding scenario, the supervisor might want to use a combination of the four roles to thoroughly process the genetic counseling incident. Finally, both members of the supervision relationship have some influence over the roles that are expressed by the other party (e.g., the student may "pull for" the evaluator role by asking for feedback about how he is doing; the supervisor may encourage the student to act as a client by asking how he feels about counseling a terminally ill individual).

The following scenario provides another illustration of different supervision roles with the student initiating a specific role.

Scenario Your student comes to you and says, "I just saw a couple who was referred for prenatal diagnosis. I went through a detailed explanation of advanced maternal age and the associated risk factors. I also explained the amniocentesis procedure, but I just don't know if they 'got it.' I mean, I asked a few times if they had any questions, but they kept saying 'no.' I really don't know if they understood any of the information." The following are examples of the different directions in which the discussion may have proceeded.

Consultant Role

Consultee: I would like to figure out a different approach.

Consultant: Let's brainstorm other approaches. What other ways have you tried to find out if clients understand information? What might you do?

Teacher Role

Student: How can I tell if they understood?

Teacher: Next time, after you've given some important information, try asking the clients what they heard you say.

Counselor Role

Client: I have no idea if I am doing any good!

Counselor: I sense this is bothering you. You seemed to be operating from an assumption that if you just ask clients if they have any questions, they'll tell you. But your intuition tells you that's not working.

Evaluator Role

Evaluatee: I really don't think they understood anything.

Evaluator: On what basis are you deciding that they didn't "get it"? What is your evidence?

STUDENT RESPONSIBILITIES

Student responsibilities in clinical supervision include developing versatility in using and responding to the supervisor's different roles of consultant, teacher, counselor, and evaluator; coming prepared to supervision; being open and responsive to feedback; following through on supervisor assignments and recommendations; disclosing important information; asking for assistance when faced with counseling issues beyond their competencies; and engaging in self-evaluation.

Just as supervisors have a responsibility to provide appropriate feedback, students have a responsibility to appropriately receive it. We offer several suggestions for receiving feedback that supervisors can ask of their students at the start of a clinical rotation.

1. *Clarify Feedback*　Let your supervisor know that you heard and understand what he said; ask for clarification until you do understand.
2. *Share Your Response*　When receiving feedback, you will have both cognitive and affective reactions. Let your supervisor know how this feedback makes you feel and tell her what you think about it.
3. *Accept Positive Feedback*　Awareness of your strengths is as important as awareness of areas for improvement; do not gloss over positive feedback too quickly. You may want to use this feedback to think of new ways to adapt your strengths to different situations.

4. *Accept Corrective Feedback*   Remember that corrective feedback is necessary for your development as a genetic counselor, that everyone has areas for improvement, and try to welcome this information rather than avoid it.

5. *Test Validity of Feedback*   Although everyone's perception is valid in their own worldview, it is not necessarily the absolute "truth." If the feedback does not fit with your perceptions, ask for clarification and check it out with others. Usually you will discover the valid part of the feedback if you persist in trying to understand it.

6. *Exercise Personal Responsibility*   As you develop as a counselor, you will become more aware of your own strengths and weaknesses. Take responsibility by asking for feedback about specific behaviors and issues that are problematic for you.

7. *Avoid Feedback Overload*   If you begin to feel bombarded with information, let your supervisor know that you have had enough and want to continue the conversation at a later time. It is important to realize that assimilating and processing feedback can take time.

Students have the ability to influence the nature of their supervision relationship, and they must know what they need and how to ask for it (Bernard, 1992). If students need support, it is most effective to assume a client role and "ask" their supervisors to be counselors. If they need advice, be students and ask their supervisors to be teachers. If they desire feedback, then become the evaluatee to the evaluator supervisor. Finally, if they want to "troubleshoot," then be a consultee and ask the supervisor to be a consultant.

METHODS IN SUPERVISION

There are different methods for conducting supervision that are clearly evident from the verbal responses used by supervisors. A study by McCarthy et al. (1994) analyzed supervisors' verbal responses, which are presented in Table 12.2, along with examples specific to supervision in the genetic counseling setting. These responses also can be classified into three categories as first described by Danish et al. (1980):

- **Continuing responses**, which encourage the student to talk and are particularly useful for initial exploration of a topic.
- **Leading responses**, which are statements in which the supervisor initiates the direction of the conversation. They are particularly helpful once the initial exploration of a topic has been completed.
- **Self-referent responses**, which involve revelation by the supervisor of autobiographic information or personal reactions to the student. Self-reference is particularly helpful for normalizing student experiences (e.g., "I have difficulty working with angry clients, too."); modeling appropriate interventions (e.g.,

TABLE 12.2   Supervisor Verbal Techniques

Continuing Responses

Encourages the student to talk.

1. Content responses to reflect student statements

   Example: "To summarize, you told me that from the family history, you feel the client's risk for cancer is low."

2. Affect responses to reflect student feelings

   Example: "You are saying that you are uncomfortable when a client becomes angry."

Leading Responses

Supervisor leads discussion.

3. Open questions to get at process issues

   Example: "Tell me how you would approach this issue?"

4. Closed questions to get at details

   Example: "What is the recurrence risk for neural tube defects?"

5. Interpretations to provide insight

   Example: "When you come to very complicated information, it seems that you start talking really fast to avoid discussing it."

6. Confrontations to point out discrepancies

   Example: "You do a better job than you give yourself credit for. Maybe you're being too hard on yourself."

7. Advice giving to suggest alternative behaviors

   Example: "Next time when a client appears upset, wait a few moments before jumping in."

8. Influence responses to try to alter student views

   Example: "It isn't usually helpful to tell clients that they should not feel bad."

9. Information giving to provide facts, resources, etc.

   Example: "Direct gene analysis for this disorder is available through many commercial and academic labs in the United States."

Self-Referent Responses

Supervisor reveals information.

10. Self-disclosure to reveal supervisor personal information

    Example: "When an adolescent is involved, I find that it works best if I take the time to talk with the teenager alone."

11. Self-involving to provide supervisor personal reactions about the student

    Example: "I think our counseling approaches are similar and I am comfortable working with you."

Source: Examples based on McCarthy et al. (1994).

"This is how I explain a balanced chromosomal translocation to a client."); dealing with supervision impasses (e.g., "I get frustrated when you say 'Yes, but . . .' to my suggestions, because I sense that you really don't want to hear my ideas."); and providing feedback (e.g., "I'm very pleased at how well you prepared for your last counseling session.").

In addition to these specific verbal responses, several other techniques can be used by supervisors during supervision. These include modeling counseling interventions (the supervisor demonstrates a skill either through a role-play with the student or in an actual genetic counseling session, which the student observes), assigning homework to the student (e.g., summarize the research findings regarding Marfan syndrome), and referring the student to other sources (e.g., contact the molecular diagnostics laboratory for more information on testing for myotonic dystrophy).

Case Discussion

Student: My supervisor did not define her role clearly, nor did she state the objectives of supervision during my rotation. She never provided any clear or specific feedback. At the end of the placement, I received a very negative evaluation. What can I do to keep this from happening in future rotations?

Response: When a supervisor is too "laid back" and not fulfilling the obligation to provide constructive supervision, the student must take a more active role in seeking feedback. The student can set a specific time for a supervision session and then come well prepared. Some of the approaches discussed below can be used by both students and supervisors to create a more focused discussion and evaluation of cases.

A comprehensive evaluation of cases can be initiated by having students keep a journal summarizing impressions of their cases. When students document their impressions immediately after a session and make further, more detailed written comments at a later time, both affective and cognitive issues are more likely to be identified. The affective issues typically emerge immediately following a session, and more cognitive details develop through later reflection and more detailed writing. Prior to meeting with the supervisor, the student can do a self-evaluation by reviewing case notes and/or tapes (if sessions are taped) and identify particular issues to be discussed.

Alternatively, the supervisor can structure informal case presentations by giving the student a list of questions and instructions to prepare for debriefing. Some examples are as follows:

1. What was the reason given for referral?
2. What were the client's expectations of this session?
3. Summarize the medical and family histories and provide a risk assessment.
4. What were the major medical and genetic issues involved in this case?
5. What were the major psychosocial issues involved in this case?

6. What options did the client have for decision making?

7. What is your follow-up plan?

Regardless of the approach selected, it is important during supervision that the supervisor do a time and process check (e.g., "I see that we have 10 minutes left. How are we doing with the questions you had about this case?"). This is actually similar to what one does in the clinical setting, where patients are made aware of the time limitations and their responsibility to raise any issues that have not yet been discussed.

Focusing Issues in Supervision

A common complaint among supervisors is, "Why don't my students get to the point? They spend so much time on trivial details that the most important information is often neglected entirely, or it comes up too late in our session to do much with it." A student might fail to present information about a case in a focused manner for several reasons:

Fishing   The student does not know how to focus; that is, he does not know which of the details are important to convey. It is a skill he has not yet learned and/or hasn't transferred to the genetic counseling and supervision relationships.

Where's the Road Map   The student lacks a framework for assessment/conceptualization of client circumstances. She does not see the bigger picture (e.g., she fails to recognize that a family history of mental retardation, in multiple generations and affecting only males, is a possible cue for fragile X syndrome or some other form of X-linked mental retardation). This failure to focus may be more likely to occur when clients present to clinic with genetic conditions for which the student has little or no prior experience.

Going with the Flow   Sometimes the supervisor derails the student's presentation because the supervisor's thinking goes in several different directions at once. The student follows right along this divergent path.

Birds of a Feather   The student's behaviors mirror those of a client: the client is very scattered in presenting information, and the student repeats this pattern in supervision. This is called "parallel process" by mental health theorists (Ekstein and Wallerstein, 1972).

Pressure Cooker   Something emotional is going on inside the student (e.g., anger, frustration, agitation). The student deals with this affect by blowing off steam or talking on and on in great detail.

I've Got a Secret   The student is experiencing anxiety. It could be due to several sources, including apprehension over being evaluated by the supervisor, client experiences that touch on personal issues (e.g., unresolved issues about death triggered by a client mourning the loss of a child), deficits in the supervisory relationship, or anticipated consequences of telling the supervisor certain information (e.g., the student avoids disclosing that he did not confront the client about her alcohol abuse during pregnancy because he does not want to do this with clients in the future; if he never focuses, the supervisor never figures out this resistance).

Shotgun Approach   The student has the irrational belief that there is one crucial piece of information which, if the supervisor only knew it, would enable her to tell the student exactly what to do with a given case or with similar cases in the future. Because the student does not know which piece of information is the crucial one, she feels compelled to provide a verbatim description of everything that transpired in the genetic counseling session.

I'm No Good   The student is motivated to convince the supervisor that he is on the verge of a genetic counseling disaster. The student magnifies his mistakes, placing the supervisor in the position of providing reassurance and minimizing errors. The student's unconscious agenda (perhaps trying to mask insecurities) prevents a focused discussion of the case.

The Laid-Back Supervisor   Students often complain, "I just don't know what my supervisor wants me to say about my cases. She always asks me what I need to discuss, but I don't even know where to begin!" Novice students are especially likely to need some structure from the supervisor. Indeed "laissez-faire" supervisors may be regarded as undesirable by both novice and more experienced students (Allen et al., 1986).

STUDENT ASSESSMENT

A great deal of supervision time is spent discussing indirect data about genetic counseling cases. Most of the time, this information is obtained via self-reports and case notes through which the student provides a verbal and written summary of what transpired in a session. This information can also form the basis for formal case presentations.

Formal Case Presentations

Formal case presentations provide a vehicle for assessing student counseling skills, giving the student feedback about genetic counseling behaviors, and generating interventions for working with a particular client. Although they can be done in individual supervision relationships, the benefits are usually greater when formal case presentations are done in small groups with a designated supervisor acting as facilitator, such as in a class setting. A typical case presentation consists of a written as well as a verbal summary of the case, followed by discussion, which includes brainstorming and feedback.

One important ground rule when giving feedback is that everyone should begin with one positive feedback statement and then one corrective statement. If this rule isn't stressed, the presentation usually proceeds in one of two ways—either everyone praises the student and no constructive learning occurs, or everyone "gangs up" and overwhelms the student with corrective feedback. The supervisor should ask group members who raise criticisms to specify what they would have done if it had been their client. Group members also can be asked to role-play important suggestions the student is receiving. Because numerous ideas can be generated quickly during case presentations and these ideas vary in their value, the supervisor should summarize the most valid points. The supervisor also needs to be sensitive to the emotional state of the student undergoing assessment and call a time-out if she becomes too anxious or discouraged.

> *Supervisor:* A student reported on a counseling session I had not observed. I questioned the way in which some information was given to the client—in a complicated way, which I felt the client did not understand. The student's response was "You were not there—you are only second-guessing." How can I better handle case reviews of sessions I did not directly supervise?
>
> *Response:* A formal case presentation is an excellent way to provide detailed supervision of a case when the supervisor was not present during the session. It requires the student to analyze the specifics of the case, present it chronologically, and to look for areas in which skills can be improved.

Because self-report is limited by the student's skills in assessing and explaining what occurred and by the student's comfort with disclosing relevant information (including mistakes), direct data also are desirable. When possible, sessions should be directly observed by the supervisor, conducted by both the supervisor and student (cocounseling), audiotaped, or videotaped. These methods provide firsthand information, which affords a more valid record of what occurred. They also allow the supervisor and student to more quickly address important issues because they can dispense with a detailed summarization of the session.

Surveying Clients

Another possible assessment approach for obtaining more direct data on student performance would involve surveying the students' clients. Some would argue that the ultimate test of whether genetic counseling is effective is the extent to which the clients' expectations are met (Michie et al., 1997). Survey results from clients served by the student could be incorporated into student feedback and supervision. Table 12.4 provides a sample client survey form.

Formative and Summative Evaluations

Evaluation is "one of the hallmarks of supervision" (Bernard and Goodyear, 1992). It is essential to the development of competent professionals and to the assurance of quality service provision. There are two major types of evaluation in supervision—formative and summative. Formative evaluation, the most common type, is the informal feedback that focuses on the student's day-to-day behaviors with specific cases

TABLE 12.3   Tips for Providing Formative Evaluation

1. **Timing**   Pick the right time, as soon as is appropriately possible after the behavior has occurred.
2. **Private**   Select the right setting; privacy is important.
3. **Balance**   Begin with one or two things the student is doing well before moving to corrective feedback.
4. **Affect**   Recognize the emotional impact your feedback is having on the student.
5. **Warn**   Avoid surprises; it is best to drop a hint of what is to come.
6. **Self-Control**   Keep your own reactions under control and in perspective. It may be a familiar mistake to you, but it's the first time for the student.
7. **Accuracy**   Be sure of all the facts involved.
8. **Behavioral**   Keep the student's personality out of the discussion. Focus on the behavior she is exhibiting or failing to exhibit that requires modification.
9. **Focus**   Keep the conversation on the student; don't compare him to other students.
10. **Rebuild**   Close with an effort to restore the student's confidence.
11. **Delay**   Think before you speak. Often when we have a strong reaction to an immediate situation, our responses tend to be more emotional and off-putting. With reflection, the feedback can be more effectively delivered. Consider telling the student that you would like to discuss the case tomorrow, taking the time to think about your supervision approach.

Source: McCarthy, unpublished.

TABLE 12.4   Examples of Questions to Be Used in a Client Satisfaction Survey[a]

1. How well do you feel your student counselor understood your situation?
2. How well did your student counselor explain information about your conditions, testing options/results, etc. to you?
3. How comfortable did you feel with your student counselor?
4. What is one thing your student counselor did especially well?
5. What is one thing your student counselor could have done better?

[a]Scale for questions 1–3: 1, not at all . . . 5, extremely well.

and clients. We have outlined some tips for providing formative evaluation in Table 12.3. These tips focus on specific behaviors and are intended to minimize some of the common difficulties encountered with this type of evaluation.

Student: As a student, I experienced a few sessions in which the genetic counselor/supervisor jumped in and started speaking with the clients regarding a topic I was just getting to.

Response: In providing formative evaluation, timing, privacy, and accuracy are extremely important (see Table 12.3). It would be most helpful if the student were permitted to complete a session with the supervisor taking note of what to "fill in" at the end. Later, at an appropriate time and place, accurate feedback could be given regarding the session. This strategy will more likely lead to a change in counseling approach than interrupting the student during a session that is, for the most part, going well.

Summative evaluation is a more formal activity in which supervisors and students look at the "whole package" and make judgments about the student's overall functioning. Because summative evaluation is based on a judgment about how well the student has done so far, it tends to be more comprehensive than formative evaluation. Therefore, summative evaluation should not focus on a single case or behavior but rather on the professional development of the student. Some guidelines for conducting summative evaluations are listed below (Bernard and Goodyear, 1992).

1. Develop minimum evaluation criteria. Articulate the standards you are comparing the student against. Students often agonize over how much importance to place on the feedback they receive when they lack these minimum standards. One approach is to review with the student the practice-based competencies established by the American Board of Genetic Counseling prior to the start of a clinical rotation. The supervisor can work with the student to identify goals for developing a set of skills for that particular rotation on which the evaluation will be based.

2. Have a strong administrative structure. Supervisors need to be supported by their employers and by the academic program directors. It is important to know who is the ultimate authority, that there are agreed on evaluation methods, and that due process will be followed if a student encounters performance difficulties.

3. There should be standardized procedures—the method and timing of evaluations should be consistent across training sites.

4. Students should be held to a reasonable number of performance criteria.

5. Supervisors should keep in mind student developmental differences. Different skills are expected from a student in her first rotation versus the last rotation.

6. Articulate clear evaluation criteria and methods.

7. Evaluate qualitatively and quantitatively.

8. Obtain input from several sources (other supervisors, the student, clients, etc.).

9. Document your evaluation in writing and provide illustrative examples.

10. Include suggestions about what the student can do to improve.

11. Be flexible—your goals for the student may need to be modified over time.

Evaluation can be a difficult activity for several reasons (Bernard and Goodyear, 1992). Students feel anxious about evaluation because it touches on their personality and intelligence. There can also be anxiety for the supervisor, who may feel as if he has opened Pandora's box (e.g., Will I have to fix what I call attention to? Is my student's deficit due to poor supervision on my part? What if my impressions are wrong?).

Evaluations at the Completion of a Rotation

At the end of a clinical rotation, it is important that the supervisor complete an evaluation summarizing the student's overall performance. If, during the rotation, the student worked with other genetic counselors, the supervisor should elicit and incorporate their feedback into the final evaluation. The student, likewise, should complete an evaluation to provide feedback about the supervisor and to communicate overall impressions about the rotation. The supervisor and the student should set aside a time to meet to exchange and review these evaluations and to formally bring the rotation to a close. These completed evaluations should be forwarded to the director of the genetic counseling training program. Sample forms for formal written evaluation of the student and supervisor are provided in Appendices 12.2 and 12.3, respectively, and can be modified to fit specific programs.

ETHICAL AND LEGAL ISSUES

A number of ethical issues may arise during the supervision relationship. Supervisors usually work with a "captive audience"; that is, students have no choice about being there (Stout, 1987). Two common ethical issues are dual relationships and confidentiality.

Dual Relationships

The nature of supervision itself is that of a dual relationship, because the supervisor is often consultant, teacher, evaluator, and counselor. The supervisor also has dual allegiance to the student and to the clientele, which can cause conflicts (Williams et al., 1990). The dual relationships that pose the greatest likelihood of problems to the supervision process are the other relationships the supervisor has or may have with the student (e.g., boss or future boss, instructor, friend, or recent classmate). If possible, the supervisor should suspend other relationships with the student during the period of supervision. This will help to maintain appropriate boundaries and reduce the potential for loss of objectivity.

Confidentiality

A related issue concerns how much of the supervision content is confidential. Since the essential nature of supervision is evaluation, students struggle with the question of how open to be with the supervisor who ultimately will evaluate them. A supervisor should never promise that everything a student says will remain confidential. In some cases, information revealed during the supervision relationship needs to be divulged to the director of the program as it may impact on client services. Confidentiality limits should be reviewed prior to discussing particular issues for which student privacy might not be maintained. A student may choose not to disclose some information, and the supervisor should respect this choice. If the information is not crucial to the student's effectiveness with clients, it can remain undisclosed.

Supervisor: I have been working for about four months with a student who initially was doing very well. One day she seemed to have uncharactertistically low energy. I asked how she was and she said "Fine, just fine." Later as her performance declined, I pressed for an explanation and she stated that she had some personal problems and did not want to discuss them with me. Her academic and clinical work are suffering, and the change in her personality is significant. What should I do if she continues to decline to talk to me?

Response: As her supervisor, you have the responsibility to address the unidentified problems because they are impacting on the student's ability to learn and to provide adequate genetic counseling. Since your offer to help has been turned down, you should respect her right to keep personal difficulties confidential. However, you have the primary responsibility of ensuring quality client care. Therefore, if the student's problems continue to affect her ability to provide patient care, it is appropriate for you to acknowledge her right of confidentiality and refer her to a professional who can help. Depending on the situation, it also may be appropriate to suggest a leave of absence from the clinical rotation.

In clinical supervision, there is the potential for inappropriate behaviors by both supervisors and students. Supervisors can fail to provide timely, accurate feedback; attend only superficially to the student's clinical work; distort evaluations; and waste time in supervision gossiping, talking about their own cases, or trying to impress the student. Students, on the other hand, can withhold critical information, waste time in supervision discussing trivial points, and act either so fragile or so hostile that the supervisor is hesitant to give feedback. Using informed consent contracts is one mechanism for decreasing the likelihood that behaviors of these types will occur.

Informed Consent for Clinical Supervision

Obtaining informed consent for clinical supervision is an optimal way to specify supervision goals and process (McCarthy et al., 1995). Informed consent in supervision, as in genetic counseling, is a communication process that involves a discussion of issues between parties. The discussion and written document should define the basic elements of the supervision process and outcomes. This contractual agreement formalizes and clarifies the relationship, educates the student about the nature of supervision, and provides the student with a model of the informed consent process for use with clients.

A written informed consent statement may include the following information: purpose of supervision, supervisor credentials, pragmatics of the supervision relationship (e.g., frequency, length, location, missed sessions), description of the supervision process (e.g., methods that will be used, roles and responsibilities of each party), administrative issues (e.g., description of evaluation procedures, due process procedures), ethical and legal issues (e.g., confidentiality limits, dual relationships, guidelines for client treatment), and a statement of agreement (indicates understanding of and adherence to the contract). Appendix 12.1 provides a sample of an informed consent statement for use in genetic counseling clinical supervision. This sample can be modified to fit varying supervisor situations.

Supervisor: A student entered into the supervisory setting being defensive and re-sistant to the process. These reactions, it seemed, stemmed from the student's own uncertainties about herself as a professional (low self-esteem), as well as personal conflicts from her life experiences. How should the supervisor deal with this behavior?

Response: Having an informed consent contract for clinical supervision can be helpful in clarifying the expectations related to the student's attitude and handling of personal issues. This format allows the supervisor to clearly disclose her expec-tations of the student, including the need for the student to develop a personal awareness of the impact of personal experiences on her ability to effectively pro-vide genetic counseling services. The informed consent contract should be com-pleted prior to the clinical rotation and a copy given to the student for reference.

OTHER ISSUES IN SUPERVISION

Cultural differences can play a role in the supervisory relationship. Culturally sensi-tive supervisors assist their students in becoming aware of their own biases, beliefs, and values; they encourage them to learn and understand the worldview of their clients; and they help to develop culturally appropriate interventions for their clients (Sue and Sue, 1981). Similarly, the supervisor becomes aware of her own biases, at-tempts to understand the worldview of her students, and works to develop and use culturally appropriate supervision methods.

Unconscious dynamics also play a large role in the supervision experience for both parties. The supervision relationship can generate strong affective reactions (Stout, 1987). Students may develop intense emotional reactions to their supervisors (adulation, resentment), and vice versa. These feelings are often based on transfer-ence and countertransference; that is, such reactions are the culmination of responses to individuals similar from one's past. Examples include students either greatly ad-miring authority figures or resenting them, and supervisors losing patience with de-pendent or rebellious behaviors. Sometimes students act out their clients' emotions on their supervisors (anger, anxiety, etc.). Such acting out can be difficult to detect and resolve because it is done unconsciously (Williams et al., 1990).

Supervisors may have conscious or unconscious agendas to impress their stu-dents with how much they know and may feel insecure as students gain competence ("You won't need me anymore"). Additional issues can involve gratification of one's own needs (e.g., having rescuer fantasies; wishing to be the ultimate authority), var-ious anxieties (anxiety over being evaluated by each other; anxiety about strong client or student emotions; performance anxiety), imposing one's values on the stu-dent (e.g., beliefs about abortion or mental retardation), fear of being disliked, or overidentification (with clients or the student). To the extent that these issues are un-recognized, they can compromise the work of supervision. Supervisors may wish to meet periodically in peer supervision groups to consult about their supervision is-sues, or they may prefer to raise the issues with the director of the genetic counseling training program.

CONSIDERATIONS FOR SUPERVISING NOVICE AND ADVANCED STUDENTS

Student developmental differences constitute another factor that can give rise to critical issues. It is a mistake to assume that all students have essentially the same competencies and the same supervision needs. There are several ways in which novices can differ from more advanced students. These differences may be viewed on continua, rather than as "all or nothing."

1. *Dependence on Supervisor*   Novices tend to be dependent on the supervisor for direction, feedback, and validation, whereas advanced students function more independently. Novices are also highly dependent on client reactions as a basis of validation (e.g., "The client was angry . . . that must mean I did a poor job of answering her questions."), while advanced students are more realistic about client reactions.

2. *Anxiety*   Novices experience more "global" anxiety about their competencies; they fear that they are "terrible" genetic counselors, while advanced students will usually have more skill-specific anxiety (e.g, "I have difficulty confronting male clients."). Advanced students may also have greater performance anxiety, because they feel they have to be expert genetic counselors.

3. *Motivation*   Although novices may be afraid of supervision, they are also eager for it and usually feel as if they can't get enough. Some authors (e.g., Kadushin, 1968) think that advanced students may be resentful of supervision, regarding it as one more "burden" in an already overloaded schedule.

4. *Personal Responsibility*   This is the "Whose problem is it?" phenomenon. Novices often feel overly responsible for all aspects of the genetic counseling relationship (e.g., if a client is unable to decide whether to have an amniocentesis, the novice believes the indecision remains because she did not explain the risk factors adequately). This is usually because the novice lacks a clear understanding of her professional role and needs more genetic counseling and self-evaluation experience. Advanced students are usually more realistic about their responsibilities as genetic counselors; when they feel overly responsible, it may be a clue that something in the case touched on a personal issue.

5. *Professional Self-Concept Crystallization*   Novices typically have poorly defined concepts of themselves as genetic counselors. They have many questions about their strengths and weaknesses, and whether they are capable of being helpful. Their concepts shift fairly easily in response to client reactions. Advanced students have more stable self-concepts that are more resistant to specific clients and genetic counseling experiences.

6. *Supervision Needs*   Novices are generally a more homogeneous group with respect to supervision needs. Beginning mental health counselors have some common concerns (Littrell, 1978): Can I learn the necessary counseling techniques? How do I meet client needs? What is my role as a counselor? How ad-

equate am I as a counselor? Do my clients like me? Advanced students are much more heterogeneous, given their varied counseling and supervision experience.

7. *Clarity of Supervision Goals* Novices are inclined to set goals that are too global and unrealistic (e.g., "I will learn everything there is to know about all metabolic diseases during my 8-week rotation in this specialty clinic."). Advanced students have a more precise idea about their needs; their goals tend to be more specific, obtainable, and individualized.

8. *Developmental Levels* Students vary as they gain counseling experience in areas such as competence, emotional awareness, autonomy, identity, respect for individual differences, purpose and direction, personal motivation, and professional ethics (Borders and Leddick, 1987). As they develop competencies for each of these areas, students will progress through stages: stagnation, confusion, and integration. Supervisors must tailor their responses accordingly.

Supervisor: I have a student who is in her final rotation and she already has a job in a pediatric setting. This rotation is in the perinatal center. She was an excellent student throughout her program but seems to be 'blowing off' this clinical experience. One obvious difficulty was motivation, but more difficult for me were the "I know it all" attitude and the disbelief of her previous supervisors that she wasn't doing well. What do I do?

Response: This student needs to be reminded of the original goals of the rotation. Problems like this will be less difficult to deal with if you have written criteria for passing the rotation and a written consent to supervision. It may also be helpful to acknowledge that supervision may have lost some of its thrill for this advanced student. Ask her to identify supervision activities that would appeal to her, and include some of these as well as the required activities in your time together.

It is important to have strategies for supervising students at different skill levels. If possible, the supervisor should assign less complicated cases to novices and save the more challenging ones for advanced students. Any case is additive for the novice who has done little or no genetic counseling. The supervisor should be more selective with advanced students who may need greater depth or breadth of experiences. Initially the supervisor should be more highly structured with novices, to diminish their anxiety and to increase productive use of time. The supervisor may also want to take an expert, teacher role with the novice, while primarily using a more collaborative, consultant role with the advanced student.

Issues faced by genetic counselors when they take on the role of supervising students at different levels of training include the following:

Identification It may be harder to identify with a novice, especially if the supervisor has been a genetic counselor for a long time.

Self-Efficacy The supervisor may feel superior to a novice, but threatened by the advanced student who has highly developed skills.

Responsibility   The supervisor may feel more responsible for the novice and may experience separation issues as the novice develops his skills. The supervisor probably feels less responsible for advanced students, believing that they can "take care of themselves."

Expectations   The supervisor may mistakenly assume that a novice has fewer competencies than she does; conversely, the supervisor may mistakenly assume that the advanced student has certain competencies that she lacks. Discussion of student experience and an assessment of student skills at the beginning of supervision can help to clarify competencies and responsibilities for goals for skill development.

CONCLUSIONS

As with the practice of genetic counseling, clinical supervision involves many complicated skills that take time and practice to develop. One should not expect to somehow automatically be able to perform the skills necessary to be an effective supervisor or student. Much effort goes into a supervisory relationship, and your preparation for a role of active responsibility in this relationship, whether supervisor or student, will help ensure a successful experience. It is important to keep in mind that investing time in becoming an effective supervisor is both a personal and a professional responsibility. Effective supervisors positively shape the future of the genetic counseling profession.

REFERENCES

Allen GJ, Szollos SJ, Williams BE (1986) Doctoral students' comparative evaluations of best and worst psychotherapy supervision. *Prof Psychol Res Pract* 17:91–99.

Association for Counselor Education and Supervision (1995) Ethical guidelines for counseling supervisors. *Couns Educ Superv* 34:270–277.

Bartels DM, LeRoy BS, McCarthy P, Caplan AL (1997) Nondirectiveness in genetic counseling: A survey of practitioners. *Am J Med Genet* 72(2):172–179.

Bernard JM (1988) Receiving and using supervision. In: Hackney H, Cormier LS (eds), *Counseling Strategies and Interventions,* 3rd ed. Englewood Cliffs, NJ: Prentice Hall, pp. 153–169.

Bernard JM, Goodyear RK (1992) *Fundamentals of Clinical Supervision.* Boston, MA: A division of Simon & Schuster.

Bloom BS (ed) (1956) *Taxonomy of Educational Objectives, Handbook I: Cognitive Domain.* New York: Longmans, Green.

Borders LD, Leddick GR (1987) *Handbook of Counseling Supervision.* Alexandria, VA: Association for Counselor Education and Supervision.

Boyd J (1978) *Counselor Supervision: Approaches, Preparation, Practices.* Muncie, IN: Accelerated Development, Inc.

Danish SJ, D'Augelli AR, Hauer AL (1980) *Helping Skills: A Basic Training Program,* 2nd ed. New York: Human Sciences Press.

Ekstein R, Wallerstein RS (1972) *The Teaching and Learning of Psychotherapy,* 2nd ed. New York: International Universities Press.

Fiddler MB, Fine BA, Baker DL, and ABGC Consensus Development Consortium (1996) A case-based approach to the development of practice-based competencies for accreditation of training in graduate programs in genetic counseling. *J Genet Couns* 5(3):105–112.

Fine BA, Baker DL, Fiddler MB and ABGC Consensus Development Consortium (1996) Practice-based competencies for accreditation of and training in graduate programs in genetic counseling. *J Genet Couns* 5(3):105–112.

Hart GM (1982) *The Process of Clinical Supervision.* Baltimore: University Park Press.

Kadushin A (1968) Games people play in supervision. *Soc Work* 13:23–32.

Littrell JM (1978) Concerns of beginning counselor trainees. *Couns Educ Superv* 18:29–35.

Loganbill C, Hardy E, Delworth U (1983) Supervision: A conceptual model. *Couns Psychol* 10:3–42.

McCarthy P, DeBell C, Kanuha V, McLeod J (1988) Myths of supervision: Identifying the gaps between theory and practice. *Couns Educ Superv* 28:22–28.

McCarthy P, Kulakowski D, Kenfield J (1994) Clinical supervision practices of licensed psychologists. *Prof Psych Res Pract* 25:177–181.

McCarthy P, Sugden S, Koker M, Lamendola F, Maurer S, Renninger S (1995) A practical guide to informed consent in clinical supervision. *Couns Educ Superv* 35:131–138.

Michie S, Mateau TM, Bobrow M (1997) Genetic counseling: The psychological impact of meeting patients' expectations. *J Med Genet* 34(3):237–241.

Stout CE (1987) The role of ethical standards in the supervision of psychotherapy. *Clin Superv* 5:89–97.

Sue DW, Sue D (1981) *Counseling the Culturally Different.* New York: Wiley.

Williams G, Cormier LS, Moline M, Morris J, Neukrug E, Wlazelek B (1990) Malpractice and ethics in supervision. Presented at the annual conference of the Association for Counseling and Development, Washington, DC.

Example Consent Form for Clinical Supervision in the Genetic Counseling Setting*

GENETIC COUNSELING CLINICAL SUPERVISION CONSENT

The purpose of this form is to acquaint you with your supervisor, to describe the supervision process and expectations, to involve you in planning your supervision experience, and to give you the opportunity to ask any questions you may have regarding supervision.

Your Supervisor I hold a master of science degree specializing in genetic counseling and am certified by the American Board of Genetic Counseling. I have worked in the university setting for approximately 8 years. My clinical work has largely involved patients with general genetics concerns, metabolic diseases, and familial cancers. I have supervised genetic counseling students in these clinical settings since 1989.

Practical Supervision Concerns We will meet weekly for one-hour individual supervision sessions. All these sessions will be held in my office. We will arrange a regular meeting time prior to your first clinical experience at this setting. Because your placement at this site is a requirement of your genetic counseling program, attendance at all sessions is mandatory. In the event that you are unable to attend a clinic or a supervision session, you are responsible for making alternative arrangements in advance, if possible. I will provide you with my phone and pager numbers.

*Adapted from: McCarthy P, Sugden S, Koker M, Lamendola F, Maurer S, & Renninger S (1995) A practical guide to informed consent in clinical supervision. *Couns Educ Superv* 35:131–138.

Practical Clinical Concerns I am ultimately responsible for the genetic counseling services provided in these clinics, including the services provided to any patient seen by students. You are responsible for coming to clinic on time and professionally attired. You are responsible for preparing the cases assigned to you in each of the clinics. This includes coming prepared to discuss the medical, genetic, and psychosocial aspects associated with the specific disorders encountered. We will discuss the cases prior to each clinic, and together we will decide on the most appropriate counseling approach. You will be expected to write clinic notes and follow-up letters to referring health professionals and the families you have seen in each clinic.

Supervision Process Clinical supervision is an interactive process intended to monitor the quality of client care, improve clinical skills, and facilitate your professional and personal growth. As a student, you have the responsibility of learning and growing as a professional through these clinical experiences. This includes coming prepared to supervision, being open and responsive to feedback, following through on all assignments and recommendations, disclosing important information, asking for assistance when you are faced with counseling issues beyond your competence, and engaging in self-evaluation. You can expect to receive timely verbal and written feedback on your clinic assignments and progress with your clinical experiences. You can also expect to have a supportive environment in which to explore client-related concerns. You will be expected to actively participate in the supervision process.

The possible benefits from this experience include improvement of your genetic counseling skills and an increased sense of your professional identity. The possible risks to you include discomfort arising from challenges to your genetic counseling knowledge, abilities, and/or skills.

Evaluation and Due Process The clinical settings in which you will be seeing patients are part of an accredited graduate program in genetic counseling, so you may use your clinical experiences here as part of your logbook of cases needed to apply for your board certification. As your supervisor, I will provide you with ongoing written and oral feedback throughout this rotation. A formal written evaluation will be conducted upon completion. Evaluation criteria include your performance in the clinical setting. I will also determine the extent to which you have met the set of objectives for this rotation and assess your progress in attaining the skills necessary to practice as a genetic counselor. A formal written evaluation of my supervision will be solicited from you at the end of the rotation.

If at any time you are dissatisfied with your supervision or the evaluation process, please discuss the matter with me directly. If we are unable to resolve your concerns, you are urged to discuss them with the director of this genetic counseling graduate program.

Legal/Ethical Issues Supervision is not intended to provide personal counseling for you. You are strongly encouraged to seek counseling if personal concerns arise. In general, the content of our supervision sessions is confidential. My evaluations of your development are shared with the director of this program and may be discussed in supervision meetings. The purpose of supervision meetings is to assist clinical supervisors in exploring teaching methods that may improve a student's progress. The formal written evaluations are kept in your file in the program director's office. You can expect that I will not discuss your progress with your fellow students. Limits to confidentiality include, but are not limited to, treatment of a client that violates the legal or ethical standards established by the institution or the genetic counseling profession.

Statement of Agreement   I have read and understand the information contained in this document.

_____   _____

(Supervisor(s) Signature)   (Date)

_____   _____

(Student Signature)   (Date)

Supervisor Evaluation of Student Skills and Performance*

GENETIC COUNSELING CLINICAL EVALUATION

Student:_____   Rotation Period:_____
Supervisor(s):_____   Rotation Site:_____

SECTION I: OVERALL PERFORMANCE IN ROTATION

Check the Appropriate Category   Additional Comments

Meeting Clinical Rotation Objectives
____   Shows skill well beyond set objectives
____   Meets and shows progress beyond set objectives
____   Meets objectives satisfactorily
____   Meets some objectives, needs help with others
____   Not able to meet objectives

Judgment in the Clinical Setting
____   Exceptional
____   Above average in making decisions
____   Usually makes the right decision
____   Often uses poor judgment
____   Consistently uses poor judgment

*Section II: Skill Level Assessment of this evaluation form is adapted from the ABGC practice-based competencies appendixed to Chapter 1 (Fine et al., 1996).

Attitude Application to Work

____ Outstanding in enthusiasm
____ Very interested and industrious
____ Average in diligence and interest
____ Somewhat indifferent
____ Definitely not interested

Relations with Staff and Patients

____ Exceptionally well accepted
____ Works well with others
____ Gets along satisfactorily
____ Difficulty working with others
____ Works poorly with others

Quality of Work

____ Excellent
____ Very good
____ Average
____ Below average
____ Very poor

Dependability

____ Completely dependable
____ Above average in dependability
____ Usually dependable
____ Sometimes neglectful or careless
____ Unreliable

Overall Evaluation and Final Rotation Grade

____ Excellent (A)
____ Very good (B)
____ Average (C)
____ Below average (D)
____ Very poor (F)

1. Please describe the major strengths of this student:

2. How could this student improve and in what areas?

3. What is your final assessment of this student's performance during your clinical rotation?

(Supervisor(s) Signature)	(Date)

(Student Signature)	(Date)

SECTION II: SKILL LEVEL ASSESSMENT

Supervisor Instructions This section is an assessment tool to help the student recognize his or her skill level in all areas of genetic counseling practice and structure personal goals for attaining these skills. These are the practice-based skills established by the American Board of Genetic Counseling that an entry level genetic counselor is expected to be able to demonstrate. Please use a scale of 1–5, where 1 is a more beginning level and 5 is more advanced, to assess the student's skill level in each area. Please use the results to help the student target areas for improvement.

Definitions of Skill Levels

- *Beginning Level* This is the level of skill expected of a beginning student with little to no genetic counseling clinical experience.
- *Intermediate Level* This is the level of skill expected of a student with some genetic counseling clinical experience. The student shows progress toward becoming proficient in this skill but needs to improve.
- *Advanced Level* This is the level of skill expected of an entry-level genetic counseling professional.

Communication Skills   ***Level of Skill***

Can establish a mutually agreed on
genetic counseling agenda with the client.   1   2   3   4   5

Can elicit an appropriate and inclusive family history.   1   2   3   4   5

Can elicit pertinent medical information including
pregnancy, developmental, and medical histories.   1   2   3   4   5

Can elicit a social and psychosocial history.   1   2   3   4   5

Can convey genetic, medical, and technical
information including, but not limited to, diagnosis
etiology, natural history, prognosis, and
treatment/management of genetic conditions and/or   1   2   3   4   5
birth defects to clients with a variety of educational,
socioeconomic, and ethnocultural backgrounds.

Can explain the technical and medical aspects of
diagnostic and screening methods and reproductive   1   2   3   4   5
options including associated risks, benefits, and limitations.

Can understand, listen, communicate, and manage a
genetic counseling case in a culturally sensitive manner.   1   2   3   4   5

Can document and present care information clearly
and concisely, both orally and in writing, as   1   2   3   4   5
appropriate to the audience.

Critical Thinking Skills   *Level of Skill*
Can assess and calculate genetic
and teratogenic risks.   1   2   3   4   5

Can evaluate a social and psychosocial history.   1   2   3   4   5

Can identify, synthesize, organize, and
summarize pertinent medical and genetic   1   2   3   4   5
information for use in genetic counseling.

Can demonstrate successful case management skills.   1   2   3   4   5

Can assess client understanding and response
to information and its implications to modify a   1   2   3   4   5
counseling session as needed.

Can identify and access local, regional, and national
resources and services.   1   2   3   4   5

Can identify and access information resources
pertinent to clinical genetics and counseling.   1   2   3   4   5

Interpersonal, Counseling, and
Psychosocial Assessment Skills   *Level of Skill*
Can establish rapport, identify major concerns,
and respond to emerging issues of a client or family.   1   2   3   4   5

Can elicit and interpret individual and family
experiences, behaviors, emotions, perceptions,
and attitudes that clarify beliefs and values.

1 2 3 4 5

Can use a range of interviewing techniques.

1 2 3 4 5

Can provide short-term, client-centered counseling
and psychological support.

1 2 3 4 5

Can promote client decision making in an unbiased,
noncoercive manner.

1 2 3 4 5

Can establish and maintain inter- and intradisciplinary
professional relationships to function as a part of
a health care delivery team.

1 2 3 4 5

Professional Ethics and Values   *Level of Skill*

Can act in accordance with the
ethical, legal, and philosophical
principles and values of the profession.

1 2 3 4 5

Can serve as an advocate for clients.

1 2 3 4 5

Can introduce research options and
issues to clients and families.

1 2 3 4 5

Can recognize his or her own limitations
in knowledge and capabilities regarding
medical, psychosocial, and ethnocultural
issues and seek consultation or refer clients
when needed.

1 2 3 4 5

Can demonstrate initiative for continued
professional growth.

1 2 3 4 5

Student Evaluation of the Clinical and Supervisory Experience

GENETIC COUNSELING STUDENT EVALUATION OF THE CLINICAL ROTATION EXPERIENCE

Supervisor(s):_____   Rotation Period:_____
Student:_____   Rotation Site:_____

SUPERVISOR'S EXPECTATIONS OF THE STUDENT

Check the Appropriate Category   Additional Comments

The supervisor's expectations for my performance for this clinical rotation period were:
_____ Far too great
_____ A little too much
_____ Very appropriate
_____ Not very high; I felt I
      was more advanced

The supervisor's expectations of my knowledge base for this clinical rotation period were:
_____ Far too great
_____ A little too much
_____ Very appropriate
_____ Not very high; I felt I was not
      given credit for what I knew

The supervisor's expectations of my clinical skills for this clinical rotation period were:

_____ Far too great

_____ A little too much

_____ Very appropriate

_____ Not very high; I felt that
I could have done more

FEEDBACK

The quality of feedback from the supervisor was:

_____ Very useful

_____ Somewhat useful

_____ Not helpful

_____ Unconstructive and harmful;
I did not feel supported

The timing of the feedback from the supervisor was:

_____ Appropriate and timely

_____ Adequate

_____ Inconsistent

_____ Inadequate; I needed more
feedback on a regular basis

Overall evaluations of my performance as a genetic counseling professional were:

_____ Very helpful; I was able
to see improvement

_____ I could have used more
structure to plan improvement

_____ An evaluation of my performance
was not discussed until the end
of my rotation

THE CLINICAL EXPERIENCE

I found the degree of independence during this rotation:

_____ Too little for my experience

_____ Appropriate for my experience

_____ I felt pushed into situations
I was not ready to handle

In preparing cases and working with patients, I found the supervisor:

_____ Helped me prepare for
patients where appropriate

_____ Helped me to think about
how to identify resources

_____ Did not help me as much
as I needed (please discuss)

I found this supervisor was:

____ Available to help me when needed

____ Occasionally available to help

____ Not available enough; I needed
more help

In working with patients, I found this supervisor:

____ Was supportive and helped
me through the sessions

____ Interrupted me during the sessions
too much; did not trust me

____ Was not available for most sessions

I found this clinical rotation to be:

____ A great experience; I learned
a great deal

____ A good learning experience but
I could have learned more (discuss)

____ Very difficult; I had a hard time
learning (discuss)

OVERALL EVALUATION OF THIS CLINICAL EVALUATION EXPERIENCE

Please discuss the major strengths of this rotation.

How could this rotation experience be improved to further encourage the professional development of the student?

What genetic counseling skills do you feel you learned or improved upon during this rotation?

_____   _____

(Student Signature)   (Date)

_____   _____

(Supervisor(s) Signature)   (Date)

13

Professional Development

Beth A. Fine and Karen Greendale

In this chapter, we discuss the concept of professional development—enhancing knowledge and skills while growing in our professional abilities and in our individual jobs. Opportunities for professional development are presented, including continuing education, specialized training, research, and involvement in institutional, regional and national activities.

The ceremony marking the completion of the requirements toward a degree is called a "commencement" because the student is considered to be at the beginning of a lifelong process of learning and professional development. The genetic counseling student begins her professional development with her first graduate courses and continues the learning process throughout her life. The evolving field of genetic counseling has led to development of the profession in response to internal forces, such as training, specialization, and certification, and external forces, such as biomedical advances and cost containment, perceived by genetic counselors to be important. Scientific and technological developments in molecular and clinical genetics, along with changes in health care delivery and policy, necessitate an ongoing approach to education to ensure high quality care of patients. The role of genetic counselors in the delivery of genetic services has expanded to fill niches created by these advances. Continuing education and related activities enable practicing genetic counselors to remain current about new technologies and testing modalities and about their relevant psychosocial, ethical, and legal implications. In essence, genetic counselors must practice at a level consistent with the standard of care even as they help to define guidelines for appropriate practice. Maintaining the status quo is not a viable alternative in a field as rapidly changing as this one!

A Guide to Genetic Counseling, Edited by Diane L. Baker, Jane L. Schuette, and Wendy R. Uhlmann. ISBN 0-471-18541-8 (cloth), 0-471-18867-0 (paper). Copyright © 1998, Wiley-Liss, Inc.

Professional development is essential for professional satisfaction. Genetic counseling is a field that encourages creativity in developing one's job description, tackling clinical research questions, and filling niches in the workplace and at state, regional, and national levels. An effective professional must be challenged and stimulated in her work.

This chapter describes the history of the genetic counseling profession and the role that professional development has played in expanding its boundaries. We will explore options for professional development within one's own institution as well as within national professional organizations.

THE GENETIC COUNSELING PROFESSION AND ROLE EXPANSION

The genetic counseling profession is relatively new, having had its start in 1969 when the Sarah Lawrence College program accepted its first class of 10 students. As this book goes to press, there are more than 1500 genetic counselors practicing in the United States and Canada. In fewer than 30 years, genetic counseling has become a unique and established profession. Kenen, a sociologist, notes that the genetic counseling profession possesses most of the characteristics used to define a profession (Kenen, 1997). These include its own body of knowledge (which combines content from genetics, medicine, ethics, and psychology), dedicated training programs, an accrediting body, and a certification process [overseen by the American Board of Genetic Counseling (ABGC)], a professional society [the National Society of Genetic Counselors (NSGC)], a journal (the *Journal of Genetic Counseling*), a newsletter (*Perspectives in Genetic Counseling*), and a published code of ethics.

To consider future directions, it is instructive to look at our past. A workshop on professional roles conducted at the first NSGC Annual Education Conference in 1981 provides a glimpse into early professional development and role expansion (Kloza, 1982). The notion of a "traditional" genetic counselor in a tertiary care, hospital-based setting, practicing prenatal and/or pediatric genetic counseling, had been established but was continuing to evolve. Innovations in service delivery led to jobs in settings such as community health clinics and biotechnology companies. In satellite and specialty clinics, genetic counselors moved into niches, providing services to families concerned about cystic fibrosis, muscular dystrophy, and Huntington disease. Genetic counselors were starting to apply their skills in interpretation and communication of genetic information in several "nontraditional" settings. Genetic counselors who were working with clients considering adoption, paternity testing, and new reproductive technologies shared their experiences in forging new ground. Genetic counselors also began to play a role in public health, administration, education of professionals at all levels, and risk screening.

In the years since the original workshop, new areas of practice have been created as genetic counselors recognized unmet needs and developed new services in response. For example, some genetic counselors work in Teratogen Information Services, offering information and counseling to the public and to health care professionals on risks associated with exposures during pregnancy. These genetic counselors ultimately formed a subspecialty group and actively participate in the Organization of Teratogen Information Services (OTIS).

Some genetic counselors have focused on educating science teachers, coordinating research projects, or acting as customer service representatives in commercial genetic testing laboratories. Others have succeeded in obtaining grant funding for clinical research studies. One advantage of writing a grant is that it provides the opportunity to define your role in new ways. Many of these expanded roles have led genetic counselors to seek additional education to accomplish their goals. Their work has often opened new doors for others in the field.

The discovery of cancer susceptibility gene mutations, especially those predisposing to familial breast, ovarian, and colon cancers, has led to the development of predictive testing and the need for genetic counseling of a large new cohort of patients. Genetic counselors responded to these discoveries by combining their knowledge about genetic principles and relevant psychosocial issues with newly acquired knowledge about cancer to develop a new subspecialty, cancer genetic counseling. In addition to counseling patients in genetics clinics about cancer susceptibility, genetic counselors moved into oncology clinics to serve this patient population. Genetic counselors working in oncology clinics provide risk assessment and genetic counseling, participate in clinical research, and coordinate genetic testing in conjunction with oncologists.

The NSGC has played a critical role in promoting and sustaining role expansion by genetic counselors. As genetic counselors have moved into new areas of specialization, the NSGC has established special interest groups (SIGs), including Pediatrics, Prenatal, Cancer, Public Health, Neurogenetics, Connective Tissue Disorders, and Infertility/Assisted Reproductive Technologies. These SIGs enable genetic counselors to network with colleagues working in the same area, fostering collaborations and professional development. The Cancer SIG, for example, uses a listserv to keep members informed about advances published in the medical literature, to advise each other on difficult cases, and to share information from relevant conferences.

The NSGC has explored professional status indicators, surveying its membership every two years regarding focus of work, responsibilities, salaries, faculty appointments, professional activities, employer support for continuing education, and job satisfaction. These data are analyzed, and a summary of the results is published by the NSGC. Many genetic counselors have utilized the survey findings to expand their roles, create new positions, and negotiate increased compensation, autonomy, and benefits. The survey results also provide a quantitative and qualitative description of the profession and demonstrate changes over time in response to economic, social, and other factors.

In summary, genetic counselors spearheaded efforts to expand professional roles, enhancing their own job satisfaction and contributing to the richness of the profession. We predict that changes in the health care marketplace, especially in the managed care arena, will lead future generations of genetic counselors to again reassess and redefine their professional roles and responsibilities.

LIFELONG LEARNING PRACTICES

The goal of any graduate program in genetic counseling is to provide students with knowledge and skills necessary to approach any particular genetic counseling case or situation. While no student comes into contact with patients or families with every known genetic condition, program directors and instructors prepare students to know where to go to access the information and resources necessary to serve a broad spectrum of clients. As new areas of practice emerge, genetic counselors must develop pertinent skills. A commitment to lifelong learning has always been valued by genetic counselors. A primary goal of the NSGC has been to support continuing education for its members (Rollnick, 1984). The mission statement, adopted in 1993, states that the "NSGC will promote the genetic counseling profession as a recognized and integral part of health care delivery, education and public policy." Practices and policies of the NSGC highlight this value. For example, an Education Committee was established under the NSGC Bylaws (NSGC, 1979) to provide continuing education for genetic counselors, other professionals, and the public at large.

The NSGC Code of Ethics describes the ethical responsibilities of its members and provides guidance in terms of genetic counselors' relationships with their clients, their colleagues, society, and themselves. Specifically, the Code states that "genetic counselors strive to seek out and acquire all relevant information required for any given situation; to continue their education and training; [and] to keep abreast of current standards of practice" (National Society of Genetic Counselors, 1992). Since members are expected to adhere to the Code of Ethics, one can therefore view continuing education and professional development as virtual requirements for the genetic counselor.

The American Board of Genetic Counseling (ABGC) was established in 1993 to serve the public and the field of genetic counseling through the establishment and maintenance of criteria and procedures for certification and recertification of genetic counselors and for the accreditation and reaccreditation of graduate programs in genetic counseling. Graduates of accredited programs are expected to achieve 27 practice-based competencies considered minimum standards for entry-level genetic counselors (see Appendix 1.1, Chapter 1). The commitment to professional development is illustrated in the following competency: "[The student] can demonstrate initiative for continued professional growth. The student *displays a knowledge of current standards of practice and shows independent knowledge-seeking behavior and lifelong learning"* (Fine et al., 1996). The requirement of this competency for an entry-level genetic counselor again highlights the value the profession places on professional development.

Certification for genetic counselors has existed since 1982 and has increased recognition of genetic counselors among members of other health professions and the public. A candidate for certification must have graduated from an accredited graduate program in genetic counseling and must submit letters of recommendation and a logbook demonstrating that the candidate played a significant role in 50 counseling sessions under appropriate supervision. Once these credentials have been approved, the candidate must pass a two-part national certification examination—one part covering general genetics, which is taken by all genetics professionals, and a second specialty examination in genetic counseling. In 1996 certification became time-limited, with recertification required every 10 years.

Clearly, lifelong learning is both valuable and essential for genetic counselors from a personal and professional point of view. Genetic counselors are obligated to remain current regarding advances in genetics, medicine, and pertinent clinical standards. Professional development depends on educational opportunities, many of which are available through genetics professional organizations and in individual practice settings. Some of these resources are discussed in the subsections that follow.

Continuing Education

"Continuing education" is a general term used to connote formalized learning after graduation. Many employers and professional organizations require documentation of continuing education activities. The number of continuing education units (CEUs) awarded is determined by the amount of time spent participating in a continuing education program. The programs must obtain prior approval for a specific number of CEUs, determined through a comprehensive review process that assesses the program's ability to meet specific national criteria.

In 1981 the NSGC developed a continuing education model to ensure quality education programs (Rollnick, 1983). Since 1996, all courses and conferences sponsored by the NSGC have had to meet established criteria for offering CEUs. This means that genetic counselors can accrue CEUs by attending NSGC-sponsored education programs and that other health professionals can attend NSGC meetings to help fulfill their own CEU requirements. Recertification of genetic counselors in the future will likely involve CEUs. Thus, while continuing education should be standard practice for genetic counselors adhering to the Code of Ethics, documentation of such activities may become mandatory as part of the recertification process.

Conferences

Conferences or professional meetings are a major forum for continuing education. The NSGC holds a national education conference each year. Generally the meeting immediately precedes the annual meeting of the American Society of Human Genetics (ASHG), to permit attendance at both conferences. The format and content of the NSGC conference is designed to meet the broad range of educational needs of the membership. More than half of NSGC members attend this conference each year. Lectures, workshops, practice-based symposia, and contributed paper and poster sessions address issues in molecular genetics, clinical genetics, and genetic counseling, as well as the psychological, cultural, ethical, and legal implications of our work. Conference faculty includes NSGC members and guest faculty from a variety of disciplines. Informal networking and sharing of ideas make this conference an extremely important forum for professional development activities. Short courses on topics of special interest, such as cancer genetics and neurogenetics, have also been organized by leaders in these fields.

The NSGC also sponsors regional meetings that focus on specific topics and encourage presentations from genetic counselors in the region. They typically have a small number of participants and promote local and regional collaborations. These meetings are often good places for students or recent graduates to make their first presentations and to consider issues of regional import.

Several other genetics societies and other organizations offer continuing education programs for genetics professionals. For example, the American Society of Human Genetics hosts an annual scientific meeting that includes lectures, workshops, symposia, and platform and poster presentations. Several thousand geneticists from around the world attend this meeting; again, there are opportunities for networking. The American College of Medical Genetics, primarily comprised of genetics clinicians, holds an annual conference that focuses on clinical genetics and the medical issues raised by genetics research.

The federally supported Council of Regional Networks for Genetic Services (CORN) holds an annual meeting; each of the 10 member regional networks holds periodic meetings as well. These conferences typically address genetic service delivery issues. The membership of CORN and the regional groups include a variety of public health and clinical genetics professionals and consumers of genetic services, such as representatives of genetic disease support groups and the Alliance of Genetic Support Groups. These meetings provide didactic and networking opportunities for genetic counselors interested in the public health implications of human genetics.

Many states have their own genetics organizations, which include all practicing clinical geneticists and genetic counselors. Ask colleagues in your own or neighboring institutions how to become involved. Other states have either formal or informal meetings where genetic counselors can discuss challenging cases, provide and receive support, and work together to influence policy making at local and state levels.

Finally, many local and national meetings address specific issues in genetics. For example, the various genetic disease support groups sponsor medical and scientific conferences. Conferences on cancer genetics have recently been organized by the American Cancer Society and by the American Society of Clinical Oncology; meetings to explore ethical issues relating to genetic testing have been sponsored by the National Human Genome Research Institute and others. Genetic counselors learn about meetings from announcements in genetics and related journals, direct mailings, the NSGC LIST-SERV, from Web sites maintained by specific institutions or organizations, and from colleagues.

Journals in Genetics and Related Fields

Literature review is probably the best method of day-to-day continuing education. Medical journals relevant to the practice of genetic counseling are available in medical libraries or by subscription. Some journals can be accessed via the Internet. It is prudent to at least review the tables of contents of the major genetics journals each month. Of course, the *Journal of Genetic Counseling* is carefully read by most members of the profession. Other journals include the *American Journal of Human Genetics,* the *American Journal of Medical Genetics,* the *Journal of Medical Genetics,*

Genetic Testing and *Prenatal Diagnosis.* Most pediatric and obstetrics and gynecology journals contain several pertinent articles per issue. The *New England Journal of Medicine,* the *Journal of the American Medical Association, Science, Nature,* and *Nature Genetics* are broader in focus and often contain articles of interest. In addition, the journals devoted to specific medical specialties such as oncology or neurology, or to public health or biomedical ethics, are important to read if your work is focused in a subspecialty area. Because there is so much to read and often so little time to do it, a journal club in your own institution can help everyone to keep up to date. Set one up if none exists! Additionally, familiarity with the Web sites relating to genetics and related topics, specifically the ones aimed at patients, is essential.

Quality Improvement in Genetic Counseling and Service Delivery

Informal and formal quality improvement (QI) programs have been set up in some genetics centers and can be seen to impact on participating genetic counselors' continuing education experience (Greendale et al., 1994; Lea et al., 1996). A QI program may stipulate specific educational requirements on an ongoing basis. Other aspects, such as periodic supervision and review of counseling sessions and clinical case review, can be considered integral to the lifelong learning process.

Advanced Education and Training

The multidisciplinary nature of genetic counseling provides many avenues for professional development. Genetic counselors can choose to pursue advanced education in bioethics, public policy, psychology, sociology, molecular genetics, and other pertinent disciplines. For example, some genetic counselors have enrolled in certificate programs in family and marital therapy as a means of expanding their practices to include long-term psychotherapy for individuals and families dealing with genetic disorders. Genetic counselors also may choose to expand their research capabilities. The application of social science research methodologies is an important but underutilized approach to answering academic questions regarding genetic counseling. Some genetic counselors may take courses or obtain advanced degrees to become equipped to apply such methodologies to genetic counseling practice.

Additional training in epidemiology, health policy, and research design are possibilities for genetic counselors working in public health settings. The involvement of genetic counselors in newborn and other screening programs and in administrative positions in state health departments has grown; formal training or self-study in this area may be required. Genetic counselors moving into specialty areas such as cancer genetics may be able to create "apprenticeships," learning from genetic counselors who have developed significant experience and expertise. For some, the expenses associated with such additional training have been covered by employers; others have taken advantage of sabbatical or fellowship funds. For example, the Jane Engelberg Memorial Fellowship, awarded through the NSGC, was created to provide an opportunity for genetic counselors to enhance skills, develop new areas of expertise, and conduct research.

In essence, advanced education and training both enhance the genetic counselor's practice and help to redefine the boundaries of genetic counseling. The genetic counselor with advanced training can share her expertise with others in the profession through consultations, presentations, and publications and act as a role model for colleagues.

OPPORTUNITIES IN ONE'S OWN INSTITUTION

Professional development is important within the individual genetic counselor's place of employment, both for personal fulfillment and to augment professional status. It is important to keep in mind that creating a new clinical service, organizing case conferences, serving on a hospital committee, or educating students and health professionals may do as much to positively affect institutional perceptions of a genetic counselor's "worth" as an increased caseload would. The skills genetic counselors possess can be applied in contexts outside the traditional genetic counseling sphere.

Professional development within one's institution can be pursued in two ways. First, one can identify opportunities for professional growth through educational offerings in the workplace. For example, one can attend grand rounds, seminars, and journal clubs. Many institutions offer skill enhancement programs in areas such as computer proficiency, grant writing, and public speaking.

Second, one can pursue opportunities for professional growth by participating in educational, clinical, and administrative activities. For example, a genetic counselor interested in ethical issues could serve on a hospital ethics committee; these multidisciplinary groups provide guidance to health care providers facing difficult dilemmas in patient care. In many medical schools, genetic counselors teach medical students in didactic genetics courses. Medical students and residents on clinical service in genetics units are often taught through more informal interactions with genetic counselors. Genetic counselors may hone their administrative skills by coordinating a multidisciplinary clinic, such as a craniofacial, hemophilia, or muscular dystrophy clinic. The genetic counselor coordinator generally oversees patient referrals to appropriate specialists, provides genetic counseling, and leads the case conference discussion on each patient, ensuring that proper follow-up is accomplished.

PARTICIPATION IN PROFESSIONAL ORGANIZATIONS

Opportunities for professional development through participation in genetics organizations exist at local, state, and regional levels. The appendix to this chapter provides contact information and an overview of the structure and function of the major professional and research-oriented genetics organizations. Both the NSGC and ASHG offer student memberships. A good first step is to join the NSGC as a student member or as a full member following graduation. Student members are welcome to join a committee, to help plan an Annual Education Conference, to submit an abstract for presentation at a meeting or to join one of the Special Interest Groups. There are so many ways to get involved that it may be difficult to choose! Consider your own background, talents, and special interests when making this decision. The benefits include the satisfaction associated with contributing to the profession and opportunities for networking, research collaboration, and professional and personal relationships with colleagues.

Because so many issues facing genetic counselors are important to colleagues in clinical genetics, some genetic counselors have chosen to channel their energies through the ASHG or ACMG. Examples of collaborative efforts include joint meetings of the Social Issues Committees of the ASHG, ACMG, and NSGC; an integrated effort to improve reimbursement for genetic counseling spearheaded by genetic counselors and clinical and laboratory geneticists through the ACMG Economics of Genetic Services Committee; and a project to develop, disseminate, and evaluate clinical guidelines in medical genetics through the ACMG Joint Practice and Clinical Guidelines Committee.

Taking on new responsibilities leads to new learning and new challenges. Participation in continuing education programs, on committees, and in national professional societies will ultimately enhance job satisfaction and the professional status of the genetic counselor, while furthering the goals of the profession. In response to recent and rapid changes in technology and health care delivery systems, genetic counselors have pushed the boundaries of the profession to meet the needs of patients and the society at large. As a consequence, personal and professional satisfaction among genetic counselors has been enhanced, making this a rewarding and gratifying profession. As the late Beverly Rollnick, second president of the NSGC and a founder of the genetic counseling profession wrote, "The future holds much promise!"

ACKNOWLEDGMENTS

The authors thank Bea Leopold, executive director of the NSGC, and Virginia Corson for invaluable help with this chapter.

REFERENCES

Fine BA, Baker DL, Fiddler MB and the ABGC Consensus Development Consortium (1996) Practice-based competencies for accreditation of and training in graduate programs in genetic counseling. *J Genet Couns* 5(3):113–122.

Greendale K, Knutson C, Pauker SP, Lustig L, Weaver DD (1994) Quality assurance in the clinical genetics setting—Report of a workshop. *J Genet Couns* 3(3):169–198.

Kenen R (1997) Opportunities and impediments for a consolidating and expanding profession: Genetic counseling in the United States. *Soc Sci Med* 45(9):1377–1386.

Kloza EM (1982) New roles for genetic counselors. *Perspect Gene Couns* 4(2):1–2.

Lea DH (1996) Emerging quality improvement measures in genetic counseling. *J Genet Couns* 5(3):123–137.

National Society of Genetic Counselors (1992) National Society of Genetic Counselors' Code of Ethics. *J Genet Couns* 1(1):41–43.

Rollnick BR (1983) Continuing education criteria and continuing education units: Policy issues. *Perspect Genet Couns* 5(1):1.

Rollnick BR (1984) The National Society of Genetic Counselors: An historical perspective. *Birth Defects Orig Artic Ser* 29(6):3–7.

Appendix 13.1

Clinical Genetics–Related Professional Societies

American Society of Human Genetics (ASHG)

Address: 9650 Rockville Pike, Bethesda, MD 20814-3998

Telephone: (301) 571-1825

Fax: (301) 530-7079

URL: *http://www.faseb.org/genetics/ashg/ashgmenu.htm*

Year founded/incorporated: 1948

Focus: human genetics research (basic and clinical) and education

Membership: human genetics professionals at all educational levels (clinicians, genetic counselors, clinical and laboratory researchers) and other interested professionals

Board of directors: 16 members; president and general board members are elected; editor of *American Journal of Human Genetics*, secretary, and treasurer are appointed

Standing committees (appointed members): Program Committee; Information and Education Committee; Social Issues Committee; Awards Committee; Nominating Committee

Publications: *American Journal of Human Genetics* (monthly); *Guide to North American Graduate and Postgraduate Training Programs in Human Genetics*

Annual meeting: held each fall in a U.S. or Canadian city

Student activities: student–mentor breakfast held at annual meeting; job search opportunities

American Board of Medical Genetics (ABMG)

Address: 9560 Rockville Pike, Bethesda, MD 20814-3998

Telephone: (301) 571-1825

Fax: (301) 571-1895

URL: *http://www.faseb.org/genetics/abmg/abmgmenu.htm*

Year founded/incorporated: 1981

Focus: accrediting body for doctoral training programs in human genetics; certifying body for doctoral providers; genetic counselors were also certified through 1990

Membership: certified doctoral level geneticists; certified genetic counselors (1981–1990)

Board of directors: 11 elected members; board of directors elects officers

Standing committees: committees are composed of elected board members (Accreditation Committee, Certification Committee, Recertification Committee, and Finance Committee); members of the ABMG are appointed to the Nominating Committee, which also includes one member of the board of directors

Publications: *Bulletin of Information/Description of Examinations Application Form*

Annual meeting: business meeting and program directors meeting held in conjunction with annual ASHG meeting

Student activities: not applicable

American Board of Genetic Counseling (ABGC)

Address: 9560 Rockville Pike, Bethesda, MD 20814-3998

Telephone: (301) 571-1825

Fax: (301) 571-1895

URL: *http://www.faseb.org/genetics/abgc/abgcmenu.htm*

Year founded/incorporated: 1993

Focus: accrediting body for genetic counseling training programs and certifying body for master's-level genetic counselors

Membership: certified genetic counselors

Board of directors: 8 elected members; board of directors elects officers

Standing committees: committees are composed of elected board members (Accreditation Committee, Certification Committee, and Finance Committee); members of the ABGC are appointed to the Nominating Committee, which also includes one member of the board of directors

Publications: *Bulletin of Information/Description of Examinations Application Form*

Annual meeting: business meeting and program directors' meeting held each fall during the NSGC Annual Education Conference

Student activities: guidance with completing certification application

American College of Medical Genetics (ACMG)

Address: 9560 Rockville Pike, Bethesda, MD 20814-3998

Telephone: (301) 530-7127

Fax: (301) 571-1895

URL: *http://www.faseb.org/genetics/acmg/acmgmenu.htm*

Year founded/incorporated: 1991

Focus: continuing education and professional issues of medical genetics professionals

Membership: medical genetics professionals at all educational levels (clinicians, genetic counselors, researchers) fit into one of several membership categories: Fellow (doctoral level), Associate (certified genetic counselors), Junior (in postdoctoral training program in medical genetics), Student (in graduate training program in genetics), Corresponding (resides outside U.S. and Canada), Emeritus (in retirement), Affiliate (demonstrated interest in medical genetics, but does not fit another category)

Board of directors: 14 elected members

Standing committees (appointed members): AMA Liaison Committee; Bylaws and Procedures Committee; Finance Committee; Public Policy Committee; Governmental and Legislative Affairs Committee; Clinical Practice Committee; Joint Committee on Professional Practice and Guidelines; Education Committee; Laboratory Practice Committee; Program Committee; Committee on the Economics of Genetic Services; Nominations Committee; Membership Committee; Publications Committee; Social, Ethical, and Legal Issues Committee

Publications: *Genetics in Medicine*

Annual meeting: held each winter in a U.S. or Canadian city

Student activities: not applicable

National Society of Genetic Counselors (NSGC)

Address: 233 Canterbury Drive, Wallingford, PA 19086-6617

Telephone: (610) 872-7608

Fax: (610) 872-1192

URL: *http://members.aol.com/nsgcweb/nsgchome.htm*

Year founded/incorporated: 1979

Focus: continuing education and professional issues of master's-level genetic counselors

Membership: full members (master's-level genetic counselors), associate members (clinical geneticists, researchers, interested others), student members (genetic counseling students)

Board of directors: 20 members; 10 elected officers: president, president-elect, secretary, treasurer, six regional representatives; appointed committee chairs:

Education, Professional Issues, Social Issues, Membership, Finance, Genetic Services; editor of *Perspectives in Genetic Counseling Newsletter* (appointed); editor of *Journal of Genetic Counseling* (appointed)

Standing committees (members volunteer): Education Committee, Social Issues Committee, Professional Issues Committee, Genetic Services Committee, Membership Committee, Finance Committee

Publications: *Journal of Genetic Counseling* (six issues/year); *Perspectives in Genetic Counseling* newsletter (quarterly)

Annual meeting: held each fall in a U.S. or Canadian city

Student activities: student LIST-SERV; mentor program; student workshops at regional and national meetings; job connection service

International Society of Nurses in Genetics (ISONG)

Address: c/o Eileen Rawnsley, Executive Director, 7 Haskins Road, Hanover, NH 03755

Telephone: (603) 643-5706

Fax: (603) 643-3169

URL: *http://nursing.creighton.edu/isong/index.html*

Year founded/incorporated: 1988

Focus: continuing education and professional issues of nurses working in genetics

Membership: nurses at all educational levels (doctoral, master's, baccalaureate) who are professionally involved with or interested in genetics, and other interested genetics professionals, fit into one of several membership categories

Board of directors: 7 elected members

Standing committees (members volunteer): Bylaws Committee; Education Committee; Ethics and Social Policy Committee; Information and Public Relations Committee; Membership Committee; Nominating Committee; Professional Practice Committee; Annual Educational Program Committee

Publications: *ISONG News*

Annual meeting: held each fall in a U.S. or Canadian city

Student activities: preconference seminar/workshop held in conjunction with Annual Education Conference; educational lectures and mentor arrangements through individual ISONG members

Council of Regional Networks for Genetic Services (CORN)

Address: Department of Pediatrics/Division of Genetics, Emory University School of Medicine, 2040 Ridgewood Drive, Atlanta, GA 30322

Telephone: (404) 727-1475

Fax: (404) 727-1827

URL: *http://www.cc.emory.edu/PEDIATRICS/corn.corn.htm*

Year founded/incorporated: 1985

Focus: forum for national coordination of genetic services among the 10 CORN regions

Membership: representatives of each regional network, national sickle cell disease programs, and the Alliance of Genetic Support Groups

Board of directors: 15 members; officers elected; member organizations elect representatives

Standing committees: committees of CORN are filled by appointment or election and include Birth Defects Surveillance; Data and Evaluation; Education; Ethics; Finance; Newborn Screening; Quality Assurance; Regional Coordinators; Sickle Cell, Thalassemia, and Other Hemoglobin Variants; and Teratogen Information Systems

Publications: newsletters produced by individual regional genetics networks; conference proceedings, national public health genetics guidelines

Annual meeting: spring

Student activities: not applicable

14

New and Evolving Technologies: Implementation Considerations for Genetic Counselors

Patricia A. Ward

The medical genetics field has been extremely dynamic during the 1980s and 1990s as a result of the mapping and identification of genes that cause human diseases. This research has received significant support from the Human Genome Project, which is anticipated to be completed by 2005. The development of new technologies for genetic testing has also steadily increased, and as a result, the biotechnology industry has rapidly grown.

These evolving technologies now provide individuals and their families with new testing options to more accurately predict their genetic risks. The potential value is immense, since these tests may assist those with increased genetic risk as they make complex personal, medical, and reproductive decisions. Therapeutic interventions for many genetic diseases are anticipated in the future as research continues to provide a better understanding of gene function and as effective strategies for intervention, including gene therapy, are developed.

A Guide to Genetic Counseling, Edited by Diane L. Baker, Jane L. Schuette, and Wendy R. Uhlmann.
ISBN 0-471-18541-8 (cloth), 0-471-18867-0 (paper). Copyright © 1998, Wiley-Liss, Inc.

As a direct result of the introduction of new diagnostic testing technologies, genetic counselors are faced with numerous professional challenges related to standards of practice and ethical issues. Commonly, genetic counselors receive requests for genetic testing for disorders for which gene identification has only recently been reported in the scientific literature. Many of these requests come from primary care providers (Whittaker, 1996), but increasingly they come directly from clients who perceive an increased genetic risk based on their family history.

It is critical that genetic counselors keep up to date on evolving technologies. Currently, there are no specific national requirements to ensure the overall quality of genetic tests. Genetic counselors must be highly educated consumers when they select laboratories for genetic testing, a task that calls for understanding the specifics of laboratory quality control and current practices. This chapter focuses on the transition of research methods to clinical application. It discusses current regulations for the accreditation of clinical laboratories, issues involved with laboratory quality, and validation of technology. Clinical cases are used to illustrate genetic counseling issues that arise when utilizing new technologies.

THE TECHNOLOGY TRANSFER PROCESS: AN OVERVIEW

Simply defined, the technology transfer process is the application of techniques developed in one setting (e.g., a research laboratory) to another setting (e.g., a clinical laboratory). The development of any new technology for clinical application must proceed from a research protocol through a process of validation to eventually become an accepted standard of care. For example, two procedures for genetic diagnosis—second-trimester amniocentesis, developed during the 1970s, and first-trimester chorionic villus sampling, developed during the 1980s—were initially carefully studied to determine their potential risks and efficacy for diagnosis of chromosomal abnormalities and a variety of single-gene disorders in utero. Both these procedures have become accepted as the standard of care for fetal diagnosis whenever the fetal risk for a particular genetic disorder equals or exceeds the potential risk for miscarriage or adverse fetal impact as a result of the procedure (NICHD National Registry for Amniocentesis Study Group, 1976; Task Force on Predictors of Hereditary Disease or Congenital Defects, 1979; Canadian Collaborative CVS-Amniocentesis Clinical Trial Group, 1989; Rhoads et al., 1989).

The utilization of linkage analysis and assays for mutations in disease genes for the purpose of genetic testing originated from protocols followed in research settings to identify or clone a specific gene. Laboratories performing clinical testing subsequently adopted these techniques. During the 1980s, laboratories performing molecular diagnostic tests worked on defining the technology transfer process and establishing guidelines under which new technologies (e.g., linkage and mutation analysis) could be clinically applied. Lebo et al. (1990) distinguished between research and clinical tests using the following definitions. For disease genes that have been cloned, they suggest that the test may be moved from the research to the clinical realm when at least 70% of matings are informative, based on published data in a peer-reviewed journal. For unknown disease loci tested by linked polymorphisms, undetected recombination had to be considered. Under these circumstances, in addition to the requirement that at least 70% of matings be informative, these authors recommend that for linkage analysis to be considered for clinical use, it should provide an accurate result within a 95% confidence interval.

The technology transfer process with regard to genetic testing has been under increasing review and revision in recent years. The ultimate goal is to ensure that genetic testing is both safe and effective. To reach this goal, there are likely to be increased regulatory requirements (Holtzman et al., 1997). The steps currently involved in the technology transfer process are shown in flowchart form in Figure 14.1.

Test Sensitivity and Specificity

The clinical use of any genetic test should not be initiated until a definite association between the gene and the disease phenotype has been established. Genetic testing, like other testing, does have limitations. It is therefore important to keep two facts in mind:

- A positive result may be a true positive (individual has inherited the gene) or a false positive (individual actually has not inherited the gene).
- A negative result may be a true negative (individual has not inherited the gene) or a false negative (individual actually has inherited the gene).

The specificity, sensitivity, and predictive value of a given test may vary depending on the specific technology and reagents used by a particular laboratory. Therefore, it is essential to accurately assess these variables whenever new tests are introduced by any laboratory performing clinical genetic studies. Table 14.1 provides definitions and a summary of the formulas used to calculate sensitivity, specificity, and predictive values.

Technology Transfer Process: What Occurs in the Clinical Arena

The research phase of the technology transfer process is initiated with the development of research protocols. Research protocols must have institutional review board (IRB) approval if specimens are utilized that can be linked to the donor, if medical records are to be reviewed, or if participants are to be contacted; individual participants must provide informed consent. Generally, IRB approval is not needed if identifiers are removed before specimens are analyzed. The careful selection and counseling of appropriate participants for the initial research protocols is key to the success of this transfer process. Individuals selected for study need to have an established diagnosis, confirmed by physical examination or medical record review. Participants should be properly informed in advance about the focus of the study, what participation entails, possible future uses of samples, and whether they will receive any results or be recontacted. During the research phase, the test results generally are not reported back to patients.

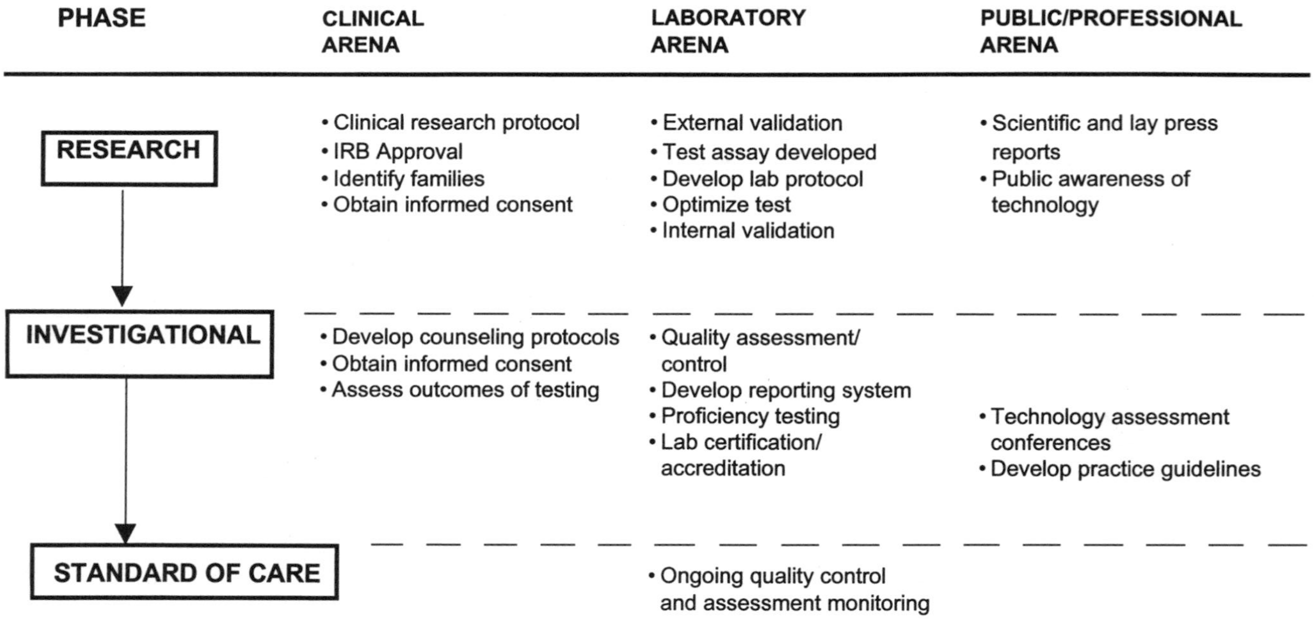

PHASE	CLINICAL ARENA	LABORATORY ARENA	PUBLIC/PROFESSIONAL ARENA
RESEARCH	• Clinical research protocol • IRB Approval • Identify families • Obtain informed consent	• External validation • Test assay developed • Develop lab protocol • Optimize test • Internal validation	• Scientific and lay press reports • Public awareness of technology
INVESTIGATIONAL	• Develop counseling protocols • Obtain informed consent • Assess outcomes of testing	• Quality assessment/control • Develop reporting system • Proficiency testing • Lab certification/accreditation	• Technology assessment conferences • Develop practice guidelines
STANDARD OF CARE		• Ongoing quality control and assessment monitoring	

Figure 14.1 *Flowchart summary of the concurrent activities in three different arenas: clinical, public/professional, and laboratory, each of which contributes to the transfer process.*

TABLE 14.1   Test Variables: Definitions and Calculations

		Disease Phenotype	
		Present	Absent
Test Result	Positive	A	B
	Negative	C	D

Clinical sensitivity	Probability that person with disease will have positive test result [A/(A + C)]
Clinical specificity	Probability that person without disease will have negative test result [D/(D + B)]
Positive predictive value	Probability that subject has the disease, given test is positive (i.e., prevalence) [A/(A + B)]
Negative predictive value	Probability that subject does not have disease, given test is negative (i.e., 1 – prevalence) [(D/(C + D)]

Source: Adapted from Holtzman and Watson (1997), p. 26.

The next phase in the transfer process is considered investigational. Under investigative protocols, researchers collect data to establish the clinical validity of genetic tests (clinical sensitivity, specificity, and predictive value). Investigational testing requires informed consent from the participants, and results are released to patients in accordance with clearly defined, IRB-approved protocols. Releasing test results would be appropriate only if the clinical validity of the test is fairly well established and there is some benefit to the participant to learn the result. Ideally, a test should be launched clinically once a good understanding of the technology, clinical validity, and clinical implications has been obtained. In some cases, however, a test is launched because it is recognized that it offers information that can be of clinical benefit in specific circumstances. For example, testing for hereditary breast cancer genes *BRCA1* and *BRCA2* was offered clinically before the detection rate and medical implications were firmly established. In addition, it may take many years to fully assess the validity and utility of some genetic tests. If a test is launched clinically on the basis of preliminary data, it is important that long-term data be collected. Assessment of a test and development of practice guidelines eventually result in the acceptance of that test as a standard of care.

Technology Transfer Process: What Occurs in the Laboratory Arena

Test validation takes place in the molecular laboratory, utilizing patient samples obtained through the protocols defined above. Informed consent must be obtained whenever there is potential for specimens collected in validation studies to be linked back to the actual participants. Test validation begins with a review of the scientific literature to ascertain clinical knowledge and history of the test, determine appropriate controls to be used, and establish a means of determining test failure. These factors are then verified by testing samples from subjects of known status who are representative of the population for whom the test is intended (Holtzman et al., 1997). Each intended use of a genetic test needs to have formal validation. In this preliminary phase, the assay is developed and optimized and a laboratory protocol is written.

The next step in the validation process is the ongoing scheduled testing of samples obtained on a research basis and review of testing data. Appropriate positive and negative controls, size standards, and other parameters are reviewed along with test data, to ensure that all criteria are validated and the test is working properly in the lab.

The final step is the verification, or internal validation, phase of the process. A minimum of five specimens per testing category are studied for various control standards for each test. These control standards serve to evaluate test accuracy and precision as well as the sensitivity and specificity of the given test. Once a test has been validated and launched, there should be ongoing quality control and assessment monitoring.

A written report of test results should clearly explain the specifics of test validity, identifying, as well, any variables that might confound the determination of the validity and reliability of the test. Some of these variables are disease prevalence, varying allele frequencies between populations, heterogeneity, penetrance, variable expressivity, ascertainment bias, mosaicism, and mistaken paternity or ethnicity. A sample laboratory report containing an explanation of test validity is shown in Appendix 14.1.

LABORATORY ISSUES: CERTIFICATION AND QUALITY

There are over 300 molecular diagnostic laboratories in the United States and Canada. Cumulatively, these laboratories now offer molecular tests for over 500 genetic disorders (personal communication with respect to HELIX database, November 1997). The medical benefits of testing for many genetic disorders are still unclear. Therefore, genetic testing for many disorders should be considered to be investigational until the medical significance and implications of results are better understood. However, the distinction between investigational and standard clinical testing is not always clearly made. In fact, most genetic testing is performed on a fee-for-service basis, and a signed consent form is not always required.

Professionals providing clinical genetic services have a significant challenge in selecting a laboratory for submission of patient samples for genetic testing. In addition to considering the testing methodology and logistics involved in sending a sample, the provider should exercise care in selecting a laboratory: to be fully informed, it is necessary to understand the current laboratory regulatory issues.

Molecular genetic laboratory testing is currently monitored by some individual states and by the Health Care Financing Administration (HCFA) of the U.S. Department of Health and Human Services, which upholds the Clinical Laboratory Improvement Act, CLIA 1988 (for *Federal Register* citation, see CLIA 88, 1992). The

CLIA act is a statutory framework for ensuring laboratory quality. Laboratories are required under CLIA to participate in proficiency testing programs, recognized by HCFA, and to take corrective action when a proficiency test is failed. Proficiency testing involves asking a laboratory to establish the identity of specimens provided. CLIA-certified laboratories are inspected every 2 years. While laboratories performing genetic testing must comply with general regulations under CLIA, CLIA has no requirements specific to genetic tests, except for cytogenetics. Currently, there are no proficiency testing programs in genetics or cytogenetics required under CLIA. Thus, CLIA certification of molecular laboratories evaluates the general laboratory operation but cannot assess the important question of molecular genetic test validity. At the present time, there is no assurance that every laboratory performing genetic testing clinically is meeting high quality standards.

The U.S. Food and Drug Administration (FDA) has authority to regulate the safety and efficacy of tests that are marketed as prepackaged "medical devices" or kits under the Medical Devices Amendments of 1976 (21 USC 321–392). Currently most assays performed in clinical molecular laboratories are developed on site and are not from kits; thus, the FDA does not assert any regulatory role. Therefore, the College of American Pathologists/American College of Medical Genetics (CAP/ACMG) Molecular Pathology Accreditation Program was established in 1995. This program uses biannual proficiency tests, including sample analysis and test interpretation (including risk calculations) components, to provide an assessment of test validity for participating molecular diagnostic laboratories. However, currently, CAP/ACMG accreditation is strictly voluntary.

Task Force on Genetic Testing (TFGT)

The Task Force on Genetic Testing (TFGT) was created in 1995 by the National Institutes of Health (NIH)/Department of Energy (DOE) Working Group on Ethical, Legal, and Social Implications of Human Genome Project. The task force was charged with making recommendations that would ensure the development of safe and effective genetic tests, their performance in laboratories of assured quality, and their appropriate use by health care providers and consumers. The working group of the TFGT was made up of 15 representatives of organizations with an interest in genetic testing, including the National Society of Genetic Counselors (NSGC), the American Society of Human Genetics (ASHG), the American College of Medical Genetics (ACMG), the Council of Regional Networks for Genetic Services (CORN), the College of American Pathologists (CAP), the American Medical Association (AMA), and the Health Insurance Association of America. In addition, the working group invited five agencies within the Department of Health and Human Services (HHS) to send liaison members to the TFGT.

In response to the deficiencies of the current CLIA certification system, the recommendation of the TFGT was as follows: "A national accreditation program for laboratories performing genetic tests, which includes proficiency testing and on-site inspection is needed to promote standardization across the country" (Holtzman and Watson, 1997, p. 48).

TFGT also recommended that a genetics advisory committee be established to assist CLIA in addressing current deficiencies in assuring the quality of genetic tests and that CLIA actively seek input from consumer groups and professional genetics societies in developing standards.

The TFGT report (Holtzman and Watson, 1997) identified specific concerns about providing sufficient protection for the public from predictive genetic tests that have not been adequately validated and whose clinical uses have not been thoroughly established. The task force recommended that the federal government consider using its acknowledged authority to ensure that *all* organizations developing new, predictive genetic tests submit protocols to an institutional review board. In addition, they recommended that third-party payers and managed care organizations reimburse only genetic tests that are performed in laboratories that can provide evidence that (1) the test has been clinically validated (based on published information or information provided by the test developer) or that the lab is participating in a systematic validation plan, and (2) the individual laboratory is qualified to provide such tests (i.e., has appropriate certification/accreditation).

As these recommendations, and potentially those of additional groups are considered, and as the number of genetic tests increases, the regulatory processes are expected to become more rigorous.

PRACTICE ISSUES: ESTABLISHING POLICY AND PRACTICE GUIDELINES

A number of elements within clinical practice influence the progression of a new technology from research protocol to acceptance as a standard of care. In determining whether a technology is ready to be established as a standard of care, it is necessary to evaluate carefully the outcomes of individual research trials and the comprehensive assessment of outcomes from related research. This review should include consideration of the medical, psychosocial, and ethical implications of all potential applications of the technology.

Technology Assessment Conferences

One effective format for the evaluation of a new technology or test is the utilization of a technology assessment conference:

> NIH Technology Assessment Conferences and Workshops are convened to evaluate available scientific information related to a biomedical technology. The resultant NIH Technology Assessment Statements and published reports are intended to advance the understanding of the technology or issue in question and be useful to health professionals and the public. (NIH Technology Assessment Panel on Gaucher Disease, 1996, p. 548)

These conferences generally include a public session during which presentations are given by experts in the field and an open discussion session is held. This is fol-

lowed by closed sessions in which the members of the expert panel address prede-
fined questions and develop a draft statement regarding the key issues of this specific
technology transfer. The draft is then critiqued by the panel participants and the statement is finalized. To date there have been two NIH Technology Assessment Conferences held that have addressed genetic technologies: Gaucher disease (1995) and
cystic fibrosis (1997).

The cystic fibrosis (CF) conference reconsidered population-based carrier screening for CF, a practice that had been discouraged in a series of professional position
statements published a few years earlier (Caskey et al., 1990; Workshop on Population Screening for the Cystic Fibrosis Gene, 1990; ASHG, 1992). The recommendations of the 1997 NIH conference were significantly different from those published
earlier. The differences in the conclusions were due, not to significant technological
improvements in CF testing nor to improved mutation detection rates, but to the results of studies addressing the interest and impact of population-based carrier screening for CF (Kaplan et al., 1991; Williamson, 1993; Bekker et al., 1994; Brock, 1996;
Loader et al., 1996). The findings of many of these research studies suggest that patients who were pregnant or planning a pregnancy were interested in CF carrier
screening and that the potential adverse implications of CF carrier testing did not outweigh the perceived benefits.

The draft of the 1997 consensus statement was made available on the World Wide
Web at the conclusion of the conference. Even before the final consensus statement
was published, some genetics centers in the country began offering CF carrier testing
to their pregnant patients (personal communication), and a large commercial laboratory began marketing this testing to medical professionals who care for patients of reproductive age. This sequence of events demonstrates the significant impact a
consensus development conference can have on the use of a technology in clinical
practice.

Practice Guidelines

Another powerful influence on the transfer of technology from research phase to accepted clinical standard of care is the establishment of professional practice guidelines such as those that were developed for presymptomatic genetic testing for
Huntington disease. In 1985 a committee composed of members from the International Huntington Association (IHA) and the World Federation of Neurology (WFN)
Research Group on Huntington's Chorea prepared guidelines for predictive testing
that were adopted by both organizations and published in the *Journal of Neurological
Sciences* (World Federation of Neurology Research Group on Huntington's Chorea,
1989) and the *Journal of Medical Genetics* (World Federation of Neurology Research
Group on Huntington's Chorea, 1990). These guidelines were revised when the identification of the *huntingtin* gene was reported in 1993, and the revisions were published in *Neurology* (Committee of International Huntington Association and World
Federation of Neurology Research Group on Huntington's Chorea, 1994).

The HD guidelines established international standards by which predictive testing
for Huntington disease is performed. A significant advantage of the guidelines is that

they were developed by an interdisciplinary group of professionals, ensuring representation from everyone involved in the care of HD patients and their families. This is crucial, since practice guidelines are often established within individual specialties. Medical professionals in other specialties, who may ultimately order these specific genetic tests, may not be aware of such practice guidelines, especially if the guidelines are not well publicized. Many of the elements of the HD guidelines have now been incorporated into clinical models for predictive genetic testing for other diseases, such as adult-onset spinocerebellar ataxias and hereditary cancer syndromes.

The approach used to develop the HD predictive testing guidelines was followed in the development of a consensus statement on the use of apolipoprotein E testing for Alzheimer disease, resulting in a joint statement by the American College of Medical Genetics (ACMG) and the American Society of Human Genetics (ASHG) Working Group on apoE and Alzheimer disease (1995). Included in the ASHG working group were representatives from the American Academy of Neurology (AAN) and the American Psychiatric Association (APA). The final draft of the statement was reviewed and endorsed by appropriate committees of the ACMG, ASHG, AAN, APA, and the NIH/Department of Energy Working Group on Ethical, Legal, and Social Implications of the Human Genome Project. The final statement concluded that available data indicated that apoE genotyping does not provide sufficient sensitivity and specificity to be used alone as a diagnostic test for Alzheimer disease. This statement sent a clear message to the medical professionals involved in the care of Alzheimer patients and their families to refrain from using apoE genotyping for diagnostic purposes until additional research outcomes are available.

How should genetic counselors use such statements and guidelines in their practice? The American College of Medical Genetics concluded its statement on population screening for *BRCA1* mutation in Ashkenazi Jewish women as follows:

> This guideline is designed primarily as an educational resource for medical geneticists and other health care providers to help them provide quality medical genetic services. Adherence to this guideline does not necessarily assure a successful medical outcome. This guideline should not be considered inclusive of all proper procedures and tests or exclusive of other procedures and tests that are reasonably directed to obtaining the same results. In determining the propriety of any specific procedure or test, the geneticist should apply his or her own professional judgment to the specific clinical circumstances presented by the individual patient or specimen. It may be prudent, however, to document in the patient's record the rationale for any significant deviation from this guideline (ACMG, 1996).

Thus the ACMG suggests that practitioners evaluate the validity of the statement or guideline with regard to the specific circumstances of each patient and use their professional judgment in adhering to or deviating from the recommendation. In addition, practitioners who choose not to adhere should carefully document their rationale for this course in the patient's medical record.

Such documentation is also important because published practice guidelines have been used in medical malpractice litigation. Hyams et al. (1995) investigated the frequency with which practice guidelines were used in medical malpractice claims. A review of 259 malpractice claims, filed from 1990 to 1992 at two insurance companies, identified 16 that involved practice guidelines. Of these, the guidelines were used to implicate the defendant physician (inculpatory evidence) in 12 cases and to exonerate the physician (exculpatory evidence) in 4 cases. Malpractice attorneys who were surveyed indicated that once a lawsuit has been initiated, practice guidelines were most likely to be used for inculpatory purposes (54% of cases vs. 22.7% of cases for exculpatory purposes). Guidelines that offer exculpatory value for the defendant physician induce attorneys not to file suits.

According to Wilson et al. (1995), for a practice guideline to be useful to a clinician, the recommendations should offer practical, unambiguous advice about a specific health problem. Moreover, a practice guideline should convince the practitioner that the benefits of following the recommendations are worth the potential harms and costs. The potential benefits of a clinical intervention, particularly if they are marginal, may be viewed less favorably if there are significant costs, discomforts, and impracticalities. Wilson et al. (1995) also stressed the importance of evaluating the "strength" or "confidence" of the recommendations in the guideline. They pointed out that these elements are influenced by a number of factors, including the quality of the research data that provides evidence for the recommendations, the magnitude and consistency of positive versus negative outcomes, and the relative value placed on different outcomes. These factors have great significance in our understanding of the differences between practice guidelines on similar topics from various specialty organizations.

Within the medical genetics community, there are two published reports regarding genetic testing of children and adolescents for adult-onset genetic disorders: a resolution by the National Society of Genetic Counselors (1995) entitled Prenatal and Childhood Testing for Adult-Onset Disorders and a report by the Board of Directors of the American Society of Human Genetics and the American College of Medical Genetics (1995) entitled Points to Consider: Ethical, Legal and Psychosocial Implications of Genetic Testing in Children and Adolescents. These position papers make somewhat different recommendations regarding genetic testing for carrier status of adult-onset disorders in children and adolescents when there are no medical benefits to be derived prior to adulthood. The ASHG/ACMG report states that "such testing generally should be deferred," while the NSGC resolution states that "pilot studies are needed to assess the medical and psychosocial risks and benefits" of such testing, and "until more data is gathered on the impact of this type of testing, extreme caution should be taken regarding the use of such tests."

Given the evolving informational nature of genetic testing and the diversity of factors used to measure the potential benefits and risks of such testing to a particular group, differences in recommended practice guidelines of different genetics societies, albeit subtle, are likely to continue. Medical genetics organizations working together will need to develop creative strategies to resolve these differences and provide consistent care to their clients. These strategies may involve establishing clinical testing policies that adhere to the most rigid practice guidelines, with the opportunity for exceptional cases to be carefully considered by an advisory committee.

The Role of an Advisory Committee

Some clinical programs that offer genetic testing have established advisory committees. For example, Baylor College of Medicine formed an advisory committee that specifically considers requests for genetic testing that are not consistent with the Huntington disease presymptomatic genetic testing guidelines (Committee of International Huntington Association and World Federation of Neurology Research Group on Huntington's Chorea, 1994). The Baylor HD advisory committee is composed of the members of the HD predictive testing clinical team (a genetic counselor, a neurologist, a clinical psychologist) in addition to a medical geneticist, a psychiatrist, a medical ethicist, a lawyer with expertise in health law and genetics, a genetic counselor with expertise in molecular testing, and a molecular genetics laboratory director. This committee has been a useful resource in a program of this type because it offers a mechanism for input from patients and their families, particularly those who do not meet eligibility criteria for participation under current clinical protocols. Additionally, the committee's meetings allow other providers of services to clients and families with HD to contribute their perspectives on genetic testing, which may be quite different from the view of medical genetics professionals. This diverse input should be taken into account as clinical providers revise their policies and protocols.

Genetic counselors need to be knowledgeable about practice guidelines, and their clinical and legal implications. Genetic counselors should be familiar with practice guidelines developed within the medical genetics profession as well as those developed by other groups of medical providers, since there may be considerable differences. Additionally, genetic counselors should actively participate in the development of practice guidelines.

COMPLEXITIES OF NEW TECHNOLOGIES IN THE GENETIC COUNSELING SESSION: ILLUSTRATIVE CASES

As providers of a significant proportion of clinical genetic services in this country, genetic counselors have played, and should continue to play, a key role in the technology transfer process. The following cases illustrate the significant role of the counselor in providing appropriate counseling for families and ensuring that informed consent is obtained. The names of the individuals, both professionals and clients, and the specific circumstances of these cases have been altered to preserve confidentiality.

Case 1: Communication About the Current Specificity and Sensitivity of a Test and Potential for Improvement as the Technology Evolves (Figure 14.2)

Background: Jennifer Valenti, a second-year genetic counseling graduate student, was doing a rotation in the university-based prenatal diagnosis clinic. She saw a couple, Paula and David Keller, who were referred because of advanced maternal age (42 years) in their current pregnancy. While obtaining the

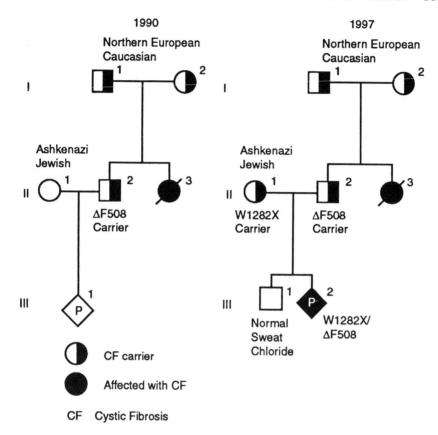

Figure 14.2 *Improvements in CF testing. Family history in 1990 reflects DNA analysis of individuals II-1 and II-2 for only the ΔF508 CF mutation. More comprehensive testing in 1997, identified individual II-1 to be a carrier for the W1282X CF mutation. In the 1997 fetal study, III-2, was found to be a compound heterozygote for ΔF508/W1282X.*

family history, Jennifer learned that during their first pregnancy, in 1990, the Kellers had elected to have an amniocentesis for prenatal diagnosis of cytogenetic abnormalities, since Paula was 35 years old at that time. During that pregnancy, Paula and David reported that they were also concerned that David's family history of cystic fibrosis (CF) increased their risk for having an affected child. David's only sister had CF and she had died in 1985 at age 32 as a result of respiratory complications.

The genetic counseling and testing for the first pregnancy was performed in another city, and Paula and David did not bring records with them. They reported that they had completed studies to determine their CF carrier status and that David was found to be a CF carrier but Paula's CF test results were negative. At the time of the 1997 appointment, the Kellers were no longer concerned about CF, having come to perceive their risk for having a child with CF to be very low. To support this perception they pointed out that their 7-year-old son was very healthy and had had a negative sweat chloride test result at age 3 months. The genetic counseling student explained the need to review the actual laboratory reports, and the Kellers signed releases so that their records could be obtained.

Review of the laboratory reports from 1990 indicated that at that time the only CF mutation being studied was the Δ F508 mutation. The reports confirmed that David was a carrier of the Δ F508 CF mutation and that Paula was negative for this mutation. Her laboratory report from 1990 indicated that her CF carrier risk had been significantly reduced, from approximately 1 in 25 to about 1 in 300. In fact, the risk modification given at that time did not take into account Paula's Ashkenazi Jewish ancestry. The differential CF mutation detection rate for individuals of different ethnic backgrounds was not appreciated in 1990. Based on current knowledge in 1997, an Ashkenazi Jewish individual with a negative family history of CF who is negative for only the Δ F508 mutation would have a CF carrier risk modification from ~1 in 29 to ~1 in 41, since the detection rate of Δ F508 for CF chromosomes in this ethnic group is ~30% (Abeliovich et al., 1992). In 1997 most laboratories offering CF molecular testing analyzed five mutations that accounted for 97% of the CF mutations in individuals of Ashkenazi Jewish background (Abeliovich et al., 1992). Such analyses, if negative, would modify the CF carrier risk for Ashkenazi Jewish individuals with a negative family history from ~1 in 29 to ~1 in 930.

Follow-Up: Jennifer conveyed the information on recent advances in CF testing to the Kellers, and Paula decided to have her blood drawn for CF mutation analysis with an expanded panel. Her results showed her to be a carrier for the W1282X CF mutation, which is identified in ~45% of Jewish CF chromosomes (Abeliovich et al., 1992). The Kellers were given a 25% risk for CF for each of their pregnancies based on their combined molecular test results. They had proceeded with amniocentesis because of the risks associated with advanced maternal age and requested that the fetal DNA be analyzed for CF mutations. The fetal results were positive for both parental mutations, indicating that this fetus had a greater than 99% risk for being affected with CF.

This case illustrates the importance of thorough genetic counseling prior to genetic testing and clear communication with the client about the current validity and predictive value of an earlier test result. The additional genetic testing provided this couple with very specific information about their carrier status and their risk for having a child with CF. Clinical scenarios such as this will occur with greater frequency as genetic testing technologies evolve and improve. Individuals seeking genetic testing should be informed of the pace with which these technologies are improving and made aware that our understanding of the validity of testing also changes with research and clinical experience.

These issues also should be clearly explained to the primary care professional who refers the client for genetic counseling and to the insurance carrier, in the event that additional genetic testing is recommended in the future. Unless insurance carriers understand how rapidly these technologies are evolving, they may deny claims for coverage of additional testing, viewing them as simply repeat analyses. Depending on the indication for a genetic test, it may be prudent for some individuals to postpone testing until a better understanding of the technology has been reached.

Case 2: Consideration of the Methodology, Comprehensiveness of the Technology, and Laboratory Differences in Light of the Specific Indication for Testing

Background: Sally Wilkins, now 12 months old, had severe hypotonia and feeding difficulties in the neonatal period. She was evaluated by neonatologist, Dr. Jorge Torres, who felt, based on the baby's clinical features, that the most likely diagnosis was Prader–Willi syndrome. As a part of the diagnostic evaluation, Dr. Torres ordered high resolution chromosome analysis in addition to fluorescent in situ hybridization (FISH) analysis for chromosome 15. The high resolution cytogenetic and FISH analyses were both negative for a deletion of 15q11.2-q13.

James Rich, a genetic counselor in a pediatric genetics clinic, was scheduled to meet with Sally's parents, Penny and Alan Wilkins. Penny and Alan were somewhat surprised that their pediatrician had referred them for genetic counseling because the problems Sally had had as a newborn were almost completely resolved and she was now a chubby one-year-old, doing well developmentally. James initially reviewed Penny's and Alan's family histories and elicited their questions and concerns. Their main question was why the pediatrician continued to think that Sally might have Prader–Willi syndrome (PWS) given that Sally's hypotonia and feeding problems had resolved. They recalled that the blood tests their daughter had had as a newborn, which were quite extensive, were negative for PWS.

James reviewed the natural history of PWS and helped Penny and Alan understand why Sally's history might still be consistent with this diagnosis. He explained that the tests that had been performed on Sally's blood looked for the most common type of mutation seen in PWS patients, a deletion in the *SNRPN* gene; but since deletions are the cause of PWS in approximately 70% of patients (Delach et al., 1994), these tests do not detect all PWS mutations. He explained that there is another test that studies the parental-specific methylation pattern of the *SNRPN* gene. The methylation test would be the best test for determining whether Sally really has PWS, since over 99% of patients with PWS will have a specific pattern showing only maternal methylation, which means that a paternal chromosome 15 has not been inherited (Glenn et al., 1993). James explained that if Sally had the methylation test and the results showed a normal methylation pattern, the diagnosis of PWS would be very unlikely. An abnormal methylation pattern, however, would confirm a diagnosis of PWS.

It is critical that genetic counselors keep up to date on evolving technologies. In addition, there are increasing instances in which multiple technologies may be utilized to provide diagnostic information for the same genetic disorder. It is often difficult to decide which technology should be utilized first. This is illustrated by the ASHG/ACMG Test and Technology Transfer Committee's report (1996) on diagnostic testing for PWS and for Angelman syndrome (AS). This report clearly defined the technologies available, including high resolution chromosome analysis, FISH studies, methylation analysis, and uniparental disomy (UPD) studies. The committee offered different strategies for consideration by the clinician facing the decision of ordering tests sequentially, noting that the approach used for a specific patient depended on local availability of testing, studies performed earlier on the patient, and the diagnostic expertise of the referring physician. In a letter to the editor, Monaghan et al. (1997) suggested that in addition to the factors cited in the ASHG/ACMG report, the cost of the diagnostic evaluations be considered in light of the percent likelihood that the patient has the PWS or AS diagnosis. They concluded:

> It is likely that, as more is learned about the etiology of genetic diseases, additional imprinting mutations will be identified. Therefore, it seems beneficial to develop a strategy for diagnosing this type of disorder. The cost savings achieved by using alternative sequential approaches rather than a single concurrent approach will increase as more imprinting mutations are identified. (Monaghan et al., 1997, p. 246)

Follow-Up: Sally's blood sample was drawn and submitted to another laboratory for *SNRPN* methylation analysis. The test results showed a maternal-only methylation pattern, confirming a diagnosis of PWS. Subsequent analysis of the family showed that Sally had maternal uniparental disomy for the PWS critical region on chromosome 15. On the basis of these laboratory data, Penny and Alan were counseled that the occurrence of PWS in their child was most likely sporadic and their recurrence risk would be approximately 1%.

The clinical scenario represented by Sally's case illustrates the complexity of selecting appropriate testing when several testing methodologies are available, as well as the limitations of testing and the significant diagnostic implications. The differences between laboratories may be extensive with regard to the methodology used, the sample and patient information requirements, the comprehensiveness of the analysis, the level of interpretation involved in reporting results, and the content of the report. Staff members who understand these variables and the indication for the particular analysis should be responsible for choosing the laboratory to be used for genetic testing. Questions to ask when selecting a laboratory for genetic testing are presented in Chapter 8. To ensure that the most appropriate analysis will be performed for a specific family, the genetic professional should play an active role in selecting the laboratory to perform the study. Additionally, patients and their family members should be fully aware of the limitations of any particular technology and the potential need to consider sequential testing to investigate the same diagnosis.

**Case 3: Limitations of Predictive Value of Genetic Test Results Must Be
Communicated and Understood Before the Technology Is Used (Figure 14.3)**

Background: Elizabeth Chang, a genetic counselor coordinating the Adult
Genetics Clinic at a university-based genetics center, had received a referral to see
a young woman, Tricia Lee, for genetic counseling about myotonic dystrophy. The
referring physician, Dr. Jordan Wilms, a local infertility specialist, explained that
he had ordered a molecular test on Tricia to check for the presence of the
trinucleotide repeat expansion mutation in her myotonin protein kinase gene
because the patient had a strong family history of myotonic dystrophy (Figure
14.3). Dr. Wilms expected the test results to be negative, since Tricia at 28 years of
age had no clinical evidence of the disease. However, the test results showed that
Tricia did have one copy of the myotonin protein kinase expansion mutation (~200
copies of the trinucleotide repeat: the normal range is up to about 35 copies of the
repeat) (Reardon et al., 1992). While Dr. Wilms explained the autosomal dominant
mode of inheritance of myotonic dystrophy and the potential risk for Tricia's future
children to inherit the mutation, he could not answer her questions about the
specific clinical implications of this result for Tricia and her future pregnancies.

Tricia had seen the wide spectrum of clinical manifestations of myotonic dystrophy in her own family. Her grandfather had late-onset and relatively mild disease and her mother and maternal aunt both had onset in early adulthood with
fairly significant physical impairment. Tricia's mother had died of cardiac complications of the disease. Tricia's brother and maternal first cousin both had congenital myotonic dystrophy. Her first cousin had died in the neonatal period. Her
brother, having survived a difficult neonatal course, was now severely physically
impaired and had moderate mental retardation. To decide whether to proceed with
her infertility evaluation and potentially use assisted reproductive technologies to
become pregnant, Tricia wished to have specific information about the clinical
phenotype that she could expect for herself and for her children, if they inherited
the mutation. Tricia was frustrated that Dr. Wilms was not able to give her answers
to these questions, and she had requested a second opinion to obtain the information she needed for future planning.

Follow-Up: Elizabeth Chang met with Tricia and explained that DNA analysis is a
highly accurate method for determining whether a myotonic dystrophy gene
mutation has been inherited and can be performed prenatally (Redman et al., 1993).
She also discussed the limitations of molecular testing for myotonic dystrophy with
respect to specific genotype–phenotype correlations. Following her appointment
with Dr. Wilms, Tricia had obtained genetic testing records on her deceased mother.
She was surprised to learn that her mother had ~ 150 copies of the trinucleotide
repeat, a lower number than had been found in her own gene expansion. Elizabeth
Chang explained that while very large expansions in the myotonin protein kinase
gene are associated with congenital myotonic dystrophy (the most severe type),
smaller expansions, as illustrated in Tricia's family, do not necessarily correlate
with clinical status. Tricia indicated that given the limitations in obtaining definite

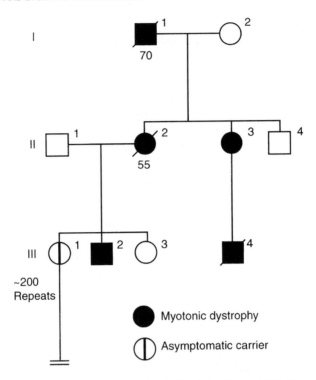

Figure 14.3 *Family history of myotonic dystrophy illustrates the variability in age of onset and severity of this disease. Individual I-1 had minimal findings beginning in his fifties, including cataracts and grip myotonia. He had also had premature frontal balding in his thirties. He died of unrelated causes in his seventies. Individuals II-2 and II-3 were diagnosed in their thirties, at about the same time as their father, following the birth and early death of III-4, who had congenital myotonic dystrophy. II-2 and II-3 both had significant muscle weakness. II-2 died at 55 years as a result of cardiomyopathy. Individual III-2 had congenital myotonic dystrophy, required ventilator support for the first few months of life, and is currently moderately mentally retarded. Individual III-1 is clinically asymptomatic at 28 years. Her DNA studies of the CGT repeat region of the myotonin pk gene identified one expanded allele (~200 CGT repeats) and one normal size allele. The normal range is up to approximately 35 repeats and the mutation range begins at approximately 50 repeats.*

information about clinical status, she was unsure about whether to proceed with her fertility workup and would need time to think over her options.

Patient and professional expectations of the utility of new technologies for genetic risk assessment are often unrealistic. The simple availability of a genetic test for a particular disease gene, even if it results in the identification of a specific mutation in the gene in question, does not ensure that the genotype–phenotype correlations will be precise enough to be used by the professional for patient management or by the client for personal decision-making purposes. These limitations should be explained to the individual considering testing prior to obtaining a sample so that a fully informed decision can be made.

Similar concerns have been expressed by many groups with respect to the use of genetic testing for inherited susceptibility to breast and ovarian cancer (Statement of the American Society of Human Genetics on genetic testing for breast and ovarian cancer predisposition, 1994; the National Breast Cancer Coalition, 1995; the National Advisory Council for Human Genome Research Statement on use of DNA testing for presymptomatic identification of cancer risk, 1994; American College of Medical Genetics Statement on population screening for *BRCA1* mutation in Ashkenazi Jewish women, 1996). Most of these groups have recommended that predictive genetic testing for cancer susceptibility be provided only under IRB-approved research protocols, until there is more information to offer clients about the predictive value of the test and the effectiveness of available management and surveillance strategies in prevention and early identification of cancer. As information about gene discoveries becomes more available, the public's demand and commercial interest in genetic testing will increase. In addition to increasing the number of genetic tests performed, these influences may result in inappropriate use and incorrect interpretation of genetic test results by professionals who do not fully understand their complexity and limitations (Giardiello, 1997).

CONCLUSIONS

To address the needs of their clients, genetic counselors must understand the process involved in the transfer of a new technology from the research laboratory to the clinical setting. They need to be aware of the rapidly evolving standards by which clinical laboratories are currently certified and the potential for involvement of other regulatory agencies in the future. Genetic counselors should devise strategies to continually educate themselves about new technologies in order to counsel their clients and obtain informed consent for testing. To appropriately select a diagnostic laboratory for their clients, a genetic counselor must be knowledgeable about test validation and laboratory variability. In addition, genetic counselors should continue to play an active role in the technology transfer process by conducting research to address the impact of utilization of new genetic technologies and by participating in the development of consensus statements and practice guidelines.

REFERENCES

Abeliovich D, Lavon IP, Lerer I, Cohen T, Springer C, Avital A, Cutting GR (1992) Screening for five mutations detects 97% of cystic fibrosis (CF) chromosomes and predicts a carrier frequency of 1:29 in the Jewish Ashkenazi population. *Am J Hum Genet* 51:951–956.

American College of Medical Genetics (1996) Statement on population screening for *BRCA-1* mutation in Ashkenazi Jewish women. *Am Coll Med Genet Newsl* 7.

American College of Medical Genetics/American Society of Human Genetics Working Group on ApoE and Alzheimer Disease (1995) Statement on use of apolipoprotein E testing for Alzheimer disease. *JAMA* 274(20):1627–1629.

American Society of Clinical Oncology (1996) Statement on genetic testing for cancer susceptibility. *J Clin Oncol* 14(5):1730–1736.

American Society of Human Genetics (1992) Statement of the American Society of Human Genetics on cystic fibrosis carrier screening. *Am J Hum Genet* 51:1443.

American Society of Human Genetics (1994) Statement on genetic testing for breast and ovarian cancer predisposition. *Am J Hum Genet* 55:I–iv.

ASHG/ACMG (American Society of Human Genetics/American College of Medical Genetics) Test and Technology Transfer Committee (1996) Diagnostic testing for Prader–Willi and Angelman syndromes. *Am J Hum Genet* 58:1085–1088.

Bekker H, Denniss G, Modell M, Bobrow M, Marteau T (1994) The impact of population based screening for carriers of cystic fibrosis. *J Med Genet* 31(5):364–368.

Boards of Directors of the American Society of Human Genetics and the American College of Medical Genetics (1995) ASHG/ACMG Report: Points to consider: Ethical, legal, and psychosocial implications of genetic testing in children and adolescents. *Am J Hum Genet* 57:1233–1241.

Brock DJ (1996) Population screening for cystic fibrosis. *Curr Opin Pediatr* 8(6):635–638.

Canadian Collaborative CVS–Amniocentesis Clinical Trial Group (1989) *Lancet* 1(8628):1–6.

Caskey CT, Kaback MM, Beaudet AL (1990) The American Society of Human Genetics statement on cystic fibrosis screening. *Am J Hum Genet* 46:393.

CLIA 88 (1992) *Federal Register* 57(40):7001–7288. Public Law 100-578 Clinical Laboratory Improvement Amemdments of 1988, 42 USC 263a et seq.

Committee of International Huntington Association and World Federation of Neurology Research Group on Huntington's Chorea (1994) Guidelines for the molecular genetics predictive test in Huntington's disease. *Neurology* 44:1533–1536.

Delach JA, Rosengren SS, Kaplan L, Greenstein RM, Cassidy SB, Benn PA (1994) Comparison of high resolution chromosome banding and fluorescence in situ hybridization (FISH) for the laboratory evaluation of Prader–Willi syndrome and Angelman syndrome. *Am J Med Genet* 52:85–91.

Giardiello FM, Brensinger JD, Petersen GM, Luce MC, Hylind LM, Bacon JA, Booker SV, Parker RD, Hamilton SR (1997) The use and interpretation of commercial APC gene testing for familial adenomatous polyposis. *N Engl J Med* 336:823–827.

Glenn CC, Porter KA, Jong MTC, Nicholls RD, Driscoll DJ (1993) Functional imprinting and epigenetic modification of the human *SNRPN* gene. *Hum Mol Genet* 2:2001–2005.

Holtzman NA, Watson MS (1997) Final Report of the Task Force on Genetic Testing: Promoting safe and effective genetic testing in the United States. Bethesda, MD: U.S. National Institutes of Health.

Holtzman NA, Murphy PD, Watson MS, Barr PA (1997) Predictive genetic testing: From basic research to clinical practice. *Science* 278:602–605.

Hyams AL, Brandenburg JA, Lipsitz SR, Shapiro DW, Brannan TA (1995) Practice guidelines and malpractice litigation: A two-way street. *Ann Intern Med* 122(6):450–455.

Kaplan F, Clow C, Scriver CR (1991) Cystic fibrosis carrier screening by DNA analysis: A pilot study of attitudes among participants. *Am J Hum Genet* 49:240–242.

Lebo RV, Cunningham G, Simons MJ, Shapiro LJ (1990) Defining DNA diagnostic tests appropriate for standard clinical care (Letter to the editor). *Am J Hum Genet* 47:583–590.

Loader S, Caldwell P, Kozyra A, Levenkron JC, Boehm CD, Kazazian HH Jr, Rowley PT (1996) Cystic fibrosis carrier population screening in the primary care setting. *Am J Hum Genet* 59:234–247.

Monaghan KG, Van Dyke DL, Feldman G, Wiktor A, Weiss L (1997) Diagnostic testing: A cost analysis for Prader–Willi and Angelman syndromes (Letter to the editor). *Am J Hum Genet* 60:244–247.

National Advisory Council for Human Genome Research (1994) Statement on use of DNA testing for presymptomatic identification of cancer risk. *JAMA* 217:785.

National Committee for Clinical Laboratory Standards (NCCLS), 771 East Lancaster Ave., Villanova, PA 19085-1596, in press.

National Institute of Child Health and Development National Registry for Amniocentesis Study Group (1976) Midtrimester amniocentesis for prenatal diagnosis: Safety and accuracy. *JAMA* 236:1471–1476.

National Institutes of Health (1997) Consensus Development Statement on Genetic Testing for Cystic Fibrosis, Draft, April 16.

National Institutes of Health Technology Assessment Panel on Gaucher Disease (1996) Gaucher disease: Current issues in diagnosis and treatment. *JAMA* 274:548–553.

National Society of Genetic Counselors (Adopted 1995) Resolution: Prenatal and childhood testing for adult-onset disorders.

Presymptomatic genetic testing for heritable breast cancer risk. Press release of the National Breast Cancer Coalition. Washington, DC, September 28, 1995.

Reardon W, Harley HG, Brook JD, Rundle SA, Crow S, Harper PS, Shaw DJ (1992) Minimal expression of myotonic dystrophy: A clinical and molecular analysis. *J Med Genet* 29:770–773.

Redman JB, Fenwick RG, Fu Y-H, Pizzuti A, Caskey CT (1993) Relationship between parental trinucleotide CGT repeat length and severity of myotonic dystrophy in offspring. *JAMA* 269(15):1960–1965.

Rhoads GC, Jackson LG, Schlesselman SE, De La Cruz FF, Desnick RJ, Golbus MS, Ledbetter DH, Lubs HA, Mahoney MJ, Pergament E, Simpson JL, Carpenter RJ, Elias S, Ginsberg NA, Goldberg JD, Hobbins JC, Lynch L, Shiono PH, Wapner RJ, Zachary JM (1989) The safety and efficacy of chorionic villus sampling for early prenatal diagnosis of cytogenetic abnormalities. *N Engl J Med* 320(10):609–617.

Task Force on Predictors of Hereditary Disease or Congenital Defects (1979) Antenatal Diagnosis: Report of a Consensus Development Conference (NICHHD):I-1-I-263.

Whittaker L (1996) Clinical applications of genetic testing: Implications for the family physician. *Am Fam Physician* 53(6):2077–2084.

Williamson R (1993) Universal community carrier screening for cystic fibrosis? *Nat Genet* 3:195.

Wilson MC, Hayward RSA, Tunis SR, Bass EB, Guyatt G (1995) Users' guides to the medical literature. VIII. How to use clinical practice guidelines. B. What are the recommendations and will they help you in caring for your patients? *JAMA* 274(20):1630–1632.

Workshop on population screening for the cystic fibrosis gene (1990) Statement from the National Institutes of Health Workshop on population screening for the cystic fibrosis gene. *N Engl J Med* 323:70.

World Federation of Neurology Research Group on Huntington's Chorea (1989) Ethical issues policy statement on Huntington's disease molecular genetics predictive test. *J Neurol Sci* 94:327–332.

World Federation of Neurology Research Group on Huntington's Chorea (1990) Ethical issues policy statement on Huntington's disease molecular genetics predictive test. *J Med Genet* 27:34–38.

Sample DNA
Laboratory Report

DNA DIAGNOSTIC LABORATORY
(address and phone/fax)

Referring Physician	**Referring Counselor**
Address	Address
Phone	Phone
Fax	Fax

Patient Name	Date of birth
Sample type	Family #
Date collected	DNA#
Date received	

METHODOLOGY
The analysis of this DNA was performed using amplification (PCR) and allele specific oligonucleotide (ASO) hybridization and/or Southern analysis and hybridization to specific radiolabeled probes. These methodologies have been used to test for the presence of the following specific mutations in the *XYZ* gene (list of mutations tested). This panel of mutations will identify approximately 90% of the *XYZ* mutant alleles in Caucasian individuals of northern European descent. The detection rates for other ethnic groups are not well-defined at this time.

INTERPRETATION (date of report)
DNA extracted from leukocytes from this patient has been analyzed using the methodology described. The DNA shows a hybridization pattern consistent with one

copy of the ABC mutation in the *XYZ* gene. These results predict that this individual is either affected with or predisposed to the *XYZ* gene disease.

DNA extracted from both direct amniotic fluid cells and cultured amniocytes from the current fetus of this patient have also been analyzed using the same methodology. The fetal DNA does not show hybridization to any of the mutations studied in the *XYZ* gene, including the *ABC* mutation, which was identified in the mother. Thus, this fetal analysis predicts, with greater than 99% accuracy, that this fetus is unaffected with the *XYZ* gene disease. Other mutations that may cause *XYZ* gene disease will not be detected by this assay.

Comparative analyses of maternal and fetal DNA have been conducted using PCR amplification of polymorphic sites in other regions of the genome to rule out severe maternal cell contamination of the fetal sample as a potential source of diagnostic error. These data distinguish between the maternal and fetal samples and show appropriate inheritance of one maternal allele.

NOTE

DNA testing is not 100% accurate. The inaccuracy may result from a number of factors, including but not limited to the following: less than complete detection of all mutations; unrecognized recombination events, if linkage analysis is used; rare variations in the DNA of an individual being tested; sample switches; mistaken paternity or other inaccuracies in the biological relationships reported in the family; incorrect disease prevalence; inaccurate reporting of ethnicity; other genetic factors (mosaicism, heterogeneity).

SIGNATURES

_____   _____   _____
Medical Director   Lab Director   Genetic Counselor

15

Computer-Based Resources for Clinical Genetics

Debra L. Collins

Genetic counseling, like other professions, has been impacted by the rapid increase in information and the need for faster and easier access to this information. Increasingly, both genetic counselors and their clients make use of computers and the Internet. This chapter provides an overview and sampling of some of the primary computer-based clinical genetics resources including diagnostic resources, cytogenetic databases, testing laboratories, teratogen databases, pedigree-drawing programs, and a broad and growing number of Internet sites. Contact information to obtain the cited programs and a partial listing of World Wide Web (WWW) addresses for Internet resources can be found at the end of this chapter, along with a glossary defining frequently used computer terms.

Many resources are available across a variety of computer platforms (e.g., Macintosh, Windows). Some are available only as commercial software, while others can be accessed free through the Internet. Sample program discs or online demonstrations are often available so that a software program can be previewed before it is purchased. To use computer programs effectively, it is important to take the time to read the descriptions that accompany the software and use the help sections.

In selecting a computer program or using an Internet site, look for the credentials of the developers. A qualified medical or scientific advisory board gives a resource more credibility. You can seek guidance in selecting genetic computer resources from your colleagues and through evaluation of published critiques, online reviews, and direct contact with the distributor. Sometimes national meetings (e.g., National Society

A Guide to Genetic Counseling, Edited by Diane L. Baker, Jane L. Schuette, and Wendy R. Uhlmann.
ISBN 0-471-18541-8 (cloth), 0-471-18867-0 (paper). Copyright © 1998, Wiley-Liss, Inc.

of Genetic Counselors Annual Education Conference, American Society of Human Genetics Annual Meeting) offer workshops that focus on computer-based genetic resources. Exhibitors at these meetings and resource rooms also have computer genetic resources available for you to try.

The continuous upgrades and changes in computer technology, hardware, Internet browsers, and other developments make it a challenge to keep pace. Internet resources will periodically change locations; new locations are likely to be prominently announced in the clinical genetics journals. Most clinical genetics centers and medical center librarians follow the online announcements regarding site address changes. Electronic searching tools using keywords specific for a resource will relocate it as well.

CLINICAL DIAGNOSTIC RESOURCES

As a genetic counselor, you need ready access to information on genetic conditions to counsel patients, perform research, and prepare educational programs. The computer resources described in the subsections that follow are frequently used by genetic counselors. These resources utilize interactive searching to locate diagnostic information, condition summaries, and current literature reviews.

Online Resources

Online Mendelian Inheritance in Man (OMIM™) is the Internet version of *Mendelian Inheritance in Man, Catalogs of Human Genes and Genetic Disorders,* which was first published by Victor A. McKusick, M.D., in 1966. OMIM is currently maintained by the National Center for Biotechnology Information (NCBI) and is continually updated by the OMIM staff through surveillance of peer-reviewed journals in genetics, molecular biology, and related disciplines. Geneticists consider *Mendelian Inheritance in Man*/OMIM to be the most authoritative reference for information on inherited conditions.

OMIM presents chronologically the critical discoveries and insights about genetic conditions. OMIM can be used both to provide a comprehensive description and overview of a genetic condition, and to establish differential diagnoses (by entering traits to generate a list of matching genetic conditions). The OMIM database can be searched to find more than 8000 human genes and genetic disorders. Included in each listing is a description of the condition, clinical features, mode of inheritance, molecular and cytogenetic findings, diagnostic criteria, clinical management, references, and the date the entry was last updated (see Figure 15.1).

The book version, *Mendelian Inheritance in Man,* is reissued every few years and is now a multivolume set. It became accessible online in 1987, through dial-up subscriptions. In 1994 an expanded online version became available through the World Wide Web. The printed version of McKusick's catalog focuses on single-gene traits, while the online version also includes multifactorial traits (e.g., spina bifida) and chromosomal conditions (e.g., cri-du-chat). The online version also includes pho-

http://www.ncbi.nlm.nih.gov/htbin-post/Omim/dispmim?306700

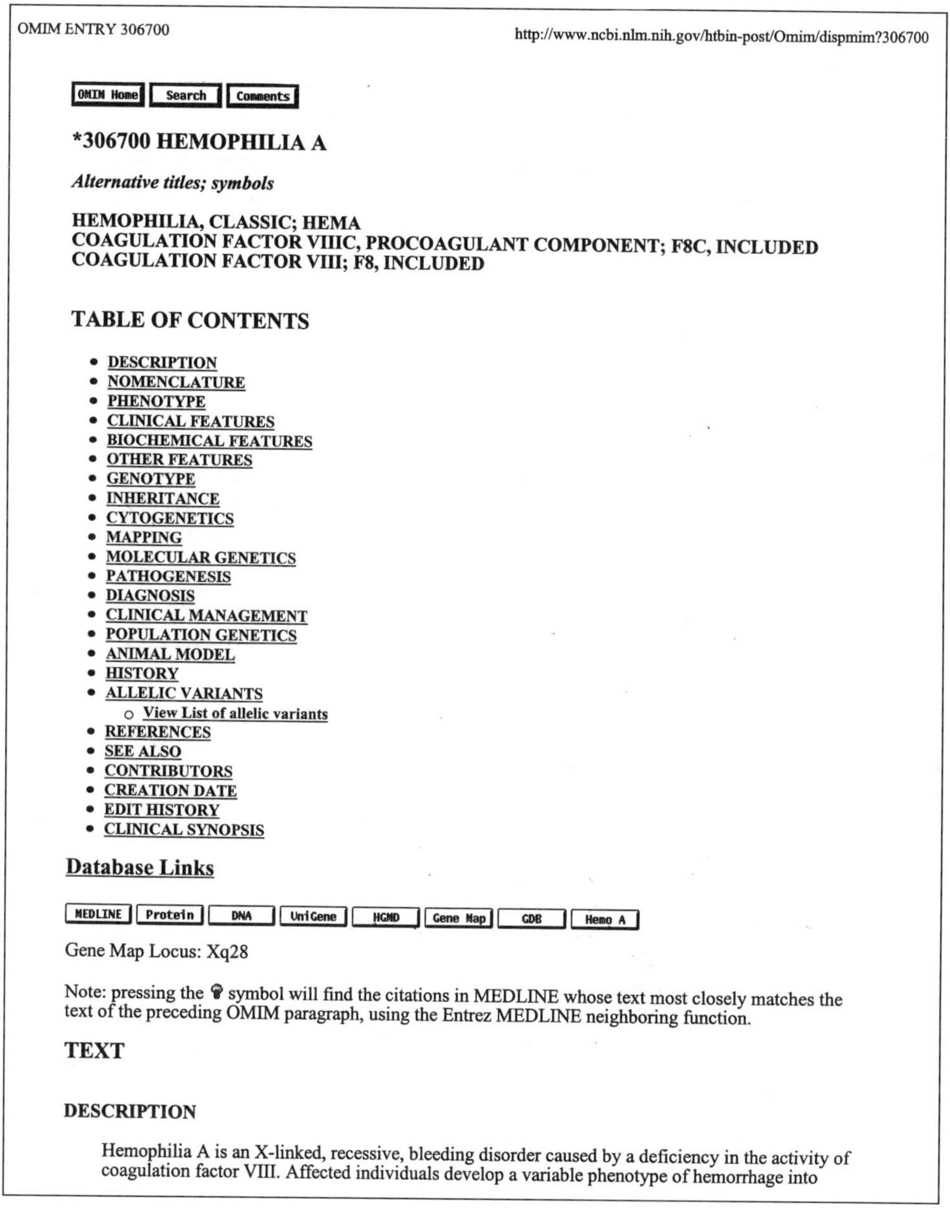

| OMIM Home | Search | Comments |

*306700 HEMOPHILIA A

Alternative titles; symbols

HEMOPHILIA, CLASSIC; HEMA
COAGULATION FACTOR VIIIC, PROCOAGULANT COMPONENT; F8C, INCLUDED
COAGULATION FACTOR VIII; F8, INCLUDED

TABLE OF CONTENTS

- DESCRIPTION
- NOMENCLATURE
- PHENOTYPE
- CLINICAL FEATURES
- BIOCHEMICAL FEATURES
- OTHER FEATURES
- GENOTYPE
- INHERITANCE
- CYTOGENETICS
- MAPPING
- MOLECULAR GENETICS
- PATHOGENESIS
- DIAGNOSIS
- CLINICAL MANAGEMENT
- POPULATION GENETICS
- ANIMAL MODEL
- HISTORY
- ALLELIC VARIANTS
  - View List of allelic variants
- REFERENCES
- SEE ALSO
- CONTRIBUTORS
- CREATION DATE
- EDIT HISTORY
- CLINICAL SYNOPSIS

Database Links

| MEDLINE | Protein | DNA | UniGene | HGMD | Gene Map | GDB | Hemo A |

Gene Map Locus: Xq28

Note: pressing the 🏮 symbol will find the citations in MEDLINE whose text most closely matches the text of the preceding OMIM paragraph, using the Entrez MEDLINE neighboring function.

TEXT

DESCRIPTION

Hemophilia A is an X-linked, recessive, bleeding disorder caused by a deficiency in the activity of coagulation factor VIII. Affected individuals develop a variable phenotype of hemorrhage into

Figure 15.1 *Sample OMIM Web page.*

tographs, videos, sound, and links to an image archive, as well as links to protein and nucleic acid sequence databases, and MEDLINE articles. In the future, this information will be expanded to include clinical photographs, sound movies, gene maps, diagrams of metabolic pathways, and gene structure. OMIM is also available for a fee on compact disc (MIM-CD™) through Johns Hopkins University Press, a consideration if your Internet access is limited. It is also possible to download the database from the OMIM site.

Tips for Using OMIM

- Take the time to read the OMIM home page to learn about the numbering system. Each entry is given a unique six-digit number, a **McKusick number**, and the first digit indicates the mode of inheritance. The presence or absence of an asterisk (*) or number symbol (#) before an entry number provides information about the locus/phenotype/gene relationship. The number system also designates allelic variants.
- To search for a specific genetic condition, enter the name of the condition.
- To establish differential diagnoses, enter keywords or traits and a list of conditions will be generated.
- If prenatal diagnosis is available for a condition, it is likely to be mentioned in the molecular genetics or diagnosis section. Use a word search of "prenatal" within the entry to find this information. The absence of a specific mention to prenatal diagnosis does not mean that none is available, since identification of the DNA mutation or known linkage may imply this capability of prenatal diagnosis.
- Check the "last updated" section to confirm when information about a condition was last revised.

Genline, which was under development at the time this book was written, contains comprehensive clinical descriptions, diagnostic criteria, management counseling and testing issues, molecular laboratories, and genetic support group links for specific genetic conditions. It is maintained by certified genetics professionals affiliated with the University of Washington. The disease profiles are expert-authored and peer-reviewed. In the future, each profile will have its own electronic commentary section and electronic bulletin board. Health care and genetics professionals who are registered users can provide feedback on errors, suggest content modifications, use the site as a discussion forum, post announcements, and share information on resources or research relating to a disease.

Commercial Software Resources

A number of commercial software and accompanying digital photographic databases are also available. All the programs mentioned here have a list of clinical features that can be searched to generate a list of matching genetic syndromes. These comprehensive programs are particularly helpful if you are seeking more information about a specific syndrome or diagnostic possibilities for a patient with complex, multisystem medical findings. Using these programs can help narrow down the diagnostic possibilities or, sometimes, broaden them. You can read about a genetic condition at different ages, which may be of assistance in diagnosis and helpful when you counsel patients regarding their concerns about how a genetic condition changes over time. Since many international cases and literature citations are incorporated, these programs can also expand your available information. These programs are continuously updated, and new versions of the programs are issued when enough updates are accumulated.

Pictures of Standard Syndromes and Undiagnosed Malformations (POSSUM) contains descriptions of more than 2500 single-gene, chromosomal, skeletal, and complex genetic syndromes compiled from the medical literature as well as from patients seen in the Genetics Clinic at the Murdoch Institute in Victoria, Australia. This database utilizes computer software coupled with a compatible laser disc. Developed in 1987 by two geneticists, Dr. Agnes Bankier and Dr. David Danks, it is available as a CD-ROM laser disc that also incorporates a skeletal dysplasia database named OSSUM (a reference to the osteogenic nature of skeletal dysplasias, with similarity to the name POSSUM). A specific entry on a genetic condition will generally provide several cases, at different ages, from which you can view photographs, X-rays, and diagnostic information. Video clips and histopathology photographs may also be included, depending on the condition. Each syndrome description includes references and commentaries and, where appropriate, is cross-referenced to McKusick's number and OMIM.

To use POSSUM to locate information about a known genetic condition, enter the specific condition. If you want to generate differential diagnoses for a patient with multiple congenital anomalies, begin by selecting the clinical features of interest from the list of features. The list of features includes all body systems in addition to life span issues. You are likely to obtain the most valuable search results if you select the most unique features and are as specific as possible. If the search returns a limited list of genetic syndromes, you can broaden the search (i.e., change "ventricular septal defect" to "congenital heart defects"). The list of syndromes, displayed as a result of your search, can then be selected individually for more in-depth information.

The online version, available for a fee, eliminates many of the memory constraints of the software version. It combines the POSSUM and OSSUM databases and list of features, as well as expands case commentaries and bibliographies. Links to OMIM are included.

Tips for Using POSSUM

- If a patient has multiple congenital anomalies, and chromosome studies are known to be normal, select "X" for "chromosomes" to eliminate chromosome anomalies in the list generated from the search.
- Select different ages for a particular condition to view a variety of clinical features.
- A separate laser disc, needed to view photographs, X-rays, and video clips is not standard for most computer systems.
- The online version can be run under Windows or on a Macintosh.

The London Dysmorphology Database (LDDB) includes more than 3000 nonchromosomal, multiple congenital anomaly syndromes. Clinical geneticists Dr. Michael Baraitser and Dr. Robin Winter initiated the database in 1984. The LDDB can be used to review clinical features of a particular syndrome or generate differential diagnoses. A detailed abstract and comprehensive reference list is available for each syndrome. The **Dysmorphology Photo Library** on CD-ROM contains over 10,000 clinical photographs and radiographs to illustrate syndromes in the dysmorphology database.

For neurogenetic conditions, a separate database by the same authors, the **London Neurogenetics Database (LNDB)**, has been available since 1988. The LNDB, which is compatible with the LDDB and Dysmorphology Photo Library, includes more than 3000 syndromes involving the central and peripheral nervous system and has information on neuroradiology, neurophysiology, and neuropathology (including nerve and muscle biopsy findings) along with references.

All the London databases have similar, user-friendly interfaces. Depending on your needs, there are two approaches for searching the databases. You can choose a specific genetic condition from an indexed master list, or you can search from a list of clinical anomalies. The latter searches can be performed using broad categories, such as "hands," or more specific features, such as "fingers, camptodactyly." There are a variety of ways to refine a search including using different selection criteria and changing the parameters on how many features are involved in a search.

Tips for Using London Databases

- Note the placement of the various clinical features by body systems. Some experience will enable you to find specific features using the author's reference system and spelling variations (e.g., *oesophagus* in the *abdomen,* or "speech delay" under *voice,* not *neurology*).
- Search for syndromes using a combination of features and levels of detail (from general to specific anomaly).
- Refined searches can be constructed, by excluding specific features, or by requesting a specific number of features from a list (e.g., find 4 of 8 features listed).
- Syndromes can be searched by chromosome location, inheritance pattern, or McKusick number.

PRIVATE DATABASES

Databases compiled by a single professional or developed for a specific professional interest may not yet be available through the World Wide Web. However, the individual responsible for the database is usually available for consultation and collaboration with genetic counselors. Examples of these databases include the Rare Disorder Network, a database that compiles information on families and laboratories for research on rare conditions, and the Metabolic Information Network, a database of metabolic conditions. Databases such as these may become accessible online in the future.

CYTOGENETIC DATABASES

Cytogenetic databases can be used to conduct a comprehensive review of the literature on a specific rearrangement, refine a prognosis for a patient, perform a risk assessment, and potentially locate other geneticists who have seen patients with similar chromosome anomalies. These databases can be particularly helpful when a patient has a rare chromosomal deletion, duplication, or rearrangement, since it is often difficult to find information about similar clinical cases in the medical literature. For cases involving a translocation or inversion, these databases can assist with determining the empiric risks for balanced and unbalanced offspring and for pregnancy loss. Some regional and national groups have compiled databases of chromosome anomalies. As these and other new databases are put online, they can be located through colleagues, or via Internet searches using keywords such as *cytogenetic, database, aneuploidy,* and/or *translocation.*

Since 1974, Digamber S. Borgaonkar, Ph.D., a cytogeneticist, has maintained an extensive database of international medical literature on chromosomal conditions, anomalies, and variants, which is published as *Chromosomal Variation in Man: A Catalog of Chromosomal Variants and Anomalies.* Purchase of this catalog includes 12 months of access to the online database, renewable annually for a fee. Chromosome anomalies at the major bands of each chromosome are described, as well as the resulting clinical features. Dr. Borgaonkar also has established a repository/registry of cytogenetic cases contributed by cytogeneticists. This resource, *Repository of Human Chromosomal Variants and Anomalies: An International Registry of Abnormal Karyotypes,* is in development for online use.

Another database, the **Human Cytogenetics Database**, was developed by clinical geneticist and cytogeneticist Albert Schinzel, M.D. It includes cytogenetic data on more than 1000 chromosome anomalies. This database describes clinical features, gives prognostic and recurrence risk information, and provides references. This database can be purchased as a software package.

GENETIC TESTING LABORATORIES

HELIX, an online international directory of clinical and research genetic laboratories, is an invaluable resource for locating laboratories offering genetic testing. This database maintains up-to-date information on genetic testing for over 500 conditions from over 300 laboratories. Only registered users have access to this private database, overseen by a board-certified clinical geneticist, Roberta Pagon, M.D., who is affiliated with the University of Washington. At publication, HELIX searches were free and supported by the National Library of Medicine.

TERATOGEN DATABASES

Access to a computerized teratogen database is invaluable if you handle inquiries from health professionals and patients about the impact of medications, environmental agents, X-rays, or viruses during pregnancy. These databases are compiled from a comprehensive review of the literature and can provide rapid access to teratogen information. These databases have medical advisory boards, are regularly updated, and include extensive resources.

Reproductive Hazard Information, Environmental Impact on Human Reproduction and Development (the **REPROTOX®** System) provides information regarding the impact of physical and chemical agents on human reproduction and development. It includes reviews of medications, recreational drugs, nutritional supplements, chemicals, and physical and biological agents on reproduction. The data presented cover all aspects of reproduction including fertility, male exposures, pregnancy, and lactation. Dr. Sergio Fabro and Dr. Anthony R. Scialli of the Reproductive Toxicology Center (Washington, D.C.) developed REPROTOX in the early 1980s. A disc, updated quarterly, can be purchased, and online access is also available for a fee. A demonstration disc is available, or a demonstration session can be reviewed online for free.

Teratogen Information System (TERIS) is a database developed by the University of Washington, TERIS PROJECT, Seattle, and has been available since 1985. It includes current information on the teratogenic effects of more than 700 drugs and environmental agents; summaries are based on clinical, epidemiological, and experimental data. TERIS includes data on teratogenicity, transplacental carcinogenesis, embryonic or fetal death, and fetal perinatal pharmacologic effects of environmental agents. Each agent summary in TERIS is based on a review of published data identified through a Toxline bibliographic record. The summaries rate the reproductive risk and explain data used to arrive at the ratings. Software, updated quarterly, can be purchased, and online access is available for a fee.

Tips for Using TERIS

- Agent summaries can be retrieved by domestic (United States), international, generic, or proprietary names.
- The computer search is based on the letters entered. If you are unsure of spelling or are unable to find the exact term, type in only the letters of which you are certain (e.g., *allyl* for *allyl alcohol,* or *aceto* for *acetonitrile*).

The **REPROTEXT**® System is a database containing information on acute and chronic exposure to over 600 industrial chemicals and physical agents and their carcinogenic, genetic, and reproductive effects. Its hazard ranking system consists of a numeric scale for general toxicity hazards and a "report card" scale for reproductive hazards. The hazard rankings from both of these systems can be combined with relative or absolute exposure estimates to obtain relative risk profiles, either for general toxicity or reproductive hazard guidelines. REPROTEXT was developed in 1987 by Betty J. Dabney, Ph.D., a biochemist with toxicology and genetics experience, and is updated regularly.

REPRORISK® is a compilation of teratogen databases that includes REPROTOX, TERIS, REPROTEXT, and Shepard's catalog. It is available as a CD-ROM or sometimes as an option on SilverPlatter (databases of professional reference information) at many medical center libraries.

The **Physicians' Desk Reference**® **(PDR)** is a source of written and pictorial drug information that is provided by pharmaceutical manufacturers. The written material for a given drug is a compilation of data and recommendations identical to those in the drug's package insert. The wording and directives included represent information that the pharmaceutical companies are permitted to present following discussion and approval by the Food and Drug Administration (FDA). PDR is available online through individual registration or arrangements with medical school libraries.

PEDIGREE-DRAWING PROGRAMS

A family history may vary from a brief, hand-drawn sketch obtained during a clinic visit to a large, complex diagram that requires numerous updates. Pedigree-drawing software programs can aid you in a variety of clinical and research applications, and the printout can be used for publications. The advantages of a computerized pedigree include clarity in presentation of family history information, ease of updating, and ability to print copies as needed. Many pedigree-drawing software programs have been developed by clinicians, epidemiologists, and researchers. Some of these are relatively simple and are available as shareware, free or for a nominal fee. Demonstration discs are generally available for free. Pedigree programs available at Internet sites may need to be expanded from their space-saving compressed versions. To find pedigree-drawing programs, and currently available downloadable demos, search the Internet using the keywords: pedigree + genetics + software.

Many pedigree-drawing programs are relatively expensive, primarily because of the integrated segregation and linkage analysis software included. Most pedigree-drawing programs can adequately accommodate the average genetics clinic patient's family, since many were adapted from software initially developed to keep track of extensive animal research pedigrees. It is generally possible to print pedigrees in portrait or landscape mode, single page or multipage, and to select different font sizes; color pedigrees may also be an option. Some programs allow you to print only a subset of the family or to print versions without names, a useful feature when confidential family information is to be published or otherwise shared. Pedigree software programs vary in the way family member information and other features are entered.

Some considerations for selecting pedigree software include the following:

• Computer system limitations, such as platform and capabilities.
• Ease of entering and updating family data; some programs require you to keep track of an individual's "pedigree number" to find a specific person.
• Automatic redrawing of pedigree as you are working, to avoid having to request this step each time a change is made.
• Automatic provision of last date pedigree was updated.
• Number of family members and data fields (names, diagnostic codes, demographic data, marriage information, twin data, comment field, etc.) the pedigree can accommodate.
• Wide selection of symbols with ability to create new symbols or add text.
• Ability to record multiple spouses, marriages within and between generations, marriage dates.

- Ability to indicate multiple pregnancies, order sibship by birthdates, or indicate "no children."
- Ability to selectively print specific branches of the family and/or data fields for specific family members and to generate pedigrees without displaying spouses.

Cyrillic is a pedigree drawing program. To compose a pedigree, you first enter the symbol for an individual. Then siblings, spouses, children, parents, and other relatives are added by means of a user-friendly format. The output can be printed without names, and in circular or other formats. An e-mail discussion group is available for problem solving and troubleshooting.

Genetic Analysis Package (GAP) is a software program for management and analysis of pedigree data, adapted from an older version named Kindred. The pedigree-drawing program is very comprehensive, with easy entry of pedigree data, color printing options, and comprehensive options for a variety of features related to marriage (adding marriage date, marriage between generations, multiple partners, etc.).

PediDraw is one of the few pedigree-drawing programs designed for Macintosh computers. It was developed by a Population Genetics Laboratory.

Prodigy, a program that allows data to be stored in a relational database, has a great amount of flexibility in managing and manipulating data. It is available both in Windows and Macintosh versions. The TreeSearch feature allows you to create a report sorting the data in the database in various ways. For example, you can ask for all individuals with colon cancer over age 50. The pedigree data can be color-coded.

INTERNET RESOURCES

The Internet connects computer users throughout the world to universities and professional resources. The Internet is accessible from most public libraries, schools, offices, and many homes. Patients often bring printed information from an Internet resource to genetic counseling sessions. The quality of this information varies from highly technical research articles to anecdotal information a family has posted on a personal home page, newsgroup, or other public area. Distinguishing between peer-reviewed professional information and subjective individual opinions may be difficult for users. Obtaining information from the Internet is like entering a library but without the structure of publishing guidelines, peer review, indexing and cataloging standards. Therefore, the ability to judge the quality of information obtained from the Internet is an important skill to develop. In assessing a Web site, it is helpful to determine whether the site is monitored by medical advisors or a scientific advisory board and whether the site is regularly updated. Most Internet browsers allow you to create "bookmarks" for sites you would like to access regularly, so that you can select them from a list without having to enter an Internet address each time.

There are many professionally developed Internet resources that are pertinent for genetic professionals. In addition to the databases discussed above, counseling aids, support group information, clinical genetics lectures, and other resources are available on the Internet. Through the Internet, it is possible to access resources from medical libraries whose catalogs are accessible online. A key resource for efficiently searching the medical literature on a topic is **MEDLINE**.

MEDLINE is an online index to medical literature. It can be accessed through PubMed, a search program designed by the National Library of Medicine (NLM). Searches can be performed by subject, language, type of publication, date, age group, or other criteria. Review articles can also be specified.

Tips for Using MEDLINE

- A search for a specific genetic condition can be limited or broadened by providing such keywords as genetics, molecular genetics, genetic testing, or genetic counseling.
- Searches can be narrowed by specifying language of publication.
- Searches can also be narrowed by specifying years. When specifying a time period, keep in mind that depending on the genetic condition, the more complete articles may have been written several years ago.
- Take the time to learn how to use the program efficiently. For example, there is a Scan command that allows you to view several references simultaneously.

To locate information on the World Wide Web, you need some familiarity with the various ways information is organized and the tools available to search for this information. Since standardization in indexing information is limited, directories (lists of resources by topic) or search engines are invaluable. Each search tool provides instructions for its use. Even if you are familiar with a particular search tool, read the help section and Frequently Asked Questions (FAQ) for that tool on a regular basis, since search features change constantly. There is no single comprehensive search tool for the WWW, and therefore you should use more than one search engine. Most search engines (e.g., AltaVista, HotBot, Lycos, Infoseek) index their databases of sites using a software program called a robot or spider. These programs roam the WWW, processing huge amounts of information and compiling results into a database accessible to a search engine at a specific WWW site. A few thousand to more than a million WWW sites or pages may be included, depending on the search engine's database and the quality of the indexing. Each search engine will produce different search results because of the differences in indexing information. Every engine indexes at least the uniform resource locator (URL) and title of the WWW sites in its database. A few good sites may be all you need, since one good site will often lead to other sites relevant to your topic. Information will not be retrieved if a site is protected by electronic security measures, such as firewalls. These sites generally prohibit robot access, and to use them you must first register and receive a login password.

All search engines provide a link to the sites retrieved. Most display the sites in order of relevance. The information you seek may be found in one of the first sites listed in the search, or it may be further down the list. Its location will depend on the keywords used, the way you requested the search, and the way the information on the page is indexed by a search engine. Additional information may be included, such as the first few lines of text, the most frequently mentioned words, or even the full text of every WWW site included. The presence of, or the length of, a resource description can vary widely among search engines. Many of these programs allow you to set the length of the description you wish to receive. Some sites are also available in multiple languages and are accessible for the visually impaired through speech translation programs.

It is important to keep in mind that new World Wide Web resources are added or updated daily as software and servers are upgraded, or as institutional and personnel changes are made. Resultant changes in the Internet locations (URL) means that you may need to utilize search engines and other mechanisms to locate and relocate resources.

Tips for Using Search Engines

- In searching, use the most unique or unusual word from your topic.
- Search for phrases using quote marks; otherwise the terms are searched for individually. For example, a search for *genetic counseling* will return thousands of WWW pages, those with either the word *genetic* OR *counseling*. A search for *"genetic counseling"* will return pages only if they contain the complete *phrase*.
- Be aware of spelling errors and variations (e.g., color/colour), and use synonyms.

Helpful Features to Consider When Selecting a Search Engine

- Allows for Boolean operators, a group of small, simple words (or symbols) used to define the relationships between the words in your query such as AND (+), OR, NOT (–), or other operators. This feature is useful when your search yielded too much, or too little, information. For example:
  - *human* AND *genetics* would yield pages containing both these words in any order anywhere on the page.
  - *human* OR *genetics* would yield all pages containing either of these words.
  - *genetics* NOT *human* would yield all pages with the first term and excluding any pages with the second.
- Allows for searches for words side by side (adjacency), words in same sentence of field (proximity), or by word roots (truncation). For example, *gene* would yield pages with genes, genetics, genetic counseling, etc.
- Allows the searcher to determine the importance of different keywords and search for Web pages, excluding specific terms.
- Allows for specific field searches (such as within the URL or title), date, media type, and/or location. For example, using the term *gene* would yield pages having this word in the URL address or in the Web page title.
- Allows "plain English" (stating a request in a sentence, which the search engine then uses to interpret what is needed), word order, case-sensitive, and field-based searching.
- Includes searches of Web sites, ftp and gopher sites, newsgroup articles, and Telnet sites, e-mail addresses, and electronic news headlines.
- Allows searches in languages other than English.

Support Groups and Information About Genetic Conditions

A variety of online resources are available which provide information on specific genetic conditions. These resources target patients, health care professionals, educators, or the general public. Individual organizations approach their online listings from their own goals and missions, so the content varies, but they often include information regarding clinical features, management and treatment, medical advances, references to professional or lay literature, and community resources (voluntary organizations/support groups). Support groups often provide the full published text of their brochures on their home pages. Generally, resources for a specific genetic condition can be accessed by searching under the name of the genetic condition (e.g., cystic fibrosis). Other resources contain information on categories of genetic conditions: for example, cancer (Cancer Family Alliance) or deafness (The Hereditary Hearing Impairment Resource Registry). A good starting point is **Genetic Conditions and Rare Conditions Information** Web page, which links users to support groups and educational and research resources. Local, regional, national, and international not-for-profit and private organizations list patient support group information. A brief list is included in the chapter appendix.

Professional Genetic Organizations

Many of the professional genetic organizations have a home page (see the appendix to Chapter 13 for URLs). Home pages generally provide information about membership requirements, focus, certification and accreditation, public policy statements, career options, conferences, and other pertinent topics. Membership directories may also be accessible from the home page, facilitating the location of colleagues and genetic centers close to where patients reside. In addition, there are several online discussion groups or listservs available for members or for a specific subgroup interested in a particular area of clinical genetics.

Information for Genetic Professionals is a World Wide Web home page coordinated by a board-certified genetic counselor at the University of Kansas Medical Center. This site is updated continuously with input from international genetic resources to provide information and links to resources for patient care, research, education, or other professional activities. Links are provided to clinical genetic centers, genome centers, international genetic organizations, and sites discussing genetic ethical and legal issues. The information presented in this chapter, in addition to information about other computer resources, is available at this site.

FUNDING AGENCIES

The National Institutes of Health (NIH), the U.S. Department of Energy (DOE), the National Science Foundation (NSF), and other federal and private agencies support genetic research, birth defects prevention, education initiatives, and disease-specific projects. Their grant announcements, requirements, applications, budget forms, and other publications are on the World Wide Web, in addition to an enormous amount of professional, consumer, education, and resource information. You can utilize a search engine to locate lists of available grants. To find an appropriate funding agency, you may need to broaden or, in other cases, refine your search. For example, genetic counseling issues could fit within the goals and priorities for funding through agencies concerned with special education or public health.

SUMMARY

The use of computer programs, databases, and Internet sites has rapidly become an integral part of preparing for genetic counseling cases and locating helpful resources for patients. Given the vast number of genetic conditions and the advances being made in molecular genetics, computer resources are critical for providing a rapid means of accessing up-to-date information. Although it takes time to become familiar with the computer resources applicable to your work as a genetic counselor, the investment is well worth the effort.

REFERENCES

Bankier A, Keith CG (1989) POSSUM, the microcomputer laser–videodisk syndrome identification system. *Ophthalm Paediatr Genet* 10(1):51–52.

Baraitser M, Winter RM (1991) *A Database of Genetically Determined Neurological Conditions for Clinicians.* Oxford: Oxford University Press.

McKusick VA (1994) *Mendelian Inheritance in Man. Catalogs of Human Genes and Genetic Disorders,* 11th ed. Baltimore: Johns Hopkins University Press.

Online Mendelian Inheritance in Man, OMIM™ (1996) Center for Medical Genetics, Johns Hopkins University (Baltimore, MD) and National Center for Biotechnology Information, National Library of Medicine (Bethesda, MD), 1996.

URL: *http://www.ncbi.nlm.nih.gov/omim/*

Winter RM, Baraitser M (1990) *The London Dysmorphology Database: A Computerised Database for the Diagnosis of Rare Dysmorphic Syndromes.* Oxford: Oxford University Press.

Appendix 15.1

Genetics Computer-Based Resources

Clinical Diagnostic Resources

Genline: University of Washington, Division of Medical Genetics, CH-25, Children's Hospital and Medical Center, Seattle, WA 98105-0317 (206) 526-2056
URL: *http://healthlinks.washington.edu/genline/*

London Dysmorphology Database (LDDB), Dysmorphology Photo Library, and London Neurogenetics Database (LNDB): Oxford University Press, 200 Madison Avenue, New York, NY 10016 (212) 726-6000
URL: *http://www.hgmp.mrc.ac.uk/DHMHD/lddb.html*
URL: *http://www.oup-usa.org/* or *http://www.oup.co.uk/*

OMIM (Online Mendelian Inheritance in Man): Johns Hopkins University School of Medicine, 725 North Wolfe Street, Baltimore, MD 21205 (301) 496-2475
e-mail: NCBI Help Desk *info@ncbi.nlm.nih.gov*
URL: *http://www.ncbi.nlm.nih.gov/omim/*

POSSUM (Pictures of Standard Syndromes and Undiagnosed Malformations): The Murdoch Institute for Research into Birth Defects, P.O. Box 1100 Parkville 3052 Melbourne, Victoria, Australia +61 3 9345-5045
URL: *http://murdoch.rch.unimelb.edu.au/possum.htm*

Cytogenetic Databases

Chromosomal Variation in Man: A Catalog of Chromosomal Variants and Anomalies: Christina Health Care System, 4755 Ogletown Stanton Road, P.O. Box 6001, Newark, DE 19718   (302) 733-3530
URL: *http://www.wiley.com/products/subject/life/borgaonkar/online.htm*

Human Cytogenetics Database: Oxford University Press, 200 Madison Avenue, New York, NY 10016   (212) 726-6000
URL: *http://www.oup-usa.org/* or *http://www.oup.co.uk/*

Genetic Testing Laboratories

HELIX: University of Washington, Division of Medical Genetics, CH-25, Children's Hospital and Medical Center, Seattle, WA 98105-0317   (206) 526-2056
URL: *http://healthlinks.washington.edu/helix/*

Teratogen Databases

Physicians' Desk Reference® (PDR): Medical Economics Company, 5 Paragon Drive, Montvale, NJ 07645   (888) 632-9998
URL: *http://www.medecinteractive.com/*

REPRORISK®: Micro Medics, 600 Grant Street, Denver, CO 80111   (800) 525-9083
URL: *http://www.mdx.com/po-rrsk.htm*

REPROTEXT®: Micro Medics, 600 Grant Street, Denver, CO 80111   (800) 525-9083
URL: *http://www.mdx.com/po-rtext.htm*

REPROTOX®: Reproductive Toxicology Center, Columbia Hospital for Women Medical Center, 2440 M St. NW, Suite 217, Washington, DC 20037-1404 (202) 293-5138 or (800) 525-9083
URL: *http://www.mdx.com/po-rtox.htm*

TERIS (Teratogen Information System): University of Washington, WJ-10, Box 38, Seattle, WA 98195   (206) 543-2465
URL: *http://weber.u.washington.edu/~terisweb/*

Pedigree-Drawing Programs

Cyrillic: Cherwell Scientific Publishing, The Magdalen Centre, Oxford Science Park, Oxford, England OX4 4GA   +44(0)865-784800; U.S. phone: (415) 852-0720
URL: *http://www.cherwell.com/ProdHome/cyrilhome.html*

Genetic Analysis Package (GAP): Epicenter Software, P.O. Box 90073, Pasadena, CA 91109 (818) 304-9487

URL: *http://icarus2.hsc.usc.edu/epicenter/gap.html*

PediDraw: Population Genetics Laboratory, Department of Genetics, P.O. Box 28147, Southwest Foundation for Biomedical Research, San Antonio, TX 78228-0147

e-mail: *paul@darwin.sfbr.org* OR *bdyke@darwin.sfbr.org*

Prodigy: Genetic Data Systems, 4702 Lincolnway E., Suite 200, Mishawaka, IN 46544 (219) 257-4000 or (800) PRODIGY

URL: *http://www.progeny2000.com/index.shtml*

Internet Resources

Alliance of Genetic Support Groups: A not-for-profit organization that maintains current information on over 300 genetic and rare conditions support groups. The Web page is regularly updated to provide accurate addresses, phone numbers, and other valuable information. Patients will find a printed brochure, list of addresses, phone numbers, and other ways to seek additional information, especially from others who share their concerns.

URL: *http://www.geneticalliance.org*

Genetic Conditions and Rare Conditions Information: Links users to support groups, educational and research resources.

URL: *http://www.kumc.edu/gec/support/groups.html*

Genetic Professional Organizations: Addresses and information on international genetics groups.

URL: *http://www.kumc.edu/gec/prof/soclist.html*

URL: *http://www.faseb.org/genetics*

Information for Genetic Professionals: clinical, research and educational resources for genetic counselors and geneticists.

URL: *http://www.kumc.edu/gec/geneinfo.html*

March of Dimes: information on a specific birth defect or infant health problem, prepregnancy, pregnancy, teen pregnancy, newborn care, drugs and environmental hazards during pregnancy (teratogens), support groups, genetic counseling and other genetic topics.

URL: *http://www.modimes.org*

MEDLINE: National Library of Medicine (NLM), Lister Hill National Center for Biomedical Communications, 8600 Rockville Pike, Building 38A, Room 707, Bethesda, MD 20894 (301) 496-4441

URL: *http://www.ncbi.nlm.nih.gov/PubMed/*

National Network of Libraries of Medicine includes resources, links to comprehensive health materials, advisory groups, and funding opportunities, links to National Library of Medicine (NLM) [*http://www.nlm.nih.gov/*], a large biomedical library, MEDLINE searches, health information, and references.

URL: *http://www.nnlm.nlm.nih.gov/*

NORD (National Organization for Rare Disorders, Inc.): education, advocacy, research, and service for voluntary health organizations serving people with rare disorders and disabilities.

URL: *http://www.pcnet.com/~orphan*

Appendix 15.2

Computer Terms

ASCII (American Standard Code for Information Interchange)   A set of characters that make up most of printed English, including capital and lowercase letters, numbers, punctuation marks, and other special symbols. Documents saved in ASCII usually can be utilized in most software or computer systems.

Browser (e.g., Netscape, Microsoft Internet Explorer)   A program that requests and interprets HTML documents on the Internet's World Wide Web and then displays them on your computer.

File Transfer Protocol (ftp)   Using the Internet, a standard that ensures the error-free transmission of program and data files. Ftp can be used in downloading programs and files regardless of make of computer, type of network, or operating system. Files may be text, executable programs, video, sound, or graphic.

Firewall   A combination of hardware and software that creates a "wall" for security purposes; can be used to protect confidential information on the Internet, or to restrict access to authorized users.

Gopher   A method of making menus of material available over the Internet. Although largely supplanted by Hypertext on the WWW (World Wide Web), there are still thousands of Gopher servers on the Internet and they probably will remain for a while.

Internet   A system of linked computer networks, worldwide, that facilitates data communication services such as remote logon, file transfer, electronic mail, and distributed newsgroups. Internet is not a computer network, but rather a way of connecting existing computer networks that extends the reach of each participating system, using the same protocols to share resources.

Keywords   A set of words to describe the content of a document.

Listserv   A type of mailing list; usually an automated system that allows people to send e-mail to one address, whereupon their message is copied and sent to all subscribers of the mail list.

Online Accessible via a computer (or terminal), rather than on paper or other medium.

SilverPlatter A company based in London that provides, by subscription, access to biomedical and other databases through libraries, with a user-friendly search engine.

Search engine One of several World Wide Web programs that uses keyword searches for information on the Internet. The search may cover titles of documents, URLs, headers, the full text, or other items. The user enters a search string [word, phrase, compound search query]. The interface is linked to a database, which includes information on Internet resources that have been accumulated by a software package, referred to variously as spiders, robots, or worms, which have downloaded documents/document titles/addresses, or some combination of these, from various hosts linked to the Internet. Search engines vary by the Internet resources their databases have indexed.

Spider (crawler, robot) A program that automatically explores the World Wide Web by retrieving a document and some or all of the links referenced in it.

Unzip a file Computer files often are compressed to save space. To extract these files from their compressed form, they must be "unzipped."

URL (Uniform Resource Locator) The standard way to give the address of any resource on the Internet that is part of the World Wide Web (WWW). The most common way to use a URL is to enter a browser program, such as Netscape or Lynx.

World Wide Web (WWW) An Internet system on which documents, graphics, video, sound, and other materials are represented in a format that different computers can interpret. Links between documents are simplified by hypertext links and URLs. A browser runs on the user's computer and provides navigational operations, either following links or sending a query to a server.

16

Putting It All Together: Three Case Examples

Vivian J. Weinblatt, Robin L. Bennett, and Elsa Reich

The preceding chapters have described the tasks, skills, and challenges inherent in genetic counseling. We have enormous responsibilities to our patients, to the institutions in which we work, to our profession, and to ourselves. The process of upholding these responsibilities while serving our patients comprises the "art" of genetic counseling. The means by which we make complex genetic concepts accessible to, and understood by, patients and their families is only the most obvious of our responsibilities. Respecting confidentiality, conducting psychosocial assessments, providing interventions, adhering to social, ethical, and legal standards, maintaining professionalism and respect for patient individuality—each of these, when woven together creates the unique "product" that we provide to patients and their families.

There is perhaps no better method to demonstrate the process of genetic counseling than by examining case summaries of interactions between a patient and genetic counselor. Observe, then, in the three cases presented in this chapter, how each genetic counselor "puts it all together" in the provision of quality care. The first case describes a family contemplating the use of prenatal diagnosis; the second demonstrates pedigree assessment, examines the impact of adult-onset disease, and notes the issues raised by presymptomatic diagnosis; and the third explores the search for a diagnosis in a child with a progressively debilitating neuromuscular disease. Each of these cases demonstrates common approaches as well as individual methods and techniques. They are not necessarily intended to be examples of ideal genetic counseling scenarios, but rather examples of the dynamics of true cases. Your own techniques

A Guide to Genetic Counseling, Edited by Diane L. Baker, Jane L. Schuette, and Wendy R. Uhlmann.
ISBN 0-471-18541-8 (cloth), 0-471-18867-0 (paper). Copyright © 1998, Wiley-Liss, Inc.

and style will evolve over time, as an amalgamation of what you observe, read, and experience.

Case Example 1
Vivian J. Weinblatt

This prenatal genetics case involves a couple, their decision-making process, and their adjustment to the identification of a balanced translocation. The couple belonged to an Orthodox Jewish community, an environment rich in tradition and strongly dedicated to the value of family and childbearing. This case demonstrates the importance of not merely understanding but supporting the patient's cultural milieu, as well as the way in which culture impacts decision making.

Rachel was a 25-year-old woman with two healthy children. She presented for genetic counseling because her first cousin, Miriam, had recently had a child, Rebecca, who was small for gestational age and had a congenital heart defect. Rebecca's chromosome analysis revealed a reciprocal translocation between chromosomes 13 and 15, with the following karyotype: 47,XY+der(13)t (13;15)(q21.2;q13). Miriam was found to carry a balanced translocation from which her child's unbalanced rearrangement was derived (Figure 16.1). This would prove to have no significance for Miriam's health or longevity; it would, however, have considerable implications for her childbearing and for her children's childbearing, as well as implications for extended family members (Summitt, 1979; Crandall et al., 1980; Boue and Gallano, 1984; Daniel et al., 1988; Stene and Stengel-Rutkowski, 1988; Gardner and Sutherland, 1996).

During meiosis, there are several different outcomes for the gametes of a reciprocal translocation carrier (Figure 16.2). If the gamete receives either both the nontranslocated or both the translocated chromosomes, one would predict a pregnancy outcome devoid of a chromosomal sequelae. If, however, the gamete includes a different combination, a monosomy 13 and a trisomy 15 or the converse would be predicted. Additionally, as is the case for Rebecca, some translocations can result in 3:1 segregation, producing partial trisomy for both the chromosomes involved in the translocation (Stene and Stengel-Rutkowski, 1988). Many of these conceptuses would be expected to miscarry before viability, but, as demonstrated by the mode of ascertainment in this family, live-born babies with significant health problems can also be the product of an unbalanced translocation (Daniel et al., 1988; Stene and Stengel-Rutkowski, 1988; Gardner and Sutherland, 1996). In fact, the mode of ascertainment aids the genetic counselor in the prediction of recurrence risks for future pregnancies. When a family is ascertained by the birth of a baby with birth defects due to an unbalanced translocation, the predicted risk for recurrence is higher than it would have been for ascertainment by chance or because of multiple pregnancy losses. Many translocations result in pregnancies too anomalous to continue to term. Ironically, the greatest risk for a liveborn child with an unbalanced translocation occurs with translocations involving smaller chromosome segments. This is because when large segments of the chromosomes are rearranged, an unbalanced conception is likely to be compromised so griev-

t(13;15)(q21.2; q13)

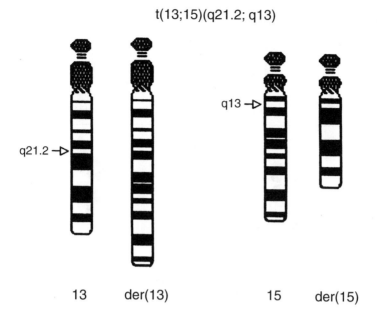

Figure 16.1 Balanced translocation identified in Miriam.

ously that the fetus would be extremely unlikely to survive early pregnancy. Conversely, when only small regions of the chromosome are affected, the fetus is more likely to survive to viability (Petrosky and Borgaonkar, 1984; Davis et al., 1985; Stene, 1989; Gardner and Sutherland, 1996).

Families who are found to carry balanced translocations have several options for addressing the recurrence risk associated with this finding. For families who wish to virtually eliminate the risk of an unbalanced pregnancy, donor gametes may be used (Brambati et al., 1994; Howard-Peebles, 1996). This procedure is simplest when the father is the carrier. Artificial insemination with donor sperm is considerably less costly and is procedurally easier than the use of donor eggs with in vitro fertilization and embryo transfer into the uterus of a carrier mother. For families who prefer natural fertilization, prenatal diagnosis by chorionic villus sampling (CVS) or amniocentesis will detect most chromosome translocations (Boue, 1979; Boue and Gallano, 1984; Pergament and Verlinsky, 1986; Jackson et al., 1992). Fluorescent in situ hybridization (FISH), molecular microdissection techniques, and chromosome painting used in conjunction with CVS or amniocentesis are useful in the detection of small translocations (Cantu et al., 1992; Powell et al., 1994; Scott et al., 1995; Engelen et al., 1996).

Clearly, when an unbalanced translocation is detected prenatally, options are limited. Providing a specific prognosis for an affected fetus is difficult and must be based on either a known proband with the same derivative chromosome or reports of similar chromosome abnormalities (deGrouchy and Turleau, 1984; Gardner and

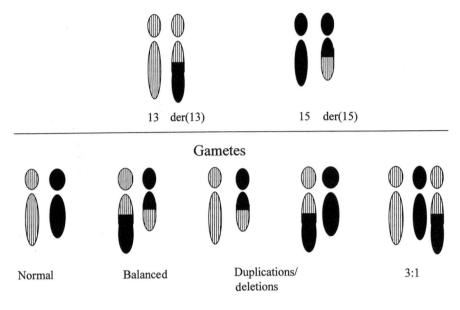

Figure 16.2 *Some of the possible chromosomal combinations in gametes of the t(13;15) carrier.*

Sutherland, 1996; Borgaonkar, 1994). Families will typically choose between continuation and termination of the pregnancy. The option of placing a child with severe disabilities up for adoption is one that some families will wish to explore.

At the time that the diagnosis of a familial translocation was coming to light, Rachel was early in her sixth pregnancy. Her obstetrician, concerned about her family history and previous pregnancy losses, referred Rachel for genetic counseling. From a detailed family history, it became clear that Rachel's history of pregnancy loss raised concern that she might also carry a translocation (Figure 16.3). Rachel reported five pregnancies prior to the current one, two of which resulted in the birth of a normal child: a son and a daughter. She also reported three miscarriages, two of which occurred prior to 12 weeks. The third loss occurred at 22 weeks. Rachel had always felt bad about these events, but because her mother and sister had also experienced miscarriages she felt that such losses were somehow expected in her family. As a result, she had refrained from sharing news of her pregnancies with friends and family until she completed her first trimester. Rachel worried that news of the losses could label or stigmatize her, in the small, intimate community in which she lived. The most recent loss at 22 weeks was the most unsettling. She had learned of the death of the fetus during a routine ultrasound exam. The late timing of the loss, as well as the need for a surgical procedure (dilation and evacuation), was significantly more traumatic for Rachel and her husband, Samuel, than the earlier losses had been. Moreover, Rachel's extended family members were aware of the pregnancy and had to be told of the fetal demise.

Being extremely anxious about the possibility that Rachel might also be carrying this translocation, the couple decided to pursue karyotyping, the results of

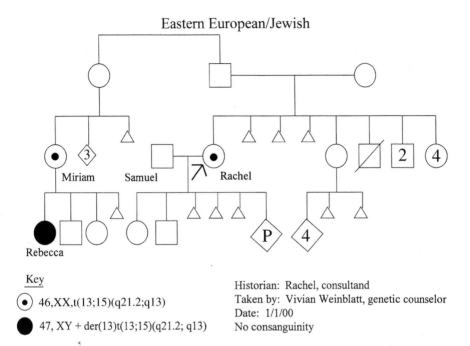

Figure 16.3   *Rachel's family history.*

which confirmed that Rachel was a translocation carrier. This diagnosis came as an acute blow to Rachel's self-esteem; she felt both guilt that she might have passed this chromosome abnormality on to her children and shame that she was somehow "tainted" by this translocation. She also expressed great concern about the potential impact of the information on the marriageability of her children.

Rachel and Samuel are members of an Orthodox Jewish community, where bearing children is extremely important and valued (Gold, 1988; Jakobovits, 1994; Ekstein, 1995). Pregnancy termination is very controversial among Orthodox Jews, with some rabbis forbidding termination under any circumstance and others considering the option, but only until 40 days postconception (Edelman et al., 1988; Gold, 1988). In light of these impediments to pregnancy termination, a prospective groom or bride and their families might hesitate to agree to a marriage involving a person with a translocation or even a history of multiple miscarriages because of the risk to offspring (Ekstein, 1995). Rachel exclaimed tearfully, "Who will marry my children, now? Our family will carry this stigma forever!"

The concern over stigmatization is so great in the Orthodox Jewish population that secrecy about disability is widespread (Ekstein, 1995). A unique, anonymous carrier screening program was pioneered as a response to this concern. Dor Yeshorim, a phrase taken from the Book of Psalms and meaning "upright generations," is a program created to introduce carrier screening to members of Orthodox Jewish communities without fostering the stigmatization of individuals and families found to carry mutations that occur with a higher incidence among people of

Jewish descent. Initially, the program offered screening for Tay–Sachs carrier status only; it has been expanded to include Gaucher disease, cystic fibrosis, and other conditions for which carrier testing is available based on individual family history. When an Orthodox couple considers marriage, they contact the program to determine whether the proposed pairing is "compatible." If both prospective spouses are carriers of the same genetic disease, the marriage is not forbidden, but some couples reconsider their engagement upon receiving such news. The particular genetic disease involved is not disclosed to the couple, protecting the privacy of the individuals and their families and reducing stigmatization (Vought et al., 1986; Ekstein, 1995). Dor Yeshorim has been enthusiastically received in the Orthodox Jewish community, and the number of disorders for which analysis is offered has increased as the ability to test for more conditions improves (Ekstein, 1996).

During the first meeting with Rachel and Samuel, we talked for some time about the randomness of genetic differences, and the lack of volition or responsibility when such a difference is discovered. Samuel was extremely supportive of Rachel, and although he expressed sadness about the results, he did not place blame. The couple decided to talk with their rabbi about the disturbing information, and discuss their options with regard to this and future pregnancies, as well as the implications for their two children.

Prior to Rachel's second visit to discuss prenatal diagnosis in further detail, she reported by telephone that she had lost her current pregnancy at approximately 30 days postconception, or just over 6 menstrual weeks. Interestingly, Rachel had come to view these losses somewhat differently in light of the diagnosis of the translocation. Paradoxically, a pregnancy that was lost through miscarriage became a somewhat reassuring event. She understood that a fetus with an unbalanced translocation was more likely to be lost than a fetus with either a balanced karyotype or no translocation at all. She began to view a loss as proof of her body's ability to "self-screen" a pregnancy and preserve only a "normal" one. She felt some struggle, in fact, when considering prenatal diagnosis because she wondered if it was really necessary. Perhaps her body was truly capable of "monitoring" itself without outside intervention. Ultimately, however, largely because of her history of a 22-week loss and because her cousin's daughter, Rebecca, had survived, she indicated that in a future pregnancy she would opt for the certainty of diagnosis afforded by prenatal testing. She did not wish to bring a multiply handicapped child into the world if such an outcome could be prevented.

Rachel discussed what she would do if she were to achieve another pregnancy. She reported that her rabbi had authorized her to undergo prenatal diagnosis by chorionic villus sampling. The rabbi also had allowed for pregnancy termination of an affected pregnancy, provided the appropriate test results could be obtained and the termination performed no later than 40 days postconception. We discussed the current controversy about CVS prior to 10 menstrual weeks (~56 days postconception), including a possible increased risk for pregnancy loss as well as the risk for limb reduction defects as a result of the procedure (theoretically greatest when CVS is performed early in the first trimester) (Brambati et al., 1992; Mastroiacovo et al., 1992; Firth et al., 1994; Froster and Jackson, 1996). Rachel understood these issues but was willing to accept the possibility of an increased procedure-related risk to preserve her option of pregnancy termination. Direct CVS results are typically available within 24–48 hours of the procedure but are somewhat less reliable than final results, which include analysis of cells obtained from tissue culture and are typically available 10–14 days after the procedure (Coleman et al., 1987; Heaton and Czepulkowski, 1987; Mikkelson, 1987). Rachel was clear about her desire to undergo early CVS and requested that our genetics division make an exception in her case and perform CVS prior to the 10-week cutoff.

This request posed an ethical dilemma. Because of Rachel's unique circumstances, including the importance of her culture and religion, I was unwilling to summarily dismiss her request. Instead, I requested a meeting with staff members of the genetics division to discuss the options. Was it ethical to deny this patient what she perceived as her only means to obtain a timely prenatal diagnosis? Was it ethical to perform a procedure when its safety and accuracy during the time frame desired had been called into question? What would the legal liability be in the event of a pregnancy loss, limb defect, or inaccurate test result in the pregnancy? I was playing the role of patient advocate, certainly not an unfamiliar role for a genetic counselor. In this case, however, I was not sure that I was in agreement with the patient's wishes. This dilemma highlights an important aspect of the genetic counseling process. It is not necessary to agree with a patient's request or decision; however if deemed reasonable, a request may be honored. It is also necessary to provide information and support during the decision-making process. Suppose, for example, that a family wants to terminate a pregnancy with an anomaly generally considered to be mild or repairable. The patient maintains the right to make that decision, regardless of what you might do were you facing the same choice. *You* are not making the decision and will not face the ramifications of its outcome. It does the family a disservice to disclose, either in words or manner, your personal opinion.

In the weeks that followed, multiple telephone contacts were made with Rachel and her rabbi (at her request). Ultimately, the prenatal diagnostic center agreed to perform the CVS before the 40th postconception day. This decision was based on the desire to respect the cultural milieu of the patient and to support the patient's decision making, since she had heard and understood the implications of this course of action. An addendum, noting the possibility of a significantly increased risk for complications, fetal anomaly, and/or incorrect diagnosis when CVS was performed prior to 10 weeks, was included on the consent form.

Three months later, Rachel called to report that she was again pregnant and to schedule an early CVS procedure. She was scheduled at 35 days postconception. At the time of the counseling prior to her procedure, the plan for reporting of results was discussed.

Rachel: My husband and I have decided that when you call with the results, we do not wish to know if the fetus is a balanced translocation carrier. We just want to know if the chromosomes are unbalanced or not.

Counselor: I am wondering what brought you to that decision.

Rachel: We want to try to avoid as much stigmatization as possible for our children. If we don't even know which ones are carriers, there should be fewer impediments to their eventual marriages.

Counselor: Your oldest child is 6. Has she asked about her cousin's mental retardation?

Rachel: We told her that Rebecca has a problem that can run in families. We plan to tell all our children about this risk before the process of finding them spouses begins. Then they will have to decide when to get their chromosomes tested—either before or after the wedding. We will also tell them about the 40-day time period, and have them discuss it with their rabbi if they so choose.

Counselor: How did your daughter respond to this news?

Rachel: I think she took it in stride. I made sure to tell her that even though a problem with a baby could happen, most of the time it doesn't, and gave her as examples herself and her brother. They are fine and healthy. I think this reassured her.

Counselor: It sounds like you gave her a very good explanation. She may come up with more questions as she grows. Children tend to process information gradually. What will it be like for you if the subject comes up again?

Rachel: I don't know. I guess I will have to just answer her questions as they come. I don't want to go over her head. Most adults don't understand about chromosomes, much less 6-year-olds.

Counselor: You're right, small pieces of information are probably the easiest to understand and process, but how will it be for you when she talks about the chance of being a carrier?

Rachel: I still feel pretty badly about it. . . . I know it's not my fault, but I still feel different, damaged, in some way. It is much easier, though, to know that there is something I can do to try to have healthy children. That is very comforting.

Counselor: I think it is very common to feel different upon learning that one carries a translocation or a gene that can cause problems. The ironic thing is that we all carry differences in our DNA. Some that we learn about in life, and some that we never uncover. No one is immune from this.

Rachel: Yes, I suppose so, but part of me wishes I never knew.

Counselor: What about the other part?

Rachel: The other part is glad to know, so that I can take steps to deal with it for the sake of my children and their children.

Rachel's CVS was performed without incident at 35 days postconception. Preliminary results, available 48 hours after the procedure, revealed a fetus with normal chromosomes; no translocation was identified. In accordance with her request, Rachel was informed only that the fetus did not have an unbalanced translocation.

It was clear that Rachel's sense of self had been altered by the diagnosis of the translocation. This is not uncommon among people facing the knowledge that there is something different about their genetic disposition (Kenen and Schmidt, 1978;

Skirton, 1995; Welkenhuysen et al., 1996). Rachel armed herself, however, with information and a plan of action that helped her regain some measure of control.

The case also raises an ethical dilemma regarding the availability of presymptomatic or carrier testing for children. Since the early days of karyotyping, parents were routinely informed of the karyotype of a child or a fetus when a familial translocation was known. Recently, with the availability of testing for Huntington disease, breast cancer susceptibility, and other adult-onset disorders, the recommendation has been to not test until majority if there is no medical benefit to be achieved by early testing: that is, defer childhood testing for these disorders and discourage prenatal testing if the family plans to continue the pregnancy in the event of a positive diagnosis [Bloch and Hayden, 1990; Sharpe, 1993; Wertz et al., 1994; Working Party of the Clinical Genetics Society (UK), 1994]. This approach has been suggested to protect children from stigmatization and from falling victim to a self-fulfilling prophesy, and to allow individuals to make competent decisions for themselves about whether they wish to be tested.

In Rachel's situation, I was initially uncomfortable withholding the information about whether her children were balanced translocation carriers. I was certain, however, that she would appropriately discuss these issues with her children in the future and would allow them to choose whether to undertake testing. Does the counselor's level of comfort with Rachel's decision imply that if a different patient wishes to learn which of her children are translocation carriers, the information should be withheld from her? There is no simple answer to this question. As genetic counselors, we must always examine the manner in which we do our work, to remain true to our patient's needs and to our professional Code of Ethics.

Over time, we have seen many of Rachel's family members. Two siblings chose karyotyping and prenatal testing. One brother and one sister, who were themselves translocation carriers, did not pursue prenatal testing. One additional sister chose not to undergo chromosome analysis. She felt that she would not be comfortable considering pregnancy termination under any circumstance. For her, positive test results would simply increase her anxiety during pregnancy, with no possible reassurance until the birth of the baby. Thus, families with strong cultural influences will still experience intra-and interfamilial differences regarding genetic testing.

Coincidentally, Rachel's cousin Miriam, as well as several of Miriam's siblings, have visited our offices to discuss and obtain early CVS testing. These families revealed knowledge of their relatives' consultations here. It is *extremely* important to preserve confidentiality. If a patient questions you about another person or asks for clarification about what may be known to the counselor, simply reply that it is inappropriate to discuss anyone else's situation and use the opportunity to explore what is behind the question. Ultimately, patients appreciate this response because it communicates that their information will likewise be kept confidential. Miriam, for example, desired information about balanced carrier status for her pregnancies. This was clearly a different decision from Rachel's. Miriam referred to Rachel's genetics visits, inquiring how many of Rachel's children were balanced translocation carriers. I did not share that Rachel preferred not to learn this information, and instead, asked Miriam why she wondered about her cousin's history. Miriam's reply made it clear that she was merely searching for reassurance that balanced carriers were normal and that some children carried no translocation at all. I was therefore able to address Miriam's concern without breaching Rachel's confidentiality. I reassured her that in the absence of an unbalanced translocation, the greatest likelihood would be the birth of normal child, who in some cases would carry no translocation and therefore would not be at an increased risk for passing the anomaly on to future generations.

This case taught me many things. I learned that advocating for a patient and concurring with her plan of action are not necessarily synonymous. Cultural differences demand more than tolerance. The counselor should educate herself to enhance her ability to accommodate individual needs. Over the course of the genetic counseling process, patients and families are educated as well. Rachel learned that for her, "knowing" a result is not always desired.

This counseling case highlights several issues:

1. The coping process of a person who has learned of the presence of a genetic alteration that poses a risk to pregnancy outcome.
2. The religion and culture of a family and its impact on decision making.
3. The importance of not making assumptions based on a patient's religion and culture. (People respond uniquely when confronting difficult choices.)
4. The evolution of a patient's view of herself after learning about the presence of genetic risk. (Rachel moved from feeling victimized by her situation to achieving a measure of control over it; she took the steps she deemed necessary to protect her family.)
5. The importance of recognizing that it is incumbent on us to reevaluate our policies in the context of an individual patient's needs. (This is not to say that every patient request should be satisfied, but all should at least be considered. Guidelines are necessary, but sometimes exceptions can be made.)
6. The decision-making related to the appropriateness of screening for genetic carrier status in children or in fetuses.

This case was a success, not because the patient was satisfied and received a normal result, but because she was supported in her cultural and religious milieu during her navigation through the medical environment. Rachel's experience exemplifies the trust that should be inherent in the relationship between patient and genetic counselor. The patient must be able to assume that the counselor will provide truthful, balanced information, support her individual differences and needs, and empower her to make a decision or choose a course of action. The counselor must, in turn, believe that when provided with the aforementioned support, the patient will make the decision or choose the course of action that is right for her.

Case Example 2
Robin L. Bennett

This case presents some of the issues that arise in providing genetic counseling for disorders that primarily have adult onset of symptoms. Julia M. was a 22-year-old pregnant medical student referred for genetic counseling because of a family history of mental retardation. Her sister had died at age 5 days, manifesting severe hypotonia and difficulties with feeding and respiration. This was a planned pregnancy for Julia and her husband John, age 23. They presented at 8 weeks' gestation. Julia and John were interested in knowing their chances for having a child with mental retardation or complications at birth, and options for prenatal testing.

Family History   The family's pedigree appears in Figure 16.4.
Julia had a 25-year-old brother, Michael, who had a healthy 5-year-old daughter. Michael was separated from his wife and not planning any more children. Julia

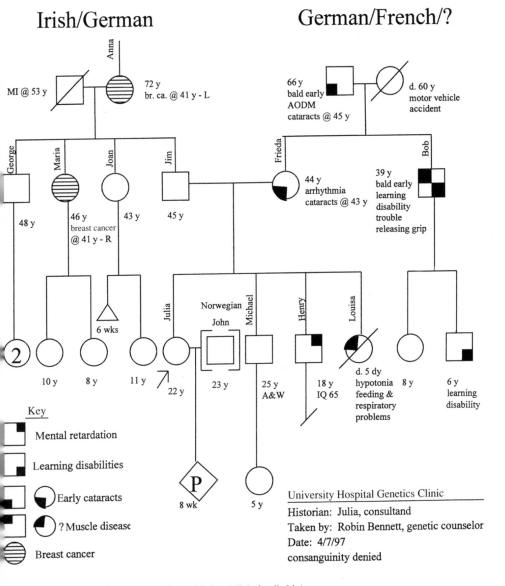

Figure 16.4   *Julia's family history.*

had a younger brother, Henry, who was 18 years old. He lived at home and had a job as a dishwasher at a sheltered workshop. Julia indicated that when Henry was evaluated for school placement, her parents were told his IQ was 65. Her sister, Louisa, had died at 5 days of age. Louisa apparently had problems that presented at birth, including very poor muscle tone. The infant also was unable to nurse and had to be ventilated. An autopsy was not performed.

Julia's 45-year-old father, Jim, was a computer analyst, and Julia stated that he had "never been sick a day in his life." Jim had a brother and two sisters. His brother, George, was 48 years old and had two daughters. His sister Maria was 46 and had two daughters. Maria had breast cancer in her right breast at age 40; she was doing well after a mastectomy and chemotherapy. The other sister, Joan, was 43 years old and had an 11-year-old daughter. Joan's first pregnancy had resulted in a miscarriage at 6 weeks of gestation. Jim's father had died of a myocardial infarction at age 53. His mother, however, remained healthy. Indeed, Julia stated that her paternal grandmother, Anna, at age 72, "wears me out, she is so active." It was interesting to note that Anna had had breast cancer at age 41, only in her left breast. When asked about the ethnic background of her paternal grandparents, Julia used the term "Heinz 57," although stating that they were mostly Irish and German.

Julia's mother, Frieda, was an attorney and was 44 years old. She was basically healthy, although she was taking medication for an irregular heartbeat. The year before, she had been diagnosed with cataracts. Frieda had only one sibling, a 39-year-old brother, Bob. He had recently left his construction job because of difficulty releasing his grip when he was climbing ladders. Julia laughed when she said "This is a real problem when you are two floors up a ladder!" Julia stated that this uncle had inherited the "Nelson curse of baldness," since Bob and his father were both bald by their mid-thirties. Bob had a healthy 8-year-old daughter. His son was 6 years old, had severe learning disabilities, and was in special classes at school. Julia remembered her mother saying that Uncle Bob also had had problems in school and had needed special education classes to graduate from high school. Frieda's father was still an active hiker at age 66, although he had been complaining of lacking the stamina he used to have. Julia's maternal grandfather had recently developed diabetes. He had early cataracts (at age 45), like Frieda. Frieda's mother, who died in a car accident at age 60, had been healthy up to that time. Julia's paternal ancestors were of German and French ancestry.

Julia did not believe that her parents' families were related. She laughed when she said "People always say my husband and I look like we could be brother and sister, but I guess I shouldn't laugh, since he is adopted and we don't know anything about his family, other than both of his parents apparently came from Norway."

Evaluation of the Family History   Julia's history of a brother with mental retardation and a maternal uncle and nephew with learning disabilities is suspicious, suggesting a genetic cause of mental retardation. At first glance, the inheritance pattern could be X-linked recessive or possibly multifactorial. It is also notable for a person to have a mother and grandfather with early cataracts. Julia's deceased baby sister had had severe hypotonia, and her maternal uncle and grandfather complained of muscle weakness. Is there a genetic disorder that would tie together all these medical conditions, or are there multiple, potentially genetic, conditions in this family? Could there be an inherited muscle disease in the family?

This family history suggests the possibility of myotonic muscular dystrophy. The features supporting this diagnosis include early cataracts, cardiac arrhythmia in Julia's mother, learning disabilities and mental retardation, the suggestion of myotonia in the maternal uncle, the presence of mental retardation and learning disabilities in Julia's brother and cousin, and early balding in males. Adult diabetes can also be a part of this form of muscular dystrophy, but it is also common among adults in the general population. In fact, except for myotonia, all these medical conditions are relatively common in the general population. This may explain why the family and their physicians had not suspected a pattern. Clearly, taking a complete family history and then considering the history as a whole (family history analysis) is extremely useful and important.

Factors consistent with the diagnosis of myotonic dystrophy in Julia's family include the following:

1. An *autosomal dominant inheritance pattern* is supported by the family history. The symptoms and signs of the condition are present in more than one generation, in both males and females. Transmission has occurred from father to son.
2. Many autosomal dominant adult-onset conditions exhibit *variable expressivity* as illustrated in this family: Louisa died shortly after birth from complications that may have been related to myotonic muscular dystrophy, yet her maternal grandfather was still hiking at age 66.
3. Autosomal dominant conditions of adult onset often exhibit age related *penetrance.* In myotonic dystrophy, penetrance is considered close to 100%, although the age of onset of symptoms is quite variable.
4. *Anticipation* is exhibited in some autosomal dominant disorders, such as myotonic muscular dystrophy. This means that the condition appears to have an earlier age of onset, with more severe symptoms presenting, in succeeding generations. Additionally there is a congenital form of myotonic muscular dystrophy that presents with severe hypotonia and difficulties with feeding and respiration. This is explained in myotonic dystrophy by the instability of a CTG triplet repeat, with larger expansions usually associated with an earlier age of onset. The transmission of congenital myotonic dystrophy seems to be exclusively through the mother, which is consistent with this family history.

Other Issues   It is not unusual at all for an unsuspected genetic condition to be unveiled in a family history taken for another indication. The appropriateness of addressing additional family history issues at the time the history is taken or deferring them to a follow-up visit needs careful consideration.

Of note in Julia's family history is the premenopausal occurrence of breast cancer in the paternal aunt and grandmother. This warrants further exploration to determine whether other individuals in the family have breast cancer or other cancers suggestive of a hereditary cancer syndrome. Ideally medical records should be obtained documenting the primary pathology of the cancer(s). Since Julia's main concern is her pregnancy, it might be best to alert Julia to the concerns raised by her family history of breast cancer, but suggest that the topic of hereditary breast cancer risks and susceptibility testing be addressed at another visit.

Confirmation of the Diagnosis   Obtaining medical records on individuals in this family is essential. Ideally Julia's younger brother should be formally evaluated to see if he has myotonic dystrophy. The issues to be resolved regarding a possible diagnosis in Henry include the following.

Has he had a karyotype to rule out a chromosomal basis for his mental retardation?

Is his mental retardation progressive (which might suggest a metabolic cause)?

Does he have dysmorphic features? (Individuals with myotonic dystrophy may have a flat facial affect, "cupid bow lips," ptosis, temporal narrowing, facial weakness, and, in males, early balding.)

Can medical records be obtained that document Louisa's hospital course from 16 years ago? Did the infant have dysmorphic features? Was her hypotonia evaluated by, for example, metabolic testing (which should be normal in myotonic dystrophy), a muscle biopsy, or an electromyogram?

Obtaining ophthalmology records on Julia's mother and grandfather would also be useful. Bilateral iridescent lens opacities and posterior cortical lens opacities are highly specific for myotonic dystrophy (Harper, 1997).

The best way to confirm the diagnosis of myotonic muscular dystrophy is by DNA testing for the CTG repeats in an affected relative. It is considered normal to have between 5 and 30 CTG repeats. Individuals who are asymptomatic or have mild symptoms usually have been 50 to 80 repeats, while those with more severe, adult onset of symptoms have a wide variability of repeat sizes ranging from 100 to 500 copies. Patients with the congenital myotonic dystrophy have large expansions in the number of repeats, with values ranging from 500 to 2000 copies (Harper, 1997).

Probably the most potentially useful people to test in this family would be Julia's mother and her brother Henry. If for some reason these family members are unavailable or not willing to be tested, DNA testing can be done directly on Julia. This, however, is less than ideal. If Julia has a repeat number in the normal range, it is extremely unlikely that she has myotonic dystrophy, although not all affected individuals have a CTG expansion (Harper, 1997). In addition, a definitive diagnosis has not been established in the family. If myotonic dystrophy isn't present, Julia may remain at risk for whatever condition is responsible for the abnormalities in her family.

Genetic Counseling   If Julia carries the DNA mutation for myotonic dystrophy, she will have the "double whammy" of being diagnosed with a chronic, progressive medical condition and the knowledge that this and future pregnancies are at

50% risk. She would also be at risk for having a child with congenital myotonic dystrophy. Julia would have the option of prenatal diagnosis through amniocentesis or chorionic villus sampling.

For her own medical care, Julia should be evaluated for symptoms of myotonic dystrophy by a medical geneticist or neurologist. Of particular importance would be a glucose tolerance test for diabetes and an EKG for the detection of any cardiac arrhythmia.

Ethical Issues   When a genetic disorder with autosomal dominant inheritance is identified in an individual, multiple family members are usually at risk. These at-risk family members may not suspect that they have symptoms of a genetic disorder. In Julia's family, for example, her mother, maternal uncle, nephew, and maternal grandfather most likely have a diagnosis of myotonic muscular dystrophy, yet they may be unaware of it. For whom would you recommend testing, and why? How is the recommendation for testing made for these at-risk family members? For example, Julia's grandfather probably has myotonic muscular dystrophy, yet he is still physically active at age 66. If he is informed of this situation, there is the potential of stigmatization by making a "healthy person sick." Could the diagnosis be used in a discriminatory manner? How would this diagnosis affect his own perception of wellness?

Myotonic dystrophy is a condition for which presymptomatic testing and childhood testing are also issues. Even though Julia's older brother is apparently healthy at age 25, he may develop symptoms of myotonic dystrophy in the future, and his child and any further children may be at risk. Should an evaluation be recommended? If Michael does carry an expansion, should his healthy 5-year-old daughter be tested? What about the issue of testing the healthy 8-year-old daughter of Uncle Bob? One or more of these individuals may have a CTG expansion, yet they may not experience appreciable problems as long as they live, other than early cataracts, or muscle weakness. The test results do not predict the course of the disorder.

Prenatal diagnosis for myotonic dystrophy is yet another complicated issue. Unless the repeat number is very high (e.g., 2000 CTG repeats, indicating that the fetus may be born with congenital myotonic dystrophy), it does not necessarily correlate with prognosis. This condition has caused significant problems in two of Julia's siblings, yet other people in her family have lived productive and active lives. Julia must wrestle with her own feelings about whether she would want to know if she and her pregnancy were affected.

Summary   Some of the issues that are often raised in providing genetic counseling for a positive family history of an inherited adult-onset disorder are highlighted by the case summary. These include the following:

- Autosomal dominant inheritance is common. Thus, multiple family members may be at risk, and a whole family may become "your patient."
- Variable expressivity may be present.
- Penetrance of the disorder may approach 100%, but the age of onset of symptoms may be quite variable.

- Anticipation may be associated with the disorder.
- The available DNA testing is unlikely to provide specific information as to age of onset of symptoms or disease severity.
- Presymptomatic/susceptibility testing may disclose the presence of a genetic disorder in a person who had thought of himself as healthy.
- Childhood testing and prenatal testing for adult-onset conditions raise unique issues. Counseling about such testing can be challenging.

The case also clearly illustrates the value of obtaining a complete three-generation family history. The information obtained pertaining to the health histories of Julia's family members provided evidence for the probable diagnosis of myotonic dystrophy. The importance of the genetic counselor's strong knowledge base and her appreciation of disease variability were critical in establishing a likely diagnosis.

Finally, the case sets the scene for anticipating psychosocial issues surrounding the potential diagnosis of an adult-onset disorder. For Julia, the question of pregnancy risk was broadened to include the risks for her own health as well as for other family members. This is not an uncommon result of the diagnosis of an adult-onset genetic disorder in a family. Genetic counselors are likely to encounter these issues more often as testing for additional disorders becomes available.

Case Example 3
Elsa Reich

The care of the patient described in this case involved a collaboration between physicians in a specialty clinic and the genetic counselor. At a time when there is an explosion of new information about the molecular basis of disease, such physicians rely more and more on the expertise of genetic counselors, who have a knowledge base that allows them to make significant contributions to the care and management of patients. The counselor was an important resource and participated actively in the diagnostic evaluation. The evaluation of this patient was reviewed as needed by the counselor with her colleagues in the regular genetics case management conference.

Background In this case, the relationship of the counselor with the specialty clinic, a neuromuscular clinic, had evolved over a long period of time. The counselor herself did not regularly attend the clinic, nor did she see all the patients. The neurologist frequently provided counseling for affected patients and carried out some genetic testing. She regularly consulted the counselor or referred patients to her when there was a question about the diagnosis that might be clarified with a molecular test, when a patient or a family member requested information about reproductive risks, or when patients specifically requested genetic counseling. When the neurologist requested the counselor's help in the evaluation of patients, the counselor usually arranged for the patient and the family to come for genetic counseling. The counselor was expected and encouraged to provide suggestions about modification of the diagnostic workup, and she frequently assumed responsibility for managing any molecular studies.

In an academic medical center, a patient with a complex condition receives the attention of a multitude of caregivers, and coordination of the care may become fragmented. The discontinuity of care may also be exacerbated when the diagnostic evaluation is carried out over a long period of time—in this case, 4 years. While the team approach may be the ideal, the reality is that the "team" is likely to be in constant flux, as proved to be the case for this patient. The child described here received her care in three separate hospital settings in one medical center, each of which had its own independent staff. No single individual consistently supervised the care of this patient, and interaction with the genetic counselor was intermittent.

Not every case is managed in an optimal way, and this case was selected because it offers the opportunity to discuss the positive aspects of counseling and case management as well as some of the ways in which more efficient and comprehensive care might have been provided. The narrative also describes the rationale for the case management.

Maria P. was an 8-year-old Hispanic girl referred to the counselor in 1993 because, for a year or two, she had been experiencing slowly progressive muscle weakness and had been walking on her toes. Maria's muscle weakness had begun in the proximal muscles of her legs and gradually involved her upper extremities in a similar distribution. She had pseudohypertrophy of her calves. Although ambulatory at 8 years of age, she fell frequently and was having considerable difficulty climbing stairs.

Maria's evaluation in the neuromuscular clinic included a physical examination, laboratory tests, electromyography, nerve conduction studies, and a muscle biopsy. Her gait was awkward, she walked with an exaggerated lordotic curve, and she demonstrated a Gower sign. The results of her medical evaluation included an elevated level of creatine phosphokinase (CPK), electromyographic findings consistent with a myopathy, normal nerve conduction studies, and abnormal muscle biopsy results with nonspecific changes, also consistent with a primary disease of muscle. At the time of the first meeting, it was believed that all the members of Maria's immediate family (a 16-year-old sister, a 6-year-old brother, and both parents) had normal CPKs and no clinical evidence of muscle disease (Figure 16.5). The child's symptoms were suggestive of a manifesting carrier of Becker or Duchenne muscular dystrophy (BMD/DMD) or one of the many forms of limb–girdle muscular dystrophy. The neurologist recommended genetic counseling and further workup if indicated.

Structure and Goals   A genetics evaluation by its very nature constitutes a limited interaction. Single sessions are frequently the norm, although a complex pediatric case might require several sessions. The charge to the counselor is challenging. The counselor must educate the patient and family members about the evaluation, including what role the genetic counselor will play, and what the family, in turn, is expected to bring to the interaction. Each aspect of the evaluation, from the information gathering to the testing, can be utilized to develop a relationship with the family, as well as to set the stage for further meetings.

The counselor formulated a plan for the care of the patient, taking into account the putative diagnosis, the family structure, and the past and potential medical care

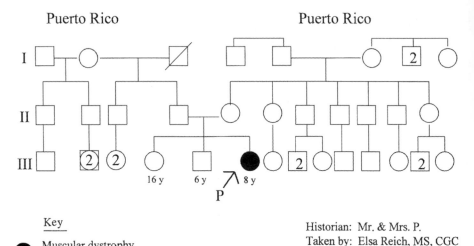

Puerto Rico   Puerto Rico

Key

● Muscular dystrophy,
  type ?

Historian: Mr. & Mrs. P.
Taken by: Elsa Reich, MS, CGC
No consanguinity

Figure 16.5   *The P. family.*

required by the patient. She considered this a tentative plan, since the unfolding of the case could indicate the usefulness of a change in approach.

First Session   At the first meeting, Maria was accompanied by her father and mother, her sister (Carmela), and a young foster child (Joe). The counselor's plan for the meeting was to learn about the parents' appraisal of the severity of Maria's condition, their major concerns, and their understanding of any possible cause. In addition, the counselor would elicit an assessment by the patient and her parents of her school adjustment, and would attempt to determine how the disability was affecting Maria's functioning both at home and at school. With the understanding that the parents might have information they preferred not to discuss in front of the children, the counselor established the parameters of the meeting.

Counselor: Because each one of you might have things that you would like to discuss with me privately, I would like to arrange an opportunity for each of you to speak with me alone. (Parents and patient note their accord.) Please tell me, Maria, what you know about why you are here today.

Maria: I don't know.

Counselor: The reason that you are here is because, as I understand it, you have been having some problems walking, and the doctors want to understand more about why this is happening. Is that something you would like to know more about?

Maria: (nods)

Counselor: Mr. and Mrs. P., what do you understand is the reason that the doctor suggested that you talk with me?

Mrs. P: She told me that Maria has muscular dystrophy and that it is hereditary and that you would explain it to us. She said that Carmela [Maria's older sister] might be a carrier.

Counselor: Is there anything else that is important for you to find out?

Mrs. P: Yes. We want to know if there is anything that can be done to treat Maria.

Counselor: At this moment, we don't know the specific cause of Maria's condition, but I should say now that most of these conditions are not currently treatable with medicine. We can do everything in our power to treat her symptoms, to make sure that she will remain active just as long as possible, that she will have the very best and most appropriate schooling available, and that she will be able to utilize all her talents.

[This is not encouraging information to parents and patients, but it allows the family to establish a realistic framework. The counselor permits the family a little time to integrate this disappointing information.]

Counselor: Can you tell me if you are planning to have other children in the future?

Mrs. P: No. We are not having other children of our own, but we have had several foster children, and we are planning to adopt Joe [the current foster child].

Commentary Any child, with the exception of a nonverbal infant, should be included in the discussion at a counseling session. Wherever possible, questions should be addressed to her, and she should be actively encouraged to participate in the session. When there are other children present, as in this case, they should be included, not only because they may have their own concerns (e.g., "Am I going to develop this condition?") but because their prior experience may have been to be treated as if they were invisible at medical meetings at which the affected sibling was the sole subject of attention. At home they may also feel neglected as the affected sibling receives more than an equal share of parental attention. A genetic counselor cannot change the experiences of a lifetime, however she may be able to give each member of the family a sense of genuine participation in this particular interaction. Such recognition positively influences the overall success of the interaction.

Establishing an agenda for the meeting as well as for the entire interaction is an important early task. Incorporating the concerns of the family into the agenda and initiating questions that may suggest issues that had not been considered forms the base on which a mutually satisfying relationship can be built. This agenda should be flexible, to encourage the development of new questions as the evaluation evolves.

Counselor: Here's what I would like to do today. Please understand that we may have to meet more than once because I want to be sure that I have done everything I can to establish an accurate diagnosis. I want to review with you what has been done thus far, what the results of the tests mean, what the possible diagnoses are, and what the implications are for all the members of the family. It is important for you to know there is a possibility that we may not be able to come to a definitive conclusion about the cause of Maria's condition, but we will make every possible effort to do so. Knowing this information may allow us to modify Maria's care and possibly set the stage for additional treatment if it becomes available in the future, although I cannot promise that this will be the case.

Case Management   One of the first tasks was to obtain a careful family history. It was specifically important to ascertain the number of males in the maternal line and their affected/unaffected status, whether there was consanguinity in the family, perhaps suggesting a recessively inherited disorder, and the ages of the father and the maternal grandfather at the time of birth of their daughters—the patient and the patient's mother, respectively. An advanced age in one of these two men could suggest the presence of a new mutation either in the patient or her mother.

The counselor learned that there were no other affected family members, and there were nine unaffected males in the maternal line (Figure 16.5). The counselor concluded that, *if* the patient had BMD/DMD, it was more likely to be the result of a spontaneous mutation in her or in her mother; since her grandmother had had four unaffected sons, she is less likely to have transmitted the gene. If necessary, the counselor could use Bayes's theorem to calculate the probability that the mother was a carrier (assuming, for the sake of this calculation, that the patient was truly affected with BMD/DMD).

One of the factors to consider in evaluating the source of a de novo mutation in a female member of a family is paternal age, which has long been known to be associated with new mutations. Maria's maternal grandfather was 19 at the time of her mother's birth, providing no support for a new mutation in her mother. However, Maria's father was 36 at the time of her birth, which is consistent with advanced paternal age and possibly related to a new mutation in Maria (Bucher et al., 1980; Crow, 1995). It has been demonstrated that in families in which there is a manifesting female carrier resulting from a de novo mutation, the mutation is invariably of paternal origin and usually a point mutation (Grimm et al., 1994). Although Maria's mother had a normal CPK, the results could not be considered completely reliable since, as a female carrier ages, her CPK may return to normal levels. In addition, only two-thirds of obligate carriers of DMD and one-half of carriers of BMD have elevated CPKs (Lange and Zatz, 1979; Sibert et al., 1979). However, the normal CPK in Maria's sister, Carmela, did reduce the probability that she was a carrier.

Although Maria's clinical presentation was compatible with being a manifesting carrier of the Becker/Duchenne type of muscular dystrophy, certain recessive forms of muscular dystrophy may be phenocopies of BMD/DMD and may account for the presence of affected females in a pedigree (Francke et al., 1989). Since the patient's parents were unrelated, there was no specific evidence favoring one of the recessive limb–girdle dystrophies. Furthermore, at the time of the first meeting, the molecular basis of the recessive dystrophies was unknown. Obviously, there remained a considerable amount to be done before a more definitive conclusion could be made about the etiology of the condition.

Further diagnostic evaluations were recommended, including chromosome studies to look for structural abnormalities of the X chromosome, Turner syndrome, or a Turner variant (which could explain the presence of symptoms of an X-linked condition in a female). The lack of phenotypic stigmata of Turner syndrome made this possibility less likely. Since a portion of the muscle biopsy sample remained, the counselor recommended that it be sent for dystrophin studies. Immunohistochemical studies of muscle dystrophin can sometimes provide evidence that a female is a manifesting carrier of BMD/DMD (Mansfield et al., 1993).

All the tests that were recommended were explained in detail, and the parents and patient agreed to proceed with testing. It was determined that the costs of the laboratory studies would be covered by the patient's insurance. (This was an important issue because the patient was covered by Medicaid and the family had a modest income.) The family was advised that the chromosome studies would be completed in 2 to 3 weeks and that results of the immunofluorescence studies, which were being done in a research laboratory, might take longer.

Commentary   It is important to prepare families and patients for the length of time required for testing to produce results. Even for tests likely to take months to complete, it is preferable to anticipate an approximate period of time within which to expect results. Families may be disappointed that it takes so long, but usually they can adjust better if they are given a time frame. The counselor is also careful to be circumspect about announcing an exact day that a result will be ready, even when informed by laboratory personnel, because results commonly are not ready at the stated time.

Case Management   Both studies were completed. The chromosomes were normal, and the dystrophin studies were inconclusive. The investigator therefore suggested X-inactivation studies. Blood was obtained from the patient, and these studies demonstrated that the patient had skewed X inactivation (>80% of X chromosomes from one parent were inactivated). Although skewed X inactivation can occur by chance, the investigator concluded that most likely the patient was a manifesting carrier of BMD/DMD (Pegoraro et al., 1994).

The counselor next pursued dystrophin mutation analysis based on the inconclusiveness of the studies performed thus far. Up to 70% of cases of DMD are associated with deletions (and 5–10% with duplications), and greater than 90% of cases of BMD have deletions (Bushby, 1993; Mansfield et al., 1993). A laboratory that would accept Medicaid was identified, and the director agreed to use intragenic CA repeats to look for deletions in the dystrophin gene. This is not the most comprehensive way of doing deletion studies, but the laboratory did not perform dosage assays. Dosage studies involve the use of a cDNA probe and Southern blot analysis to identify a dosage difference between controls and carriers for deletions and for duplications (Prior, 1991). However, the only state-licensed laboratory performing these studies was expensive and required payment prior to initiating the analyses, which was not an option for this family. As a preliminary, the counselor sent blood samples from the patient and both parents for these limited deletion studies.

Commentary   Although arrangements for molecular studies as well as other tests can be made for patients covered by Medicaid, each test must be justified in detail. This is frequently quite time-consuming. Counselors and others are often called on to be creative in searching out resources to do molecular studies that are not considered "standard" by third-party payers. This necessity, which varies from state to state and among different institutions, may delay testing significantly.

Case Management Some weeks after the blood samples had been obtained, the counselor learned from the laboratory director that nonpaternity had been identified. In addition to analyzing the intragenic markers in the dystrophin gene, the laboratory directory had also analyzed several unrelated markers that further confirmed the nonpaternity. In the best determination that could be made, no deletion was identified. Already a significant period had elapsed, and an exact diagnosis remained elusive.

Commentary The counselor considered whether it was necessary to reveal her knowledge of nonpaternity to any one else and concluded that it was not. It is sometimes propitious to approach the mother and discuss the findings with her privately, but since, at that point in the investigation, knowledge of the biologic paternity was not essential to obtaining an accurate diagnosis, the counselor decided not to bring the matter up. The issue could be revisited in the future if necessary.

These studies were not completed until almost a year after the original meeting. The patient had been seen several times in the neuromuscular clinic, always accompanied by her father, not her mother. When the counselor telephoned the home, it was Mr. P. who took responsibility for discussing the case. Mrs. P. seemed somewhat distant and removed. As a result of multiple contacts with the neurologist, and periodically with the family, both in person and by telephone, the counselor learned that during the year just past the father had devoted most of his nonwork time to Maria, taking her places, picking her up after school, and neglecting the other children in the family. The mother seemed to have immersed herself in the care of the foster child. The clinic social worker repeatedly asked to see Mrs. P., and although she agreed to come to clinic, she failed to do so for most of the year. It was difficult to reconcile the original picture of the mother, which was that of a parent completely invested in her daughter, with the emerging picture of a woman who was pulling back. It seemed even more essential not to further disrupt this family's relationship by disclosing the knowledge of nonpaternity that had been obtained inadvertently.

The counselor next recommended that dosage studies be carried out. After several months, the clinic was able to come up with payment to cover the costs of testing. (No sample was sent on the father.) While waiting for the results to be completed, the counselor called the family from time to time to report on the status of the analysis. Four months later, the counselor learned that the studies had not revealed evidence of either a deletion or duplication in the dystrophin gene. This negative result did not rule out the possibility that Maria's condition was BMD/DMD, since deletions or duplications are not detected in all affected individuals and dosage differences may not always be demonstrated by current technology.

Following the completion of these studies, with the agreement of the medical geneticist (now 16 months after the initial meeting), the counselor recommended to the neurologist that another muscle biopsy be done to repeat the dystrophin studies. With the exception of the X-inactivation study results, hard data to support the diagnosis of BMD/DMD were still lacking. The counselor arranged for the family to come for a follow-up appointment and to be seen in the neuromuscular clinic on the same day. Prior to the appointment, she conferred with the social worker in the neuromuscular clinic, who had attempted to provide emotional support for the family. The social worker reported that Maria was severely depressed and, according to Mrs. P., frequently talked about dying. Maria was particularly anxious to change schools because she was in a regular school that was not adequately equipped for physically disabled children and there were few, if any, other disabled children in the school. In addition, the family was having other problems, which remained undefined. The patient came in for this appointment accompanied by both parents.

Session   Maria's disease had progressed rapidly since the counselor had last seen her. Her calves were very large, and she was not only using braces but was almost completely confined to a wheelchair. Her upper extremities were extremely weak, and she could no longer dress herself. She seemed quite depressed. Because of the child's depression, the counselor requested the opportunity to speak with her parents alone at the outset. The counselor made an effort to talk with Maria, but she was noncommunicative, and it seemed preferable to work toward enabling a referral for more in-depth therapy.

Counselor: Tell me what's happening. Susan [the social worker] has told me that Maria is very depressed.

Mrs. P: (Eyes tearing) Yes. She is always talking about dying, about wanting to die. She doesn't like her school. She feels out of place.

Counselor: What is it that is making her depressed? Is it specifically her condition, or is it her school situation?

Mrs. P: I think it's the school. She has fallen at school several times. They don't have the proper facilities for her. She wants to be in a school where there are more children like her.

Counselor: Susan told me that she has been talking with Maria when you bring her in, Mr. P., but she hasn't seen you, Mrs. P., for a long time.

Mrs. P: He [referring to Mr. P.] does everything for her. He doesn't let her do anything for herself. I am also busy taking care of Joe. The adoption is almost final. I am trying to make up for [my husband] with the other children.

Counselor: Are you spending time with the other children, Mr. P.?

Mr. P: Well, not very much. I know I should.

Counselor: I know that this is very difficult and painful for both of you, but your other children need attention as well, and it isn't necessarily good for Maria to know that she commands all your attention.

Mr. P: Yes, I know. Carmela [the older daughter] makes a lot of comments about everything being for Maria. I guess I spoil her.

Counselor: It sounds as if it would be helpful for the whole family to change the way you handle this situation for the benefit of everyone. The other children will get some needed attention from you, Mr. P., Maria will achieve more independence, and perhaps the two of you [the parents] can have some time together, too. From what you've been telling me, it seems that the whole family needs some help to get through this difficult period. Susan has said that she will help you identify another school for Maria, and also find a therapist who can work with all of you.

Following the discussion about the family interactions, the counselor reviewed the results of all the testing and then related to the parents that it was the recommendation of the specialists involved in caring for Maria that another muscle biopsy be done. The parents were reluctant to give their consent, particularly in view of Maria's emotional status, and the father also expressed the fear that the muscle biopsy could worsen her condition. He was assured that this was not the case. The parents were also discouraged because a diagnosis had not been reached and because they felt that Maria had gone through enough. It was difficult to disagree with them. The session ended with the parents stating that they would consider repeating the biopsy.

Commentary This account of the session touches on only some of the concerns of the parents and the patient, but it demonstrates the importance of addressing the care of the patient and how the involvement of the parents in the care interferes with their relationships and interactions with the other children in the family. This was also a situation that called for ongoing psychotherapy. Most genetic counselors are not trained to provide these services, nor do they have adequate time. However, they do facilitate referrals for these services and encourage families to participate. Susan, the social worker, told the family that she would arrange for them to meet with a family therapist, but, unknown to the counselor, the arrangements were delayed.

The continued pursuit of an accurate diagnosis for Maria's condition may seem a less important issue at this point, but depending on the family, such knowledge may remain a priority. It is usually important to review again with families the reasons for establishing a diagnosis (e.g., it provides an explanation for the course of the condition; it may yield prognostic information; it may help to plan future medical care; it may provide accurate information about the risks for family members). In this case, the family had become frustrated with the unsuccessful attempts to establish a diagnosis and had also developed concerns regarding the stress of such studies on Maria. Renegotiating goals and agendas with the family may help to clarify the direction of the case. For example, arranging to perform procedures such as obtaining biopsy samples and additional blood samples when the patient is scheduled to have other tests may alleviate the concern about adding to Maria's stress.

Case Management Some three months later, the counselor learned that the P. family had yet to go to their first therapy appointment, but they were due at the neuromuscular clinic the next day. They had missed two prior appointments. The counselor called the family to urge them to attend their appointment at the neuromuscular clinic. After the appointment, the family stopped in to see the counselor and stated that although they had decided to go ahead with the biopsy, they now had problems with insurance coverage, since Maria had been dropped from Medicaid and their private insurance did not cover the testing contemplated. They asked the counselor to write a letter to Medicaid stating the importance of restoring coverage. This was done.

Over the next few months, primarily through the efforts of the parents and the social worker, Medicaid benefits were restored. Just at this time, the counselor learned at a genetics meeting of a reported association between the dystrophin-associated proteins, the sarcoglycans, and recessively inherited limb–girdle muscular dystrophy (McNally et al., 1994; Roberds et al., 1994; Romero et al., 1994; Piccolo et al., 1995; Van den Berge et al., 1995). The counselor reminded the neurologist, who had assumed responsibility for having the second muscle biopsy done, of the importance of this test and found one of the investigators interested in the newly identified proteins who was willing to accept a specimen. The counselor hoped to successfully repeat the dystrophin studies and to have immunofluorescent studies of adhalin (α-sarcoglycan) done; but despite repeated requests, the repeat biopsy was not performed.

Gradually, over the next two and half years, the counselor's contacts with the family became attenuated. The counselor called and left messages that usually were not returned. She spoke repeatedly with the neurologist but learned very little about the progress of the patient. Approximately 4 months prior to the end of this narrative, a medical geneticist, new to the group, had a referral from an affiliated hospital to see a 12-year-old girl with muscular dystrophy. Of course, it was Maria. The counselor related the story of Maria, and her frustrating attempts to have the muscle biopsy repeated. It was particularly distressing to learn that Maria had just undergone a tendon-lengthening procedure and could have had a muscle biopsy at that time. The counselor went to see Maria and her mother in the hospital and also discovered that she had undergone spinal surgery the preceding year. The biopsy had never been done.

The counselor learned that Maria was not going home, but was to be transferred directly to an affiliated rehabilitation hospital. The counselor took advantage of the further hospitalization to finalize arrangements for the biopsy. Two and a half years after the initial recommendation, the biopsy was finally repeated. The dystrophin studies were repeated as well, and the adhalin studies were done. After the biopsy was done and prior to learning the results, the counselor spoke numerous times with Mrs. P., apprising her of the progress of the tissue study.

Commentary   Since reliable genetic counseling is based on the diagnosis and the implications of a specific pattern of inheritance, the importance of an accurate diagnosis cannot be underestimated. In the future, treatment of genetic disease may be based on knowledge of an exact molecular etiopathogenesis. Therefore, it becomes incumbent on the genetics professionals to assure that an optimal effort is made to establish a diagnosis. Whatever the degree of effort, however, it remains possible that no diagnosis will be arrived at and that the counseling of families in such cases will remain problematic (Hoffman and Clemens, 1996).

Last Session   Maria, accompanied by her mother, came to learn the results of the studies. She was completely wheelchair bound and had severe generalized weakness. The counselor explained the results to Maria and her mother, telling them that despite the skewed X inactivation, Maria did not have a dystrophinopathy (i.e., she was not a manifesting carrier). The dystrophin was normal in distribution and molecular weight. Maria did, however, show evidence of one of the sarcoglycanopathies (the exact type still undetermined). The probability of finding a mutation was unknown, and it remained possible that no mutation would be found (Duggan and Hoffman, 1996; Duggan et al., 1996; Hoffman and Clemens, 1996; Sewry et al., 1996). The counselor had obtained slides from the biopsy sample demonstrating the normal immunofluorescence of the dystrophin and the paucity and irregularity of the adhalin on the muscle fibers, and the counselor showed the slides to the patient and her mother. Maria had become so weak that she was unable to lift the small slide viewer to her eye, but had to support one arm and hand with the other and bend her head way over to look through the viewer.

One important implication of the results was that Maria's sister's risk for having an affected child was now known to be very small, since a rare recessive rather than an X-linked condition was at issue. Despite the rapid progression of her disease, Maria's general mood was better. She was not markedly depressed.

Counselor: Is there anything else that you want to know?

Mrs. P: Yes. Is there any treatment? Is there anything we can do?

Counselor: Unfortunately, no. I'm sorry. I know that you had hoped, once we had a diagnosis, that there would be a successful treatment, but at least for now, there is no successful treatment. On the other hand, we don't know exactly what we can expect in the future, either in terms of how she'll do or what may become available in terms of treatment (Stec et al., 1995; Mahjneh et al., 1996; Morandi et al., 1996; Passos-Bueno et al., 1996; Eymard et al., 1997). All we can do is hope. I know this isn't what you had hoped, but it's the best we can do.

Mrs. P: I am glad at least that we have an answer.

Commentary   This case demonstrates the many facets of a genetic counselor's role in the care of a patient. Although the management could have been improved (viz., the diagnosis could have been facilitated more effectively), it is not unusual for a diagnostic evaluation to extend over a long period of time. There are time constraints, financial constraints, and emotional and physical constraints. There are conflicting priorities on the part of both the professionals and the patients and their families. The families have other burdens that may interfere with their willingness and ability to follow recommendations. Parents may be overwhelmed by the many decisions that must be made with respect to the care of the patient, and frequently they must stage their decisions.

Physicians and counselors care for many patients requiring complex management, and when families continue to be unresponsive, as the parents in this case were at one point, the professionals may back off, waiting for the family to resume contact. When a counselor makes several overtures that are ignored, it is difficult to know whether to continue to contact the family or to allow a "breather." When many professionals participate in the care of a patient, it is not always clear who should coordinate the care. It was somewhat difficult for the counselor to provide optimal care for Maria because she was not always informed of the patient's status. Alternatively, in the absence of hearing from the patient or the physician, the counselor should have initiated more consistent contact to track the patient's progress and perhaps made a greater effort to review the medical records; however, there were numerous impediments to doing this successfully. Despite the prolonged elapse of time from the beginning of the evaluation until the end, the family was pleased that the diagnosis had been pursued, that they had been recontacted, and that the testing and counseling had been completed.

Conclusion   In structuring any genetic counseling session, it is important to integrate the information gathering and giving with an understanding of the emotional impact of the medical consequences on the parents, the patient, and extended family members. The concerns of these parents included the unknown prognosis of the disease for their daughter, the possibility that other children in the family might also develop the disease, the cost of medical care, securing appropriate education for the child, the psychological well-being of the child and other family members, and reproductive issues for themselves, their children, and other family members.

This case is illustrative of several aspects of the work of a genetic counselor, providing a picture of the rich and varied role that such a professional can play. The counselor was required to draw on all the skills developed through training and experience, addressing both the medical and psychosocial aspects of the case. The case demonstrates the contributions a genetic counselor can bring to the diagnostic evaluation of a patient, the quality of the relationship between a counselor and a family, the importance of pursuing a diagnosis, and the counselor's responsibility to stay current with respect to new information and scientific developments.

CONCLUSION

It is clear upon review of these cases that genetic counseling is comprised of complex, discrete tasks, all of which serve to educate and support the families with whom we interact. Although the field of genetic counseling has become more specialized over time, with some genetic counselors working in prenatal settings, others with adults or children, and yet others within oncology, all of us contribute to the body of work which is the "art" of genetic counseling. I suspect that no genetic counselor has performed the "perfect" genetic counseling session. One can always improve skills or polish technique. Just as we are learning more about the human genome, we are learning different ways in which to approach our patients. The dynamic nature of our profession is our greatest challenge; it captures our interest, challenges our skills, and feeds our imaginations.

REFERENCES

Ackerman TF (1996) Genetic testing of children for cancer susceptibility. *J Pediatr Oncol Nurs* 13(1):46–49.

Bloch M, and Hayden MR (1990) Opinion: Predictive testing for Huntington Disease in childhood. Challenges and implications. *Am J Hum Genet* 46(1):1–4.

Boards of Directors of the American Society of Human Genetics and the American College of Medical Genetics (1995) ASHG/ACMG report, points to consider: Ethical, legal, and psychosocial implications of genetic testing in children and adolescents. *Am J Hum Genet* 57:1233–1241.

Borgaonkar DS (1994) *Chromosomal Variation in Man. A Catalog of Chromosome Variants and Anomalies.* New York: Alan R. Liss.

Boue A, Gallano P (1984) A collaborative study of the segregation of inherited chromosome structural rearrangements in 1356 prenatal diagnoses. *Prenatal Diagn* 4:45–67.

Boue A (1979) European collaborative study on structural chromosome anomalies in prenatal diagnosis. Group report. In: Murken JD, Stengel-Rutkowski S, Schwinger E (eds), *Prenatal Diagnosis.* Stuttgart: Enko, pp. 34–46.

Brambati B, Formigli L, Mor M, Lului L (1994) Multiple pregnancy induction and selective fetal reduction in high genetic risk couples. *Hum Reprod* 9(4):746–749.

Brambati B, Simoni G, Travi M, Danesino C, Tului L, Privitena O, Stioui S, Tedeschi S, Russo S, Primignani P (1992) Genetic diagnosis by chorion villus sampling before 8 gestational weeks: Efficiency, reliability and risks on 317 completed pregnancies. *Prenatal Diagn* 12(10):789–799.

Bucher K, Ionasescu V, Hanson J (1980) Frequency of new mutants among boys with Duchenne muscular dystrophy. *Am J Med Genet* 7:27–34.

Bushby KMD, Gardner-Medwin D, Nicholson LVB, et al. (1993) The clinical, genetic and dystrophin characteristics of Becker muscular dystrophy. II. Correlation of phenotype with genetic and protein abnormalities. *J Neurol* 240:105–112.

Cantu ES, Khan TA, Pai GS (1992) Fluorescent in situ hybridization (FISH) of a whole-arm translocation involving chromosomes 18 and 20 with alpha-satellite DNA probes: Detection of a centromeric DNA break? *Am J Med Genet* 44(3):340–344.

Coleman DV, Heaton DE, Czepulkowski BH, Tyms AS (1987) Cytoculture of chorionic villi. In: Liu DTY, Symonds EM, Golbus MS (eds), *Chorion Villus Sampling.* Chicago: Year-Book Medical Publishers.

Crandall BF, Lebherz TB, Rubenstein L, Robertson RD, Sample WF, Sarti D, Howard J (1980) Chromosome findings in 2500 second trimester amniocentesis. *Am J Med Genet* 5:345–356.

Crow JF (1995) Spontaneous mutation as a risk factor. *Exp Clin Immunogenet* 12:121–128.

Daniel A, Hook EB, Wulf G (1988) Collaborative USA data on prenatal diagnosis for parental carriers of chromosome rearrangements: Risk of unbalanced progeny. In: Daniel A (ed), *The Cytogenetics of Mammalian Autosomal Rearrangements.* New York: Liss, pp. 73–162.

Davis JR, Rogers BB, Hagaman RM, Thies CA, Veomett IC (1985) Balanced reciprocal translocations: Risk factors for aneuploid segment viability. *Clini Genet* 27:1–19.

Duggan DJ, Hoffman EP (1996) Autosomal recessive muscular dystrophy and mutations of the sarcoglycan complex. *Neuromuscular Disord* 6:475–482.

Duggan DJ, Fanin M, Pegoraro E, Angelini C, Hoffman EP (1996) α-Sarcoglycan (adhalin) deficiency: Complete deficiency patients are 5% of childhood onset dystrophin–normal muscular dystrophy and most partial deficiency patients do not have gene mutations. *J Neurol Sci* 140:30–39.

Edelman C, Heimler A, Stamberg J (1988) Prenatal diagnosis by chorion villus sampling (CVS): Acceptability in light of Orthodox Jewish views on abortion. *Am J Hum Genet* 43(3):A166.

Ekstein J (1995) Premarital and anonymous screening for recessive genetic diseases: Recent experience with Gaucher disease. *NIH Technology Assessment Conference on Gaucher Disease*, pp. 75–76.

Ekstein J (1996) Personal communication.

Engelen JJ, Loots WJ, Albrechts JC, Motoh PC, Fryns JP, Hamers AJ, Geraedts JP (1996) Disclosure of five breakpoints in a complex chromosome rearrangement by microdissection and fluorescent in-situ hybridization. *J Med Genet* 33(7):562–566.

Eymard B, Romero NB, Leturcq F, Piccolo F, Carrie A, Jeanpierre M, Collin H, Deburgrave N, Azibi K, Chaouch M, Merlini L, Themar-Noel C, Penisson I, Mayer M, Tanguy O, Campbell KP, Kaplan JC, Tome FM, Fardeau M (1997) Primary adhalinopathy (α-sarcoglycanopathy): Clinical, pathologic, and genetic correlation in 20 patients with autosomal recessive muscular dystrophy. *Neurology* 48:1227–1234.

Firth HV, Boyd PA, Chamberlain PF, Mackenzie IZ, Moriss-Kay GM, Huson SM (1994) Analysis of limb reduction defects in babies exposed to chorion villus sampling. *Lancet* 343(8905):1069–1071.

Francke U, Darras BT, Hersh JH, Berg BO, Miller RG (1989) Brother/sister pairs affected with early-onset, progressive muscular dystrophy: Molecular studies reveal etiologic heterogeneity. *Am J Hum Genet* 45:63–72.

Froster UG, Jackson L (1996) Limb defects and chorion villus sampling: Results from an international registry 1992–94. *Lancet* 347(9000):489–494.

Gardner RJM, Sutherland GR (1996) *Chromosome Abnormalities and Genetic Counseling.* New York: Oxford University Press.

Gold, M (1988) *And Hannah Wept. Infertility, Adoption and the Jewish Couple.* New York: Jewish Publication Society.

deGrouchy J, Turleau C (1984) *Clinical Atlas of Human Chromosomes,* 2nd ed. New York: Wiley.

Grimm T, Meng G, Liechti-Gallati S, Bettecken T, Muller CR, Muller B (1994) On the origin of deletions and point mutations in Duchenne muscular dystrophy: Most deletions arise in oogenesis and most point mutations result from events in spermatogenesis. *J Med Genet* 31:183–186.

Harper PS (1997) Myotonic dystrophy. In: Rimoin DL, Connor JM, Pyertize RE (eds), *Emery and Rimoin's Principles and Practice of Medical Genetics,* 3rd ed. New York: Churchill Livingstone, pp. 2425–2443.

Heaton DE, Czepulkowski BH (1987) Chorionic villus and direct chromosome preparation. In: Liu DTY, Symonds EM, Golbus MS (eds), *Chorion Villus Sampling.* London: Chapman & Hall.

Hoffman EP, Clemens PR (1996) HyperCKemic, proximal muscular dystrophies and the dystrophin membrane cytoskeleton, including dystrophinopathies, sarcoglycanopathies, and merosinopathies. *Curr Opin Rheum* 8:528–538.

Howard-Peebles PN (1996) Successful pregnancy in a fragile X carrier by donor egg (Letter). *Am J Med Genet* 64(2):377.

Jackson LG, Zachary JM, Fowler SE, Desnick RJ, Golbus MS, Ledbetter DH, Mahoney MJ, Pergament E, Simpson JL, Black S, Wapner RJ (1992) A randomized comparison of trans-cervical and transabdominal chorion villus sampling. *N Engl J Med* 327(9):594–598.

Jakobovits Y (1994) The longing for children in a traditional Jewish family. In: Grazi R (ed), *Be Fruitful and Multiply. Fertility Therapy and the Jewish Tradition.* Spring Valley, NY: Genesis Jerusalem Press.

Kenen RH, Schmidt RM (1978) Stigmatization of carrier status: Social implications of het-erozygote genetic screening programs. *Am J Public Health* 68(11):1116–1120.

Lange K, Zatz M (1979) A new method for the analysis of age trends in CPK levels with ap-plication to Duchenne muscular dystrophy. *Hum Hered* 29:154–160.

Mahjneh I, Bushby K, Pizzi A, Bashir R, Marconi G (1996) Limb–girdle muscular dystrophy: A follow-up study of 79 patients. *Acta Neurol Scand* 94:177–189.

Mansfield ES, Robertson JM, Lebo RV, Lucero MY, Mayrand PE, Rappaport E, Parrella T, Sar-tore M, Surrey S, Fortina P (1993) Duchenne/Becker muscular dystrophy carrier detection using quantitative PCR and fluorescence-based strategies. *Am J Med Genet* 48:200–208.

Mastroicovo P, Botto LD, Cavalcanti DP, Lalatta F, Selicarni A, Tozzi AE, Barociani D, Cigolotti AC, Giordano S, Petroni F (1992) Limb anomalies following chorion villus sam-pling: A registry based case-control study. *Am J Med Genet* 44(6):854–864.

McNally EM, Yoshida M, Mizuno Y, Ozawa E, Kunkel LM (1994) Human adhalin is alterna-tively spliced and the gene is located on chromosome 17q21. *Proc Natl Acad Sci USA* 91:9690–9694.

Mikkelson, M (1987) Chromosome analysis on chorionic villi. *Bailliere's Obstet Gynecol* 1(3).

Morandi L, Barresi R, Di Blasi C, Jung D, Sunada Y, Confalonieri V, Dworzak F, Mantegazza R, Antozzi C, Jarre L, Pini A, Gobbi G, Bianchi C, Cornelio F, Campbell KP, Mora M (1996) Clinical heterogeneity of adhalin deficiency. *Ann Neurol* 39:196–202.

Passos-Bueno MR, Moreira ES, Marie SK, Bashir R, Vasquez L, Love DR, Vainzof M, Iughetti P, Oliveira JR, Bakker E, Strachan T, Bushby K, Zatz M (1996) Main clinical features of the three mapped autosomal recessive limb–girdle muscular dystrophies and estimated propor-tion of each form in 13 Brazilian families. *J Med Genet* 33:97–102.

Pegoraro E, Schimke RN, Arahata K, Hayashi Y, Stern H, Marks H, Glasberg MR, Carroll JE, Taber JW, Wessel HB, et al. (1994) Detection of new paternal dystrophin gene mutations in isolated cases of dystrophinopathy in females. *Am J Hum Genet* 45:989–1003.

Pegoraro E, Schimke RN, Garcia C, Stern H, Cadaldini M, Angelini C, Barbosa E, Carroll J, Marks WA, Neville HE (1996) Genetic and biochemical normalization in female carriers of Duchenne muscular dystrophy: Evidence for failure of dystrophin production in dys-trophin-competent myonuclei. *Neurology* 46:1189–1191.

Pergament E, Verlinsky Y (1986) Fetal karyotyping by chorionic tissue culture. In: Brambati B, Simoni G, Fabro S (eds), *Chorion Villus Sampling. Fetal Diagnosis of Genetic Disease in the First Trimester.* New York: Dekker, pp. 119–129.

Petrosky DL, Borgaonkar DS (1984) Segregation analysis in reciprocal translocation carriers. *Am J Med Genet* 19:137–159.

Piccolo F, Roberds SL, Jeanpierre M, Leturcq F, A2zibi K, Beldjord C, Carrie A, Recan D, Chaouch M, Reghis A, et al. (1995) Primary adhalinopathy: A common cause of autosomal recessive muscular dystrophy of variable severity. *Nat Gen* 10:243–245.

Powell CM, Taggart RT, Drumheller TC, Wangsa D, Qian C, Nelson LM< White BJ (1994) Molecular and cytogenetic studies of an X; autosome translocation in a patient with prema-ture ovarian failure and review of the literature. *Am J Med Genet* 52(1):119–126.

Prior WT (1991) Genetic analysis of the Duchenne muscular dystrophy gene. *Arch Pathol Lab Med* 115:984–990.

Roberds SL, Leturcq F, Allamand V, Piccolo F, Jeanpierre M, Anderson RD, Lim LE, Lee JC, Tome FM, Romero NB, et al. (1994) Missense mutations in the adhalin gene linked to autosomal recessive muscular dystrophy. *Cell* 78:625–633.

Romero NB, Tome FM, Leturcq F, el Kerch FE, Azibi K, Bachner L, Anderson RD, Roberds SL, Campbell KP, Fardeau M, et al. (1994) Genetic heterogeneity of severe childhood autosomal recessive muscular dystrophy with adhalin (50 kDa dystrophin-associated glycoprotein) deficiency. *C R Acad Sci Sê III Sci Vie* 317:70–76.

Scott JA, Wenger SL, Steele MW, Chakravarti A (1995) Down syndrome consequent to a cryptic maternal 12p;21q chromosome translocation. *Am J Med Genet* 56(1):67–71.

Sewry CA, Taylor J, Anderson LV, Ozawa E, Pogue R, Piccolo F, Busby K, Dubowitz V, Muntoni F (1996) Abnormalities in alpha-, beta- and gamma-sarcoglycan in patients with limb–girdle muscular dystrophy. *Neuromuscular Disord* 6:467–474.

Sharpe N (1993) Presymptomatic testing for Huntington disease: Is there a duty to test those under the age of 18 years? *Am J Med Genet* 46:250–253.

Sibert JR, Harper PS, Thompson RJ, Newcombe RG (1979) Carrier detection in Duchenne muscular dystrophy. Evidence from a study of obligatory carriers and mothers of isolated cases. *Arch Dis Child* 54:534–537.

Skirton H (1995) Psychosocial implications of advances in genetics. 1. Carrier testing. *Prof Nurs* 10(8):496–498.

Stec I, Kress W, Meng G, Muller B, Muller CR, Grimm T (1995) Estimate of severe autosomal recessive limb–girdle muscular dystrophy (*LGMD2C, LGMD2D*) among sporadic muscular dystrophy males: A study of 415 males. *J Med Genet* 32:930–933.

Steel CM (1997) Cancer of the breast and female reproductive tract. In: Rimoin DL, Connor JM, Pyeritz RE (eds), *Emery and Rimoin's Principles and Practice of Medical Genetics,* 3rd ed. New York: Churchill Livingstone, pp. 1501–1523.

Stene J, Stengel-Rutkowski S (1988) Genetic risks of familial reciprocal and Robertsonian translocation carriers. In: Daniel A (ed), *The Cytogenics of Mammalian Autosomal Rearrangements.* New York: Liss, pp. 3–72.

Summitt RL (1979) Cytogenetic disorders. In: Jackson LG, Schimke RN (eds), *Clinical Genetics, A Source Book for Physicians.* New York: Wiley, pp. 35–84.

Van den Bergh PY, Tome FM, Fardeau M (1995) Etiology and pathogenesis of the muscular dystrophies. *Acta Neurol Belg* 95:123–141.

Vought L, Schuette J, Bach G, Grabowski GA, Desnick RJ (1986) Tay–Sachs disease compatibility screening in the Orthodox Jewish community. *Am J Hum Genet* 39(3):A183.

Welkenhuysen M, Evers-Kiebooms G, Decruyenaere M, van den Berghe H, Bande-Knups J, Van Gernen V (1996) Adolescents' attitude towards carrier testing for cystic fibrosis and its relative stability over time. *Eur J Hum Genet* 4(1):52–62.

Wertz D, Fanos J, Reilly P (1994) Genetic testing for children and adolescents. Who decides? *JAMA* 272(11):875–881.

Wolowelsky JB (1994) New ethical issues. In: Grazi R (ed), *Be Fruitful and Multiply. Fertility Therapy and the Jewish Tradition.* Spring Valley, NY: Genesis Jerusalem Press.

Working Party of the Clinical Genetics Society (UK) (1994) Report on the genetic testing of children. *J Med Genet* 31:785–797.

Index

Accreditation, of training programs, 16–17,
    77, 295, 334
Accreditation Counsel for Graduate Medical
    Education (ACGME), 77
Acculturation, 176
Adult learners, 100–102
Affective dimensions, in counseling, 8
Alliance of Genetic Support Groups, 111,
    219, 387
    support group directory, 219
American Board of Genetic Counseling
    (ABGC), 13, 17, 295–296, 332, 334, 342
American Board of Medical Genetics
    (ABMG), 13–14, 16–17, 77, 342
American Board of Medical Specialities, 76
American College of Medical Genetics
    (ACMG), 16, 76, 339, 343, 356
American Hospital Association, 246
American Medical Association, 76
American Medical Record Association, 246
American Psychiatric Association (APA), 356
American Society of Human Genetics
    (ASHG), 5, 269, 338–339, 341, 353, 356
Americans with Disabilities Act (ADA),
    240, 269
Analogies, in interviews, 67
Anger, 143

Assimilation, 176
Association, defined, 88
Association for Counselor Education and
    Supervision (ACES), 297
Associative countertransference, 137
Audiovisual materials, patient education,
    116–117
Audit, of medical records, 243–244
Autonomy, 132–133, 254

Bad news:
    giving, 139–142
    reactions to, 142–145
Bandura, Albert, 105
Battery, litigation and, 264
Bayesian analysis, 209–211
Beneficence, 253
Bereavement, cultural differences, 182
Biochemical tests, 93–95
Biomedical ethics, 252
Bloom's Taxonomy, 298–299
Body language, 57
Body proportions, assessment of, 87, 90
Boundaries, professional conduct, 279–281
Breach of contract, 265
Brochures, patient education, 117
    assessing, 220–221

Canadian College of Medical Genetics
(CCMG), 13
Care ethics, 252
Case discussion, psychosocial:
goal of, 155
preparation for, 155–156
Case examples, overview, 391–417
Case management:
case preparation, 205–207
clinic visit, plan formulation, 221–222
communicating results, 225–226
communicating with specialists and researchers, 226–227
functions of, 199–200
genetic testing, coordination of, 211–219
initial intake, 200–202
genetic conditions, seeking information
on, 207–208
medical records, obtaining, 202–205
patient resources, support groups,
219–221
risk assessment, 208–211
session management, 222–225
Causation, litigation and, 265
Certification, of providers, *see* Credentialing
credentialing 18, 76, 334
clinical medical geneticists, 76–77
genetic counselors, 334–335
professional development and, 334
requirements, 17–18
technological advances and, 352–354
types of, 12–15
Change model, stages of, 104
Chart note, 232
Chief complaints, in medical history, 80
Children, *see* Pediatrics
breaking bad news to, 141–142
informed consent and, 255–256
minors, medical decision making,
262–263
presymptomatic testing in, 268–269
role of, cultural differences, 182
Chromosomal Variation in Man: A Catalog
of Chromosomal Variants and Anomalies, 377, 386
Client-centered approach, 147
Clinical rotations, evaluations of students,
312–313, 323–330
Clinical diagnostic resources:

commercial software, 374–376, 385
online, 372–374
Clinical geneticists, role of, 13.
tools and resources of, 95–96
training of, 76–77
Clinical Laboratory Improvement Act (CLIA
1988), 352–354
Clinical supervision, defined, 296–297
consent form example, 320–322
Clinic visit, case management, 221–222
Closed-ended questions, 60–61, 72
Closing statements, in interviews, 67–68
Colleagues:
as clients, 281–282
relationships with, 288–289
Communication, generally:
counselor-client dynamics, 68–69
cultural barriers to medical care, 190
in interviews, 64
of results, 225–226
with specialists and researchers, 226–227
with team members, 222–223
style, *see* Communication style
Competence, defined, 255
Computer-based resources:
clinical diagnostic resources, 96, 372–376,
385
computer terms, 389–390
conferences, genetics 335–336
cytogenetic databases, 376–377, 386
funding agencies, 383–384
genetic testing laboratories, 96, 377, 386
internet resources, 380–383, 387–388
pedigree-drawing programs, 379–380,
386–387
private databases, 376
teratogen database, 377–379, 386
Confidentiality:
ethical issues, 254–255
family history, 31
issues of, generally, 8–9
legal issues, 267
in medical records, 239–240
pedigree construction, 42–43
psychosocial counseling, 131
student supervision, 313–314
Conflict of interest, 282
Consanguinity, 39–40
Consequence-based utilitarianism, 251

Consolidated Omnibus Reconciliation Act of 1986, 246
Consultants, medical genetics evaluation, 96–97
Consumer information processing theory, 104–105
Context, of genetic counseling, 11–12
Coping styles, counselees', 145–146
Cost containment, impact of, 115
Council of Regional Networks (CORN):
  Code of Ethical Principles for Genetics Professionals, 270
  as information resource, 344–345
Counseling aids, location and development of, 116, 221–222
Counseling Aids for Geneticists, 116
Counselor-client dynamics:
  agendas, 69
  assumptions, 69–70
  client communication style, 68–69
  counselor's self-awareness, 146
  duration of interaction, 112–113
  identifying with client, 71
  power, 70–71
  professional conduct and, 282–283
  responsibilities, 70
Countertransference, 137
Credentialing, 18, 76, 334
Culture, defined, 175
Current Procedural Terminology (CPT) billing codes, 242
Cyrillic, 380, 386
Cytogenetic databases, function of, 376–377, 386

Damages/awards, litigation and, 265
Deformation, defined, 88–89
Denial, 142–143
Depression, 144
Dermatoglyphics, 90
Despair, 144
Diagnosis:
  in different stages of life, 286
  establishment/verification of, 9–10
  in genetic counseling session, 107
Diagnostic studies, medical genetics evaluation, 93–95
Difficult behavior, dealing with, 287
Difficult issues, discussion of, 139–145

Directory of National Genetic Voluntary Organizations and Related Resources, 219
Discrimination, use of genetic information for, 269–270
Discrimination Model, 298
Disruption, defined, 88
DNA banking, 218–219
DNA laboratory report, sample, 368–369
DNA tests/testing:
  diagnostic, 76
  disclosure, 7
  generally, 93, 95
  verification of pedigree information, 40
Dual relationships:
  professional conduct, 281–284
  student supervision, 313
Duties, defined, 250–251
Duty, defined, 264
Duty to recontact, 272–273
Duty to warn third parties, 270–271
Dysmorphology examination, 85–90
Dysmorphology Photo Library, 376
Dysplasia, defined, 89

Emery and Rimoin's Principles and Practice of Medical Genetics, 207
Emigration, 176
Empathic attunement, 132–133, 139–140, 151
Empathy, psychosocial counseling:
  break in, 138–139
  role of, 133–136
Equality, of care, 256
Ethical issues, see Ethics
  case analysis, 257–260
  duty to recontact, 272–273
  duty to warn third parties, 270–271
  ethical theories, 251–252
  discrimination, use of genetic information for, 269–270
  friends, as clients, 281–282
  presymptomatic testing in children, 268–269
  sex selection, 267–268
  student supervision, 313–315
  unexpected findings, 271–272
Ethics:
  of care, 252
  defined, 250–251
  function of, 249

influence on genetic counseling, 252–256
moral behavior, defined, 250
NSGC, Code of Ethics, 257, 291–293
principle-based, 252–256
Ethnic background
as barrier to patient education, 113–114
pedigree construction, 38–39
significance of, 11
Ethnicity, defined, 174–175
Eugenics movement, 2–3, 147
Evaluation and Management (E/M) CPT
codes, 242

Family, counselor's relationship with, 283–284
Family history:
accuracy, 41–42
basics, 29–32
efficiency, 41–42
importance of, 27, 29, 38, 50, 129
interpretation of, 43–48
medical history, 84
negative, 44, 48
obtaining, generally, 202
questionnaire, 31
pedigree construction and, 27
psychosocial aspects of obtaining, 48–49
sample, 52–54
targeted, 30, 38, 202
time constraints, 41–42
Family life cycle, 180
Family systems theories:
-illness model, 152–153
overview, 151–152
systematically based psychotherapeutic
technique, 153–154
Family therapy, major models of, 168–170
Field defect, defined, 88
First-degree relatives, pedigree construction,
35, 37
Focused questions, 59–61
Fraud, 265

Galton, Francis, 28
Gender, generally:
pedigree symbols for, 33
roles, cultural differences in, 181
Genetic Analysis Package (GAP), 380, 387
Genetic conditions, seeking information on,
207–208, 383, 387

Genetic counseling, generally:
context and situations, 11–12
defined, 5–6
goals of, 5–6
historical perspective, 1–2
interaction, components of, 9–11
landmarks, professional and educational,
15–18
models of, 2–4
philosophy and ethos of, 6–9
professional growth, 18
providers of, 12–15
sessions, *see* Genetic counseling session
skill acquisition, 18
Genetic counseling resource book, 115
Genetic counseling session:
case management, 222–225
client support, ongoing, 110–111
communication of risk, 108–110, 211
components of, generally, 105–106
diagnosis, 107
Genetic counselors, role of, 13
information gathering, 106–107
information giving, 107–108
teachable moment, 106
Genetic Testing, 337
Genetic testing, coordination of:
advisory committees, 358
complexities, 358–363
DNA banking, 218–219
laboratories, selection of, 212–213
linkage analysis, 216–218
logistics of sending specimens, 214–216
prenatal samples, 218
research laboratories, 213–214
Geneticists, role of, 12–13 *See also* Clinical
Geneticist
Genetics nurse clinicians, 13
Genetics subspecialists, 14
Genetics team, 12
Genline, 374, 385
Genogram, 129, 153–154
Gilligan, Carol, 149
Grief, 144
Group case discussion, 155
Guilt, 143

*Handbook of Normal Physical Measure-
ments*, 96

Health, cultural differences, 189
Health belief model, 104
Health care delivery, 6
Health Care Financing Administration (HCFA), 241
Health education, 102–105
Health/Illness Beliefs and Attitudes Genogram, 153–154
Health Insurance Portability and Accountability Act (HIPAA), 241
Health promotion, 102–105
HELIX database, 96, 212, 352, 377, 386
Human Cytogenetics Database, 377, 386
Human Genome Project, 267

Ideals, defined, 250–251
Illness, cultural differences, 189
Imaging studies, 95
Immigration, defined, 176
Informational materials, design of, 117–118
Information for genetic professionals, 383, 387
Information gathering:
  generally, 9
  in genetic counseling session, 106–107
  in interviews, 58
  pedigree construction, 32, 35–38
Information giving:
  generally, 10
  in genetic counseling session, 107–108
Information resources, computer-based, *see* Computer-based information resources
Informed consent:
  accuracy, 256
  authorization, 256
  for clinical supervision, 314–315, 320–322
  defined, 255–256
Initial intake, case management, 200–202
Instructional aids, 115–117
Insurance, malpractice, 267
Interactive videodisc (IVD), patient education, 118
Intergenerational beliefs, 153
International Society of Nurses in Genetics (ISONG), 13, 344
Internet:
  information on genetic conditions, 383
  as information resource, 116

professional genetic organizations, 383, 387
  search engines, 382
  support groups sites, 220–221, 383
  World Wide Web (WWW), 381–382
Interpreter, selection of, 191
Interviewing techniques:
  listening skills, 65–68
  purpose of, 55–56
  questioning, 59–62
  redirecting, 63–64
  reflecting, 63
  rephrasing, 62–63
  shared language, promotion of, 64
  silence, 65
Interviews, *see* Interviewing techniques
  family history, 30–31
  getting started, 57–58
  goals of, 56
  listening skills and, 65–68

Journals in genetics, 336–337
Joint Commission on the Accreditation of Health Care Organizations (JCAHO), 241
Journal of Genetic Counseling, 336
Journal of Medical Genetics, 336
Journal of the American Medical Association, 337
Justice, defined, 256–257

Kessler, Seymour, 132
Klinefelter syndrome, 142
Kohut, Heinz, 148–149

Laboratories:
  logistics of sending specimens, 214–216
  research, 213–214
  selection factors, 212–213
  technological advances, 351–352, 361–362
Leading responses, 305
Learning problems, documenting, 91
Learning styles, types of, 100
Legal issues:
  duty to recontact, 272–273
  duty to warn third parties, 270–271
  ethics and law, relationship between, 260–261
  genetic counseling, generally, 263–267

malpractice, 263–264, 257, 267
medical documentation, 240, 244–245
minors, medical decision making, 262–263
professional protection from litigation, recommendations for, 266–267
sources of law, 261–262
student supervision, 313–315
Legal rights, defined, 261
Legislation and regulations, medical documentation, 240–241
Letter writing, case management, 228
to referring provider, 234
Liberty, defined, 256
Life history/narrative model, 102
Lifelong learning:
advanced education and training, 337–338
conferences, 335–336
continuing education, 335
importance of, 334–335
journals, 336–337
quality improvement in genetic counseling and service delivery, 337
Line of descent, 32, 35, 38
Linkage analysis, 216–218, 348–349
Listening skills, in interviews:
analogies, 67
client closing statements, 67–68
client's narrative, 65
gaps, omissions, and inconsistencies, 65–67
recurrent references, 67
Litigation, medical documentation and, 245. *See also* Legal issues
London Dysmorphology Database (LDDB), 375–376, 385
London Neurogenetics Database (LNDB), 376

McKusick, Victor A., 207, 372
Magical thinking, 142
Major anomalies, defined, 86
Majority group, defined, 175
Malformation, defined, 88–89
Managed care, impact of, 115
March of Dimes, 387
Marriage:
between biological relatives, 31
cultural differences, 181
Medicaid, 216, 243

Medical documentation: *see also* Medical Records
case management, 227–228
external review of, 242–243
family history, 31–32
governmental and fiduciary requirements, 241–242
importance of, 231, 234–235
information disclosure, 239–242
legal issues, 240, 244–245
legal issues and, 266–267
medical genetics evaluation, 91–93
medical records audit, 243–244
pedigree construction, 40
recommendations for, 235–239
retention of medical records, 245–246
timing of, 237
types of, 232–234
Medical genetics evaluation:
clinical case conferences, 96–97
clinical medical geneticist, 76–77, 95
diagnostic studies, 93–95
documentation of, 91–93
historical perspective, 75–76
medical history, basic components of, 79–84
outside expert consultants, 96–97
physical examination, 85–90
purposes of, 77–79
types of, 90–91
Medical history, medical genetic evaluation:
chief complaint, 80
family history, 83
history of present illness, 80, 82
importance of, 9
past medical history, 82–83
review of systems (ROS), 83–84
social history, 83
special considerations in, 84
Medical literature, search of, 207–208
Medical records: *See also* Medical documentation
audits, 243–244
information disclosure, 239–243
obtaining, 202–205
retention of, 245–246
sample, 244
signing/countersigning, 239
writing guidelines, 235, 238–239

Medicare, 216, 241, 243, 246
MEDLINE, 96, 207, 373, 381, 387
Mendelian Inheritance in Man: Catalogs of Human Genes and Genetic Disorders (McKusick), 207, 372
Metabolic and Molecular Bases of Inherited Disease, The, 207
Migration, defined, 176
Miller, Jean Baker, 149–150
Minor anomalies, defined, 86
Minority group, defined, 175
Molecular Genetics Laboratory Questionnaire, 214
Moral behavior, defined, 250
Multiculturalism:
  barriers to medical care, 190–191
  bereavement, 182
  case example, 172–174, 192–196
  children, role of, 182
  culture, defined, 175
  development of, 177
  ethnicity, defined, 174–175
  family structure, 180–181
  gender roles, 181
  health, 189
  illness, 189
  majority group, 175
  marriage, 181–182
  minority group, 175–177
  multiculturalism defined, 175
  race, defined, 175
  religions, 183–188
  rituals, birth and death, 182–183
  self-awareness, 178–180
  superstitions, 183, 189
  Western medicine culture, 177–178
Mutual participation model, 102

Narrative:
  in interviews, 65
  psychosocial counseling, 130
Narrow questions, 59–60
National Center for Biotechnology Information (NCBI), 96
National Commission of Quality Assurance (NCQA), 241
National Institutes of Health (NIH), 383
National Network of Libraries of Medicine, 388

National Organization for Rare Disorders, Inc. (NORD), 111, 220, 388
National Society of Genetic Counselors (NSGC):
  Code of Ethics, 240, 257, 277–278, 291–293, 334
  functions of, generally, 13, 15–16, 29, 213–214, 240, 332–333, 353
  goal of, 334
  membership in, 338–339
  overview, 343–344
  practice guidelines, recommendations for, 356–357
Nature, 337
Nature Genetics, 337
Negligence, 263–264
New England Journal of Medicine, 337
Nondirective counseling, 8, 132, 147
Nongeneticists, 14–15
Nonmaleficence, 253–254

Occupational Safety and Health Act, 240–241
Office of Maternal and Child Health (MCH), 15
On-line Mendelian Inheritance in Man (OMIM), 96, 207, 213, 372–374, 385
Open-ended questions, 59, 61

Parental needs, 110–111
Patient education:
  adult learners, 100–102
  audiovisual materials, 116–117
  barriers to client understanding, 113–115
  brochures, 117
  consumer preferences, 111–113
  CD-ROM, 118
  design and development model, case example, 122–126
  future directions, 118–119
  genetic counseling session, 105–111
  health education and promotion, 102–105
  importance of, 99–100
  informational materials, design of, 117–118
  instructional aids, application of, 115–117
Patient letters:
  medical documentation, 234
  patient education, 116
Patient resources, 219–221

PediDraw, 380, 387
Pedigree, generally:
  accuracy, 41–42
  affected status, documentation of, 40
  analysis of, 43–48
  adoptive families, 34–35
  confidentiality issues, 42–43
  consanguinity, 39–40
  construction and symbols, 32–35
  cousins, 35
  deceased relatives, 34, 37
  ethnic background and, 38–39
  evolution of, 28–29
  family history and, 27
  identifying information, 40
  information gathering, step-by-step
    process, 32, 37–38
  legend, 35
  medical genetics evaluation, 92–93
  sample, 52–54
  standard information in, 35–37
  updating, 42
  verification of information, 40
Pedigree-drawing programs, 379–380,
    386–387
Pedigree Standardization Task Force (PSTF),
    29
Pedigree symbolization:
  common, 33
  genetic evaluation/testing information, 41
  line definitions, 32, 34–35, 38
  miscarriage, 33
  pregnancy, 33
  sibshipline, 32
Personal professional conduct code, develop-
    ment of, 285–288
Physical examination, in medical genetic
    evaluation: See also Medical Genetics
    Evaluation
  documentation, 91–93
  dysmorphology examination, 85–90
  general examination, 85
Physician's Desk Reference (PDR),
    378–379, 386
Pictures of Standard Syndromes and Undiag-
    nosed Malformations (POSSUM), 375,
    385
Pluralism, defined, 176
Poverty cycle, medical care and, 190–191

Power, counselor-client dynamics, 70–71
Practice-based competencies:
  assessment skills, interpersonal, counsel-
    ing, and psychosocial, 24–25
  communication skills, 21–23
  critical-thinking skills, 23–24
  professional ethics and values, 25–26
  student supervision, 296
Practice guidelines, technological advances
    and, 355–356
  Alzheimer Disease, 356
  Breast cancer, 356
  Huntington disease, 355–356
Prenatal counseling, 11
Prenatal diagnosis, 11, 110, 112
Prenatal Diagnosis, 337
Present illness, history of, 80, 82
Presymptomatic testing, 107, 225
Principle-based ethics, defined, 251–252
Principles, defined, 250–251
Privacy issues:
  generally, 8–9
  legal issues, 271
  in medical records, 239–240
Professional conduct:
  appropriate relations with self and others,
    284–285
  boundaries, 279–281
  dual relationships, 281–284
  NSCG Code of Ethics, development of,
    277–278
  personal professional code, development
    of, 285–288
  relationships with colleagues and society,
    288–289
  responsibilities, 278–279
Professional development:
  genetic counseling profession and role ex-
    pansion, 332–333
  importance of, 331–332
  lifelong learning practices, 334–338
  opportunities in one's own institution, 338
  professional organizations, participation
    in, 338–339
Professional review organizations (PROs),
    243–244
Professional societies:
  American Board of Genetic Counseling
    (ABGC), 13, 17, 295–296, 332, 334, 342

American Board of Medical Genetics (ABMG), 13–14, 16–17, 77, 342
American College of Medical Genetics (ACMG), 16, 76, 339, 343, 356
American Society of Human Genetics (ASHG), 5, 269, 338–339, 341, 353, 356
Council of Regional Networks for Genetic Services (CORN), 344–345
International Society of Nurses in Genetics (ISONG), 13, 344
Internet listing, 387
National Society of Genetic Counselors (NSGC), 13, 15–16, 29, 213–214, 240, 332–334, 338–339, 343–344, 353
participation in, 338–339
Projection, 137, 145
Projective identification, 145
Providers, types of, 12–15
Psychosocial aspects, of family history, 48–49
Psychosocial assessment, 129–130
Psychosocial counseling:
case example, 157–161
counselees' coping styles, 145–146
counselor's self-awareness, 146
discussing difficult issues and giving bad news, 139–145
empathic break, 138–139
empathy, 133–136
general, 10–11
genetic counselor as psychotherapist, 128–129
ongoing genetic counseling practice, stages of, 156–157
patient's narrative, 130
psychosocial assessment, structure of session, 129–130
psychosocial case discussion, 155–156
theories, overview of, 146–155
transference and counter transference, 136–138
working relationship, 131–139
Psychosocial dimensions, in counseling, 8
PubMed, 207

Quality, technological advances and, 352–354
Quality improvement (QI) programs, 337
Questioning, in interviews, 59–62
Questions, as emotional reaction, 144

Race, defined, 175
Reasoned action theory, 105
Recall, patient education, 113
Recurrent references, in interviews, 67
Redirecting, in interviews, 63–64
Reduced penetrance, 45–48
Reed, Shelton, 2
Referrals, 9, 30, 111, 135
Reflecting, in interviews, 63
Relationships, professional conduct, 281–285, 288–289
Religion, cultural differences, 183–188
Rephrasing, in interviews, 62–63
Repository of Human Chromosomal Variants and Anomalies: An International Registry of Abnormal Karyotypes, 377
Reproductive Hazard Information, Environmental Impact on Human Reproduction and Development (REPROTOX System), 377–378, 386
REPRORISK, 378, 386
REPROTEXT System, 378, 386
Research, generally:
laboratories, genetic testing at, 213–214
protocols, clinical services *vs.*, 286–287
studies, 227
Responsibilities:
counselor-client dynamics, 70
professional conduct, 278–279
Results, of tests:
communication of, 225–226
technological advances and, 363–365
unexpected findings, 271–272
Review of systems (ROS), in medical genetic evaluation, 83–84
Rights, defined, 250–251
Risk, defined, 108
Risk assessment, 10, 12, 208–211
Rituals, cultural differences, 182–183
Rogers, Carl, 147
Rolland, John, 152–153
Rules, defined, 250–251

Science, 337
Scientific literacy, patient education and, 114–115
Self-awareness:
counselor's, 146
multiculturalism and, 178–179

Self-efficacy, 105
Self-empathy, 133, 149–151
Self in relation theory, 149–151
Self objects, 148–149
Self psychology, 148–149
Self-referent responses, 305, 307
Self-understanding, 149
Sequence, defined, 88
Shame, 143
Shared decision-making (SDM) programs, 118
Silence, in interviews, 65
Skin examination, 89–90
Small talk, in interviews, 57–58
Smith's Recognizable Patterns of Human Malformation, 207
SOAP note, 232–234
Social learning theory (SLT), 105
Society, genetic counselor relationships with, 288–289
Specimens, logistics of sending, 214–216
Standard of care, 255
Standard three-generation pedigree, 29–30. *See also* Pedigree
Statutes of limitations, 246
Student assessment:
client surveys, 310
evaluations, 310–313
formal case presentations, 309–310
Student supervision:
agenda, 315
case discussion, 307–308
cultural differences, 315
discussion topics, 300–301
ethical and legal issues, 313–315
example consent form for clinical supervision in the genetic counseling setting, 320–322
feedback, 304–305
focusing issues, 308–309
goals of, 298–300
genetic counselors, responsibilities of, 297–298
methods of, 305–307
novice and advanced students, 316–318
practice-based competencies, 296
student assessment, 309–313
student evaluation of the clinical and supervisory experience, sample, 328–330

student responsibilities, 304–305
student roles, 301–304
supervision, defined, 296–297
supervisor evaluation of student skills and performance, sample, 323–327
supervisor responsibilities, 297–298
supervisor roles, 301–304
Subjective risk estimate, 109
Summative evaluations, 310–312
Superstitions, cultural differences, 183, 189
Supervision, student, *see* Student supervision
Support groups, 111, 219–221, 383, 387
Supportive counseling, 110–111
Symbols, pedigree construction, 32–35
Syndrome, defined, 88

Task Force on Genetic Testing (TFGT), 353–354
Teachable moment, 106
Technological advances, implementation considerations:
complexities of, illustrative cases, 358–365
laboratory issues, certification and quality, 352–354
practice issues, policy and practice guidelines, 354–358
technology transfer process, 348–352
Technology assessment conferences, 354–355
Technology transfer process:
in the clinical arena, 349, 351
development of, 348–349
in the laboratory arena, 351–352
test sensitivity and specificity, 349
Templates, pedigree construction, 31–32
Teratogen databases, 377–379, 386
Teratogen Information System (TERIS), 378, 386
Test sensitivity and specificity, 349, 351, 358–361
Third cousins, pedigree construction, 35
Time constraints:
in counseling session, 223–224
pedigree construction, 41–42
Tort claims, 264–265
Training programs:
accreditation, 16–17, 77
clinical medical geneticists, 76–77
types of, 15–16

Transference, 136–137, 149
Truth telling, 254

Uniform Health Care Information Act,
   240–241
U.S. Department of Energy (DOE), 383
Utilitarianism, consequence-based, 251
Utilization of services, voluntary, 6–7

Values, defined, 250–251
Variable expressivity, 44–45
Verification, in pedigree construction, 40
Virtue ethics, 251
Virtues, defined, 250–251

Voluntariness, 256

Working relationship, psychosocial counseling:
   boundaries, 131–132
   confidentiality, 131
   countertransference, 138
   disruptions in, 136–139
   empathic breaks, 138–139
   empathy, 133–136
   patient autonomy, 132–133
   transference, 136–137
World Wide Web (WWW), 381–382. *see
   also* Internet